WITHDRAWN

MAKING
AND USING
ANTIBODIES

A PRACTICAL HANDBOOK

EDITED BY
GARY C. HOWARD
MATTHEW R. KASER

CRC Press
Taylor & Francis Group
Boca Raton London New York

CRC Press is an imprint of the
Taylor & Francis Group, an informa business

CRC Press
Taylor & Francis Group
6000 Broken Sound Parkway NW, Suite 300
Boca Raton, FL 33487-2742

© 2007 by Taylor & Francis Group, LLC
CRC Press is an imprint of Taylor & Francis Group, an Informa business

No claim to original U.S. Government works
Printed in the United States of America on acid-free paper
10 9 8 7 6 5 4 3 2 1

International Standard Book Number-10: 0-8493-3528-0 (Softcover)
International Standard Book Number-13: 978-0-8493-3528-0 (Softcover)

Library of Congress Cataloging-in-Publication Data

Making and using antibodies : a practical handbook / edited by Gary C. Howard and Matthew R. Kaser.
    p. cm.
Includes bibliographical references and index.
ISBN 0-8493-3528-0 (alk. paper)
    1. Immunoglobulins--Handbooks, manuals, etc. I. Howard, Gary C. II. Kaser, Matthew R.

QR186.7.M35 2006
571.9'67--dc22
    2006047535

Visit the Taylor & Francis Web site at
http://www.taylorandfrancis.com

and the CRC Press Web site at
http://www.crcpress.com

Dedication

This book is dedicated to

Rebecca and Amanda Howard
G.C.H.

and to

Michael and Elizabeth Kaser,
Margery Ord, and Lloyd Stocken
M.R.K.

Table of Contents

Preface

Antibodies are perhaps one of the most extraordinary species of protein ever to evolve during the history of life on Earth. These "magic bullets" have become an indispensable tool in the study of biology and medicine. In biology, they have been a key component of the surge in fundamental knowledge that has occurred in the last quarter century. In the practice of medicine, multiple vaccines have led to the control (at least in the developed world) of many infectious diseases, such as polio, mumps, measles, chicken pox, and the almost total eradication of smallpox.

We hope this book will be useful to biomedical researchers and students. Although new methods for making and using antibodies will certainly be found, their current applications—ELISAs, Western blotting, immunohistochemistry, and flow cytometry—are so powerful that they will remain critical to biomedical science for a considerable period.

We want to thank the contributors to this volume. Their professional knowledge, excellent writing, and enthusiastic support made the book possible. We also owe great thanks to our editor at CRC Press, Judith Spiegel, for her valuable help and great patience with this project.

Matthew R. Kaser and Gary C. Howard
Castro Valley, California

Editors

Matthew R. Kaser, D.Phil. earned his D.Phil. in biochemistry from Oxford University (UK) in 1988. After postdoctoral positions at the University of California, the University of Texas, and at REI Harbor-UCLA Medical Center, he was appointed to a faculty position at the University of California, San Francisco, Department of Pediatrics and then served as a scientist and patent agent at Incyte Genomics in Palo Alto, California. Dr. Kaser has been practicing as a patent agent since 1999, was associate director of intellectual property at Mendel Biotechnology, and is now a senior partner at Bell & Associates in San Francisco. He has presented research papers at a number of regional, national, and international conferences and coauthored more than a dozen publications.

Gary C. Howard, Ph.D. earned his Ph.D. in biological sciences from Carnegie Mellon University in 1979. He completed his postdoctoral training at Harvard University and The Johns Hopkins University and was a research assistant biochemist at the University of California, San Francisco. He then joined Vector Laboratories in Burlingame as a biochemist and Medix Biotech (a subsidiary of Genzyme) in Foster City, California, as chemistry manager and operations manager. Currently, he is principal scientific editor at the J. David Gladstone Institutes, a private biomedical research institute affiliated with the University of California, San Francisco.

Contributors

Paul Algate, Ph.D.
Issaquah, Washington

Jory Baldridge, Ph.D.
GSK Biologicals
Hamilton, Montana

Lee Bendickson
Iowa State University
Ames, Iowa

Joseph P. Chandler, Ph.D.
Maine Biotechnology Services, Inc.
Portland, Maine

John Chen, Ph.D.
BioCheck, Inc.
Foster City, California

Frederic Fellouse, Ph.D.
Department of Protein Engineering
Genentech, Inc.
San Francisco, California

David A. Fox, M.D.
Hybridoma Core Facility
University of Michigan
Ann Arbor, Michigan

Kristi R. Harkins, Ph.D., M.B.A.
BioForce Nanosciences, Inc.
Ames, Iowa

Gary C. Howard, Ph.D.
The J. David Gladstone Institutes
San Francisco, California

David N. Howell, M.D., Ph.D.
Department of Pathology
Duke University Medical Center
and
Veterans Affairs Medical Center
Durham, North Carolina

Matthew R. Kaser, D.Phil.
Bell & Associates
Castro Valley, California

Lon V. Kendall, D.V.M., Ph.D.
Center for Laboratory Animal Science
School of Veterinary Medicine
University of California
Davis, California

M. Elaine Kunze
Flow Cytometry and Imaging
Huck Institutes of the Life Sciences
Penn State University
University Park, Pennsylvania

Sara E. Miller, Ph.D.
Department of Pathology
Department of Molecular
    Genetics and Microbiology
Duke University Medical Center
Durham, North Carolina

Sally Mossman, Ph.D.
Seattle, Washington

Marit Nilsen-Hamilton, Ph.D.
Iowa State University
Ames, Iowa

José Ramos-Vara, D.V.M., Ph.D.
Animal Disease Diagnostic Laboratory
School of Veterinary Medicine
Purdue University
West Lafayette, Indiana

Julie Ackerman Saettele, M.B.A.
Trinity Biotech USA
Kansas City, Missouri

Kathleen C. F. Sheehan, Ph.D.
Center for Immunology
Department of Pathology and Immunology
Washington University School of Medicine
St. Louis, Missouri

Sachdev Sidhu, Ph.D.
Department of Protein Engineering
Genentech, Inc.
San Francisco, California

Elizabeth M. Smith, M.Sc.
Hybridoma Core Facility
Department of Internal Medicine
University of Michigan School of Medicine
Ann Arbor, Michigan

George P. Smith, Ph.D.
University of Missouri
Columbia, Missouri

1 Antibodies

Matthew R. Kaser and Gary C. Howard

CONTENTS

1.1 A VERSATILE MOLECULE

Antibodies are perhaps one of the most extraordinary species of protein ever to evolve during the history of life on Earth. If a "magic bullet" exists in biomedicine, antibodies may very well be it. In mammals, these molecular agents augment and complement the innate immune system and are synthesized as an adaptive response to a challenge from outside the organism (e.g., a bacterial or viral infection or an exogenous organic compound).

We often cite Edward Jenner's work in 1796[1] as the first practical use of antibodies. He elicited an immune reaction to an attenuated virus viral antigen that subsequently protected the organism from a closely related antigen. Although Jenner was the first to perform such a vaccination under controlled conditions, the first use of viral material to provide immunity from future infections, variolation, also known as inoculation, appears to date from many centuries, if not millennia, earlier in China.[2,3]

Today, immunology is an integral part of science and medicine. Antibodies are used in ways unheard of 40 years ago when Edelman and Porter first isolated an immunoglobulin molecule. As one measure, between 1998 and 2003, the market for antibody-based drugs has experienced an almost explosive growth (53%) in use for diagnostics and therapeutics. Monoclonal antibodies, first developed by Kohler and Milstein,[4] now have a global therapeutic market of more than $7 billion, and hundreds of potential products are at the preclinical stage (for a review, see Ramachandra[5]).

Antibodies are used in proteomics and diagnostics to detect tumor and bacterial antigens; they are being synthesized in goat milk and in plants to be used as vaccines and antitumor agents; and as drug-delivery vehicles for treatment of viruses, bacterial infections, and, of course, cancer.

The choice of which form of antibody to be used can depend on the ultimate goal of the project. In general, monoclonal antibodies are most useful for purification and analysis of epitopes upon the surface of native proteins and otherwise hidden epitopes in denatured proteins; polyclonal antibodies are of use in Western blotting and purification of distinct protein orthologues; chimeric antibodies are used for study of xenogenic molecules in an experimental model of the same idotype as part of the chimeric antibody.

Each of our contributors has made recommendations based on his or her experience. New approaches are bound to be tried and some will be successful. For example, a combination of monoclonal and polyclonal antibodies may be found to be most effective for detecting large molecules on a biochip. If in doubt, consult with experienced colleagues and perform several pilot assays.

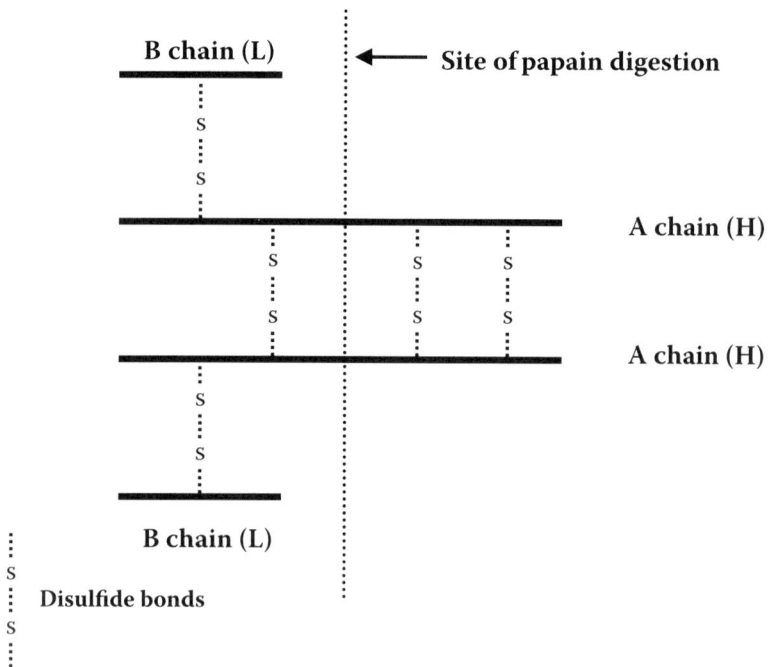

FIGURE 1.1 Antibody structure. Antibodies are Y-shaped tetrameric molecules with two heavy (H) chains and two light (L) chains.[6,7] Digestion with the plant protease papain yields 45-kDa Fab fragments which bind antigen (the antibody combining site, in Porter's terminology). A 55-kDa Fc fragment does not combine with antigen and can be crystalized. The two Fab fragments are connected together via disulfide bonds. Using a different protease, pepsin, from calf stomach, they noted that a slightly different proteolysis resulted, thereby giving rise to the terminology of Fab, equivalent but slightly larger than the Fab fragment. Pepsin digested more peptide bonds present on the Fc fragment and actually created a number of small antigenic fragments, whose significance was not realized until much later. (Reprinted with permission from Porter (1963) *Br. Med. Bull.* 19: 197–201.)

IgG

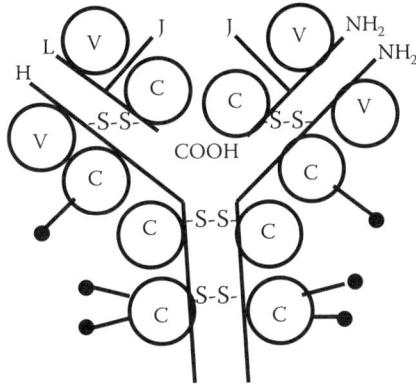

FIGURE 1.2 The various regions of a generalized immunoglobulin molecule. The molecule consists of four polypeptide chains, each chain having a variable (V) region that binds antigen, a joining (J) region, and a constant (C) region. The chains are further subcategorized as light (L) or heavy (H) chains, based on their size under dissociating conditions in 8M urea gels.

1.2 ETHICAL CONSIDERATIONS

Much of the work on antibody structure and mechanisms of action has involved the use of animal experimentation. However, use of animals for such testing has become of concern to society and scientists. For many lines of experimentation, alternative methods, such as primary or immortalized cell lines, whole plants or leaves, bacterial and bacteriophage propagation, and biochemical or chemical synthesis reaction, can replicate or approximate the tissue or organism under study.

However, some answers require the use of animals. In those cases, it is important to adhere carefully to the all applicable regulations on the use and health of experimental animals. Universities, research institutes, and other organizations have ethical oversight committees to approve protocols that use animals.

In any case, the number of animals used should be kept to the absolute minimum that may be required to obtain statistically significant answers. Power calculations, research experience,and expert input can help to ensure an appropriate choice.

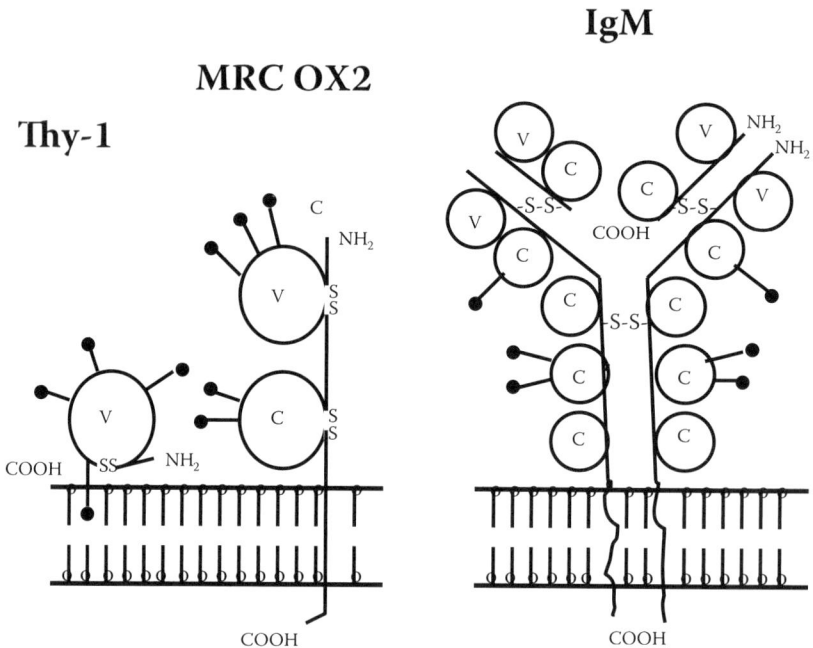

FIGURE 1.3 Antibodies of the immunoglobulin superfamily are part of the larger protein family of immunoglobulins (Ig), proteins that are synthesized by lymphocytes that bind to particular sets of epitopes usually found on, or produced by invasive or "non-self" organisms. Most of the other members of the Ig family, such as IgF and IgA, are synthesized and presented at the surface of lymphocytes where they bind to and are activated by antigens in the immediate cellular or tissue environment. Transmembrane domains of the molecule then transduce the binding signal to subsurface or cytosolic proteins, thereby establishing a second messenger-pathway response throughout the cytoplasm and, often, through to the genome. Antibodies such as IgG or IgE, on the other hand, are soluble proteins and freely migrate through the circulatory system acting as immunologic "scouts" for the organism. Members of the immunoglobulin superfamily (IgSF) have been found in every vertebrate species studied and appear to have arisen during the early Ordivician Period around 470 million years ago.[8] With respect to the human immune system, observations suggest that DNA rearrangement of the V-J-D genes account for most of the antibody diversity in primates and rodents; however, somatic hypermutation and somatic gene conversion are also of significance in other vertebrates.[9] This observation correlates with the current understanding of mammalian evolution whereby primates and rodents shared a common ancestor in northern Laurentia subsequent to the split from the Laurasiatheres around 85 million years ago.[10,11] The variable (V) and constant (C) domains of the IgSF are typically approximately 100–110 and 90 amino acid residues in length, respectively.[12] There is generally about 20–30% identity of these domains between members of the IgSF. The similarity and homology of residue sequences between molecules of the superfamily either within or between species are consistent with the thesis that the V-J-C regions of the immunoglobulins themselves arose by sequence duplication, divergence, and deletion (at a number of times) of a primordial Thy-1–like sequence in a simpler organism. Thy-1 itself is most likely involved in regulation and modulation of cell proliferation by another cell type; the ancestral molecule may have had a similar function in cell-cell interactions in primitive eukaryotes. Also shown are immuno-globulin superfamily members MRC OX2 and IgM.

1.3 SAFETY

A modern research laboratory has many potential safety hazards (e.g., chemicals, equipment, biologicals, radiation, electrical, animals). Although safety is extremely important, no handbook can describe all the hazards that can be encountered in any given laboratory.

Cell culture, isolation of proteins, analysis, and disposal or reagents frequently use chemicals that are regulated by government agencies. Be sure that your research protocols have been vetted and conform to local regulations and practices. Use of radioisotopes and carcinogens (such as ethidium bromide) has been significantly reduced over the past 10–15 years with the development of fluorophores and other means for detecting protein molecules *in vivo, in situ*, and *in vitro*. In addition, many assays that used to compose toxic or harmful reagents have now been supplanted by kits that contain either smaller amounts or safer alternatives.

The most important overall items involve training of laboratory personnel and adherence to all regulatory guidance. Everyone who works in a laboratory must be given appropriate instruction on the equipment, chemicals, biologicals, and protocols that will be used. That instruction should also include general laboratory safety, including how to deal with accidents (e.g., chemical or radioactive spills, sharps, fires).

1.4 ORGANIZATION OF THE HANDBOOK

We have organized this handbook with the laboratory user in mind; in principal, methods are sometimes presented in two chapters, but we have included cross-referencing of some methods to other chapters where appropriate. We have also tried to keep protocols on one or two pages, side by side, so that turning a page is not required when at the bench. In some chapters, several material and methods appears as an appendix to the chapter.

The goal of this handbook was to provide both an introduction as well providing the student with a robust review of the methods currently available. We hope that the research community has agreed with us!

REFERENCES

1. Jenner, E., *The Origin of the Vaccine Innoculation,* D.M. Shury (printer), Soho, London, 1801.
2. Xie, S. and Zhang, D., 30, 133, 2000.
3. Ma, B., Zhonghua Yi Shi Za Zhi., 25:139, 1995.
4. Kohler and Milstein, *Nature*, 256, 495, 1975.
5. Ramachandra, *BioSci. Technol.*, May (suppl.), 27, 2005.
6. Porter, *Br. Med. Bull.*, 19, 197, 1963.
7. Tonegawa, *Nature* 302, 575, 1983.
8. Rast et al., *Immunogenetics* 40, 83, 1994.
9. Sitnikova, T. and Su, C. *Mol. Biol. Evol.*, 15: 617, 1998.
10. Murray et al., *Science*, 294, 2348, 2001.
11. Springer et al., *Proc. Natl. Acad. Sci. U. S. A.*, 100, 1056, 2003.
12. Barclay et al., *Biochem. Soc. Symp.*, 51, 149, 1986.

2 Antigens

Paul Algate, Jory Baldridge, and Sally Mossman

CONTENTS

2.1   CHOOSING AN ANTIGEN

Molecules that can be recognized by a specific immune response are referred to as antigens. However, not all antigens are immunogens. Immunogens are molecules that elicit a humoral or a cell-mediated immune response. Some smaller molecules (haptens) are unable to stimulate an immune response, unless they are coupled to a larger reactive substance (carrier). Thus *immunogen* and *antigen* are distinct, yet related, terms. For an antigen to elicit an antibody response, it also must be an immunogen.

The first step in making any antibody is to choose an appropriate antigen for use as the immunogen. The ability to successfully raise an antibody with the required specificity and utility is directly related to the choice and quality of the antigen that is used as the immunogen. Before initiating any antibody development, one should carefully consider for what the antibody is to be used. Choosing the appropriate antigen will maximize the chances of producing an antibody with the required properties. An antibody that is to be used as a Western blot reagent or for immuno-histochemistry must recognize denatured protein and will therefore tend to be specific for linear epitopes. In contrast, an antibody detecting cell-surface proteins in which conformational epitopes are important, such as in flow cytometry, should by necessity recognize native proteins. The ability to recognize antigen expressed in its native form is essential if the antibody is to be functional with potential therapeutic use.

A second consideration should be how to screen for the desired antibody. The ability to generate multiple sources of antigen greatly increases the immunization and screening strategies that can be employed. This chapter will discuss the pros and cons of choosing a particular antigen, whether a peptide, a prokaryotic or eukaryotic recombinant protein antigen, a whole-cell antigen, plasmid DNA, or an adenovirus-expressed antigen. Methodologies and strategies are presented and discussed that enable one to generate different antigens that can be used as immunogens and for screening for antibodies with appropriate properties.

The sequencing of the human genome has made the nucleotide and protein sequences for many potential antigens readily available. Publicly available databases and search engines are available through the National Center for Biotechnology

Information (http://www.ncbi.nlm.nih.gov/). They provide a rich source of information that can be used to identify and analyze potential antigens. One of the key elements in defining an antigen as a potential immunogen is to determine the homology it may have to its ortholog protein in the species in which the antibody is to be raised.

Antibodies to any given antigen should be raised in a species where homology to the endogenous ortholog protein is minimized to increase foreignness and thus increase its immunogenicity. Several species of animal have been used to raise antibodies, including rabbits, mice, rats, guinea pigs, goats, sheep, donkeys, and chickens. The most commonly used animals are rabbits and mice/rats because they are easy to maintain and generally give good antibody responses. Rabbits tend to require more immunogen, but yield a large amount of sera. Mice and rats yield small amounts of sera, but can be used to produce monoclonal antibodies.

A further consideration is to determine the potential for homology with related proteins or family members to minimize the potential for developing antibodies with unwanted cross-reactivity to similar antigens. HomoloGene (http://www.ncbi.nlm.nih.gov/entrez/query.fcgi?DB=homologene) is a system for automated detection of homologs among the annotated genes of several completely sequenced eukaryotic genomes. Analysis of homologs and orthologs, using systems such as HomoloGene, allows for the educated design of an antigen to maximize immunogenicity and minimize cross-reactivity. Where strong homologies are identified (>90% at the amino acid level), truncated proteins or peptides can be used that correspond to regions of the least homology. Bioinformatics is also useful to identify and define functional protein domains that can be specifically targeted when functional antibodies are required. The TMpred program (http://www.ch.emb-net.org/software/TMPRED_form.html) makes a prediction of membrane-spanning regions and their orientation. The algorithm is based on the statistical analysis of TMbase, a database of naturally occurring transmembrane proteins[1] and allows for the topography of cell-surface molecules to be determined such that extracellular domains can be identified and targeted to produce antibodies that have utility for flow cytometry. Another program that is useful for subcellular localization prediction is PSORT II and related programs and data sets (http://www.psort.org/).

2.2   PEPTIDES AS ANTIGENS

Peptides provide arguably the quickest way to generate an antigen necessary to begin an immunization protocol to generate antibodies. Many institutes and companies find it cost effective to maintain core peptide synthesis capabilities. However, where not available, several commercial companies offer custom peptide synthesis services that are reliable, cost-effective, and deliver high-quality peptides within a couple of weeks in quantities suitable for an immunization schedule. For a typical peptide immunogen, 10 mg of a 15–amino acid peptide at >80% purity can be purchased for about $500 (circa 2005) and is enough material for immunizing several animals and providing reagent for screening the antibodies by enzyme-linked immunosorbent assay (ELISA).

Peptides are particularly useful for raising antibodies specific to regions of an antigen, such as a novel domain or regions of least homology, or when other antigen sources are not available. In many instances, recombinant proteins may not be available as a source of antigen because of their inability to be expressed and purified. This is often the case with large membrane spanning proteins, such as G protein–coupled receptors, a family of pharmaceutically important molecules. Peptides corresponding to the short extracellular domain protein loops have proven to be viable immunogens for generating antibodies to these complex molecules.

Peptides as short as six amino acids have been used to raise antisera, but peptides in the order of 10–12 amino acids are generally more immunogenic.[2] An example of a shorter epitope is the Flag epitope tag, which is widely used in the purification, identification, and functional analysis of proteins. This is an octapeptide sequence (DYKDDDDK) to which polyclonal and monoclonal antibodies are readily available. We have found that peptides in the range of 12–15 amino acids make very good immunogens in most instances. Longer peptides, as long as 30–35 amino acids, and limited only by the chemistry of the peptide synthesis itself, also make good immunogens and offer the advantage that the peptides can potentially form secondary structure that may be relevant. Peptide immunogens are limited in that they present short, linear, epitopes that may not be recognized in a whole protein antigen. Thus antibodies raised against peptides tend to recognize linear epitopes and in general work very well in Western blots and other applications in which antibodies recognize denatured proteins. However, antibodies raised to peptides generally do not recognize protein conformations and therefore are less likely to react with native molecules. Such antibodies tend not to be functional and may not work for flow cytometry. However, it should be noted that in some instances peptide immunogens do result in antibodies that recognize native protein—for example, when a linear peptide epitope is not masked by the tertiary structure of the antigenic protein.

Peptide antigens, because of their short length, should be considered haptens that require linkage to a carrier protein to increase their immunogenicity. Such coupling is usually considered necessary for polypeptides less than 3 kDa and is probably beneficial for any not greater than 10 kDa. Typical carrier proteins include keyhole limpet hemocyanin, bovine serum albumin, and ovalbumin. The Imject Immunogen Preparation Kits (Pierce Biotechnology, Rockford, IL) are quick, easy to use, and available for the carriers described. To use the chemistry in these kits, peptides should be synthesized with a terminal cysteine residue to provide a free sulfhydryl group (–SH) to which the maleimide-activated carrier can conjugate in an easy one-step reaction. Carrier conjugated peptides are purified by desalting or dialysis and frozen in aliquots until used. Animals are injected initially with peptide mixed with an adjuvant such as complete Freund's adjuvant (CFA), "boosted" initially with peptide in incomplete Freund's antigen (IFA) and subsequently without adjuvant.

2.2.1 IMMUNIZATION SCHEDULE FOR PEPTIDE ANTIGENS

2.2.1.1 Mice: BALB/c, Female Mice, 6–8 Weeks of Age

Prime with 50 μg peptide, 10 μg CFA, adjust volume to 200 μl with phosphate-buffered saline (PBS), deliver by intraperitoneal (i.p.) route; boost (at least twice) every 3–4 weeks with 50 μg peptide, 10 μg IFA, to 200 μl with PBS, i.p.; final boost with 100 µg peptide without adjuvant, i.p. (or intravenous [i.v.] if in a small enough volume), 3–4 days before harvesting the spleen.

2.2.1.2   Rabbits: New Zealand White, Female Rabbits, 8 Weeks of Age

Prime with 0.4 mg peptide + 0.1 mg muramyl dipeptide (MDP adjuvant), to 1 ml with PBS, + 1 ml CFA, subcutaneously (s.c.) (at four sites: two inguinal, two axillary); first boost (+ 4 weeks) with 0.2 mg peptide, to 1 ml with PBS, + 1 ml IFA, s.c. (deliver to four sites: two inguinal, two axillary); second boost and subsequent boosts every 4 weeks, 0.1 mg peptide, in 50–100 µl PBS, no adjuvant, i.v.; ~20 ml of sera production bleed 5–7 days after boost.

Antibody responses to the peptide should be measured on a preimmunization bleed and on sera harvested between boosts by an ELISA with plates coated with non–carrier-conjugated peptide to differentiate a response to the peptide from a response to the carrier. Be sure to only conjugate an aliquot of your peptide to leave some for this purpose!

2.3   PROTEIN ANTIGENS

Protein antigens have historically been isolated from tissues or cells with classical protein purification chemistry methodology that was time consuming, highly specialized, and often resulted in low yields of poorly characterized protein. Recombinant protein technology, developed in recent years, has allowed for the relatively easy production of high yields of pure protein. Purity of the protein is important because it ensures that any immune response is specific to the protein of interest and not to an immunodominant contaminant. Many reagents, expression systems, and kits are commercially available, enabling proteins to be produced from many heterologous sources, including prokaryotes (e.g., *Escherichia coli*), insect cells (baculovirus), yeast (*Saccharomyces cerevisiae* and *Pichia pastoris*), and various mammalian cell systems. Protein quality, speed, and yield are often the most important factors to consider when choosing an appropriate expression system. Overall cost and time are lowest with prokaryotic expression, but are inversely related to the probability of expressing functional protein, which tends to be higher with protein expressed in eukaryotic systems. Recombinant DNA techniques allow for the construction of fusion proteins in which specific affinity tags are added to the protein sequence of interest. These affinity tags simplify purification of the recombinant fusion proteins by affinity chromatography methods.

Affinity tags include Flag, c-myc, HA, glutathione S-transferase (GST), green fluorescent protein, and 6xHis. In addition, antibodies to these tags have enabled fusion proteins to be detected by Western blot, flow cytometry, and immunohistochemistry, while often having little effect on the functional properties of the protein. Purification of Flag-, MYC-, and HA-tagged proteins requires an antibody-capture affinity column that can make large-scale purification costly. Protein purification based on the GST tag relies on its strong reversible affinity for glutathione-covered matrices, which makes large-scale production practical. However, the GST tag itself is highly immunogenic and can dominate an immune reaction to such a tagged protein. The GST can be cleaved from the fusion protein but this can be incomplete and is costly. The 6xHis tag purification is based on the ability of the six histidine residues to chelate metal ions, such as nickel, allowing purification by affinity chromatography on an immobilized nickel column. Purification is scalable, cost effective, can be performed in a single step, and carried out under native and denaturing conditions. The 6xHis is relatively immunologically inert, and its small size generally has no effect on the structure and biologic function of a fusion protein, making it a good tag for the purification of a protein that is to be used as an immunogen. Moreover, many vectors are commercially available that allow for the production of 6xHis tagged protein in *E. coli*, insect cells, and mammalian cells.

2.3.1 PROKARYOTIC PROTEINS

Expression and purification of tagged fusion proteins from *E. coli* arguably represent the easiest and quickest way to generate protein antigens in amounts necessary for immunization. The downside to prokaryotic-derived antigens is that they will not be subjected to posttranslational modifications as is the native eukaryotic protein; they will lack glycosylation that may be important in native protein conformation and availability of antigenic epitopes. With general molecular biology cloning methodologies[3] and some basic column purification technology, it is possible to identify an antigen of interest, clone it, express it, and purify milligram quantities in as little as 4–6 weeks. Vectors and purification systems are available from major commercial sources, including Novagen (EMD Biosciences, Madison, WI), Qiagen (Valencia, CA), Invitrogen (Carlsbad, CA), and Stratagene (La Jolla, CA) and come with excellent manuals and protocols. Vectors are many and varied, employing different promoters, multiple cloning sites and are designed to have the tag at the amino- or carboxy-end of the expressed protein.

2.3.1.1 Construction of Expression Vectors

The gene encoding the antigen of interest is isolated as a cDNA sequence by polymerase chain reaction (PCR) with a polymerase that contains proof-reading capabilities (e.g., *Pfu* polymerase) and appropriate primers containing restriction sites, and cloned into an appropriate 6xHis expression vector with these restriction sites. Primers should be carefully designed to provide restriction sites that are compatible with the vector into which the fragment is to be cloned and result in the amplified fragment cloned in the same reading frame as the 6xHis tag in the vector regardless of whether you choose to place the tag at the amino- or carboxy-end of the gene. A strategy should be employed to design an expression construct for optimal protein production and purification as well as for ensuring a good immune response. Protein transmembrane regions should be avoided where possible. These hydrophobic regions often cause problematic expression in *E. coli* and problems with protein refolding and purification. These potential problems tend to increase with the number of transmembrane regions. In our experience, single-spanning membrane proteins pose little problem, whereas multispanning membrane proteins, such as the seven-spanning G protein–coupled receptors, are difficult, if not impossible to express and purify. In addition, signal peptides should be avoided. They are cleaved during translocation and should not be relevant epitopes for the mature protein and will result in cleavage of any tag fused to it. Fragments should be a size that is easy to amplify without introducing mutations and easy to ligate into a plasmid vector (<1,000 amino acids; 3.0 kb). Although larger fragments can be used, they may be more difficult to work with and may not be necessary to provide a protein with conformational epitopes. As discussed previously, fragments should be chosen that are as unique as possible, when compared with other proteins, to avoid cross-reactivity by antigens that may be generated. Finally, production of partial proteins will decrease the risk of generating a functional protein with toxic effects in *E. coli* that could prevent expression. It is prudent to make several constructs of varying length and content and to determine empirically which expresses the best. All constructs should be isolated as clones and confirmed by sequencing to ensure that sequence integrity and reading frames are maintained. Plasmid vectors are produced by standard overnight culture of transformed *E. coli* and purification from bacterial lysates. Plasmid purification kits from Qiagen are routinely used to purify plasmid of a suitable quality for all subsequent use.

Several *E. coli* host strains are available and should be matched to the vector and promoter system used. It is often useful to transform several different strains and perform pilot scale expression to determine which strain works best for your protein. Sometimes a promoter can be leaky in a general purpose expression host (e.g., BL21(DE3)), a problem if the expressed protein is toxic, so that a more stringent expression host should be used (e.g., BL21(DE3)pLysS). Other host strains have been engineered to allow disulphide-bond formation in the *E. coli* cytoplasm (e.g., AD494 host series; Novagen, Madison, WI). Furthermore, bacterial strains have been developed to avoid human codon usage bias (e.g., BL21 CodonPlus; Stratagene, La Jolla, CA), thus minimizing both the risk of truncated protein translation from inefficient recognition of some mammalian codons and the potential for a high frequency of frame shifts, each being conditions that could adversely affect the expression of antigen.

2.3.1.2   Expression and Purification of 6× Histidine-Labeled Recombinant Antigens

Reagents

Sonication buffer	20 mM Tris, pH 8.0, 500 mM NaCl (store at 4°C)
Wash buffer	20 mM Tris, pH 8.0, 500 mM NaCl, 10 mM imidazole
Elution buffer	20 mM Tris, pH 8.0, 500 mM NaCl, 250 mM imidazole
CHAPS wash	Add 0.5% CHAPS to wash buffer
DOC wash	Add 0.5% DOC to wash buffer (always run at room temperature)
Pellet binding buffer	20 mM Tris, pH 8.0, 8 M urea
Pellet wash buffer	20 mM Tris, pH 8.0, 500 mM NaCl, 8 M urea
Pellet elution buffer #1	20 mM Tris, pH 8.0, 500 mM NaCl, 8 M urea, 300 mM imidazole
Pellet elution buffer #2	Same as #1 but at pH 4.5

2.3.1.2.1   Small-Scale Expression Cultures

Small-scale expression (100 ml) and purification pilot experiments should be performed to identify clones, hosts, and induction times that are optimal for a given
protein antigen. Aliquots (10 ml of culture) of induced cells can often be lysed
directly in sample buffer and analyzed directly by sodium dodecyl sulfate-polyacrylamide gel electrophoresis (SDS-PAGE) and Western blotting with a suitable antibody if available (e.g., anti-tag antibody) to identify optimal culture conditions and
determine the solubility of recombinant protein. It is also prudent to run small-scale
purification by nickel-affinity chromatography on your optimized conditions before
proceeding with a large-scale preparation. Small-scale column purification is a
100-ml version of the large-scale purification described below.

2.3.1.2.2   E. Coli Culture and Protein Induction

1. Inoculate 20 ml of LB broth containing appropriate antibiotic (e.g., 100
   μg/ml carbenicillin or ampicillin). Grow at 37°C overnight with vigorous
   shaking.
2. Inoculate 1 liter cultures (with antibiotic) at 1:50 with the non-induced
   overnight culture. Grow at 37°C with vigorous shaking until absorbance
   (600 nm) of 0.6 is reached.
3. Take 1-ml sample immediately before induction.*
4. Induce expression by adding isopropyl-β-D-thiogalactopyranoside to a
   final concentration of 1 mM.
5. Incubate the culture for an additional 4–5 h (determined by small pilot-
   scale experiment). Collect second 1-ml sample.*
6. Harvest cells by centrifugation at 4,000*g* for 20 min.
7. Freeze and store cell pellet overnight at –20°C.

2.3.1.2.3   For the Frozen Bacterial Pellet

1. Thaw bacterial induction cell pellets on ice.
2. Add 25 ml of sonication buffer per liter of induction culture.
3. Add 1 complete protease inhibitor tablet (Roche, Indianapolis, IN) and 2
   mM phenylmethylsulfonyl fluoride to sonication buffer/pellet mix.
4. Completely resuspend pellet with pipette.
5. Add 0.5 mg/ml lysozyme (made fresh from lyophilized lysozyme stored
   at –20°C).
6. Decant into a glass beaker and stir bar, gently stir at 4°C, 30 min.
7. Sonicate 8 × 20 sec on high (20 watts) or 2× French press at 1,100 psi
   (sonicate 4 × 20 sec if needed); keep on ice.
8. Once sonicate* has low viscosity, spin at 11,000 RPM, 30 min, 4°C.
9. Save supernatant and pellet.*

2.3.1.2.4   For the Supernatant (Soluble Protein)

1. Keep supernatant at 4°C until use.
2. Add imizadole to a final of 10 mM.
3. Equilibrate Ni++NTA (Qiagen) resin with wash buffer, spin down, and decant wash (use 0.5 ml resin/l pellet).
4. Resuspend resin in small volume of wash buffer and add to supernatant and slowly stir for 1 h at 4°C.
5. Prepare column and buffers. Rinse column with wash buffer.
6. Pour supernatant/Ni resin into column, collect flow through.*
7. Wash column with 20 ml of wash buffer or until absorbance returns to baseline.*
8. Elute with 6 × 4 ml fractions of elution buffer, or until absorbance returns to baseline.*
9. Elute with 3 × 4 ml fractions of elution buffer containing 8 M urea.*
10. Save fractions at 4°C.* (Take a sample of the resin to run on a gel.)

2.3.1.2.5   For the Sonication Pellet (Inclusion Body)

1. Keep pellet at –20°C until use.
2. Wash pellet with 25 ml of 0.5% CHAPS wash* by sonicating 1 × 20 sec (or as needed) at 15–20 watts.
3. Spin at 11,000 RPM for 25 min. Repeat four times or until supernatant is clear.*
4. Repeat above steps 3–5 times with 0.5% DOC wash.
5. Resuspend pellet in pellet binding buffer, with sonication if necessary.
6. Equilibrate Ni++NTA resin with pellet binding buffer, spin down, and decant wash (use 0.5 ml resin/l pellet).
7. Add resin to resuspended pellet, stir at room temperature for 45 min.
8. Prepare column and buffers. Rinse column with pellet binding buffer.
9. Pour pellet/Ni resin into column, collect flow through.*
10. Wash column with 20 ml of pellet binding buffer.*
11. Wash column with 20 ml of binding buffer with 0.5% DOC.*
12. Wash column with 20 ml of pellet wash buffer, or until absorbance is at baseline.
13. Elute with 6 × 4 ml fractions of pellet elution buffer #1.*
14. Elute with 3 × 4 ml fractions of pellet elution buffer #2.*
15. Save fractions at 4°C.* (Take a sample of the resin to run on a gel.)

---

* Always save an aliquot at each purification step to check on SDS-PAGE. Aliquots of all steps should be run on SDS-PAGE to track the induced recombinant protein. If an antibody is available, the sample gels should be assayed by Western blot to determine the recovery and purity of recombinant protein during purification.

2.3.1.3   Ion Exchange Chromatography

Recombinant fusion protein that still shows evidence of some contaminant proteins can be purified further by ion exchange chromatography. Although this additional purification is not usually required, it can sometimes help remove residual contaminants and endotoxin. Anion exchange chromatography can be performed using Q Sepharose fast flow resin and cation exchange chromatography can be performed with SP Sepharose fast flow resin (both from Amersham Biosciences, Little Chalfont, UK).

Antigen eluted from either the Ni-column or from an ionic exchange column should be dialyzed in 10 mM Tris, pH 8.0 (four changes), and concentrated using a centrifugal protein spin column concentrator with the appropriate molecular weight cutoff (Vivaspin; Vivascience AG, Hannover, Germany). The amount and concentration of the recovered recombinant protein should be determined using a protein assay reagent such as BCA Protein Assay reagent (Pierce Biotechnology, Rockford, IL). N-terminal sequencing, where available, will confirm the nature and purity of the final purified recombinant protein.

With prokaryotic-derived protein, it is important to ensure that the final product is endotoxin free, particularly when it is to be used as a mouse immunogen, to ensure that the mice are not killed by a pyrogenic response to this toxin. An aliquot of the final antigen should be analyzed for bacterial endotoxin contamination with a Limulus Amoebocyte Lysate assay (LAL; QCL-100, Cambrex Bio Science, Walkersville, MD).

2.3.2   EUKARYOTIC PROTEINS

Recombinant proteins derived from eukaryotic cells have the potential advantage of being subjected to the same posttranslational modifications, such as glycosylation, as the native protein. Such a recombinant antigen may present itself in a more native conformation and lead to antibodies that are useful as reagents for flow cytometry and cell capture. Moreover, antibodies that recognize native antigen are more likely to be capable of blocking or stimulating the normal activity of the antigen *in vivo* and therefore have potential therapeutic value. For these reasons, recombinant proteins derived from eukaryotic cells are an attractive source of protein immunogen.

Vectors are now available that enable tagged fusion proteins to be expressed and purified in a manner similar to that described for prokaryotic recombinant proteins. However, differences in promoter activity and vector copy number often result in expression at much lower levels than those seen in *E. coli*. Furthermore, culture of eukaryotic cells, especially baculovirus and mammalian cells, tends to be more expensive and time consuming, even when eukaryotic batch fermentation is an option. In spite of these negatives, eukaryotic recombinant proteins should be considered in any strategy that requires recognition of a native protein, especially epitopes that are known to be influenced by posttranslational modification.

One problem associated with the expression and purification of eukaryotic recombinant protein is that cells often have to be lysed for protein to be isolated and purified. Costs of producing large quantities of eukaryotic-derived protein are reduced considerably if the cells secrete the recombinant protein into the culture medium, where it can be continually collected and purified. Systems have been developed where immunoglobulin (Ig) fusion proteins are expressed and secreted continually into the culture supernatant, mimicking hybridoma antibody production. Various cell-surface receptors, adhesion molecules, growth factors, and orphan ligands/receptors have been expressed as fusion proteins by linking them to the Fc region of human, or mouse, Ig heavy chain by PCR and cloning into a mammalian vector using a promoter such as that from cytomegalovirus (CMV).[4] Fc-fusion proteins can be purified by protein A or protein G chromatography from supernatants of transfected cells adapted into protein-free medium.[5] Fc-fusion proteins, also called immunoadhesins, have provided useful tools for both *in vitro* and *in vivo* studies to dissect molecular interactions that control biologic functions and have provided molecules for the treatment of human disease. The drug Enbrel (Amgen, Thousand Oaks, CA) is made from the fusion of two naturally occurring soluble human 75-kDa tumor necrosis factor receptors linked to an Fc portion of an IgG1. This genetically engineered soluble tumor necrosis factor receptor supplements the body's regulatory process by inhibiting tumor necrosis factor and is used in the treatment of rheumatoid arthritis and other immune diseases. Efficient mammalian cell expression, relative ease of purification, and maintenance of a biologically active molecule make immunoadhesins an attractive antigen source for use as an immunogen.

2.3.2.1   Immunization Schedule for Recombinant Proteins

2.3.2.1.1   Mice: BALB/c, Female Mice, 6–8 Weeks of Age

Use 25–50 μg protein i.p. in 10 μg of CFA (0.2 ml total volume in PBS); at least two boosts, 3–4 weeks apart, with the same dose of protein i.p. in 10 μg IFA (0.2 ml) or other suitable adjuvant (see "Adjuvants" chapter in this volume). Spleens should be harvested for fusion 3–4 days after the final boost.

2.3.2.1.2   Rabbits: New Zealand White, Female Rabbits,
8 Weeks of Age

Prime with 0.4 mg protein + 0.1 mg muramyl dipeptide (MDP adjuvant), to 1 ml with PBS, + 1 ml CFA, s.c. (four sites: two inguinal, two axillary); first boost (+ 4 weeks) with 0.2 mg protein, to 1 ml with PBS, + 1 ml IFA, s.c. (four sites: two inguinal, two axillary); second boost and subsequent boosts every 4 weeks, 0.1 mg protein, in 50–100 μl PBS, no adjuvant, i.v.; ~20 ml sera production bleed 5–7 days after boost.

2.4   WHOLE-CELL IMMUNOGENS

Whole cells present an effective alternative for generating antibodies to an antigen in its native form if it is usually expressed on the cell surface. Exogenous expression of a membrane protein, in a cell line that is of the same species (and strain if a mouse) used for antibody production, should result in a recombinant cell that is immunologically unreactive apart from the expressed recombinant antigen. Such a cell line would be expected to posttranslationally modify the recombinant antigen like the native protein and present the antigen at the cell surface in a conformation that is relevant.

Vectors pCEP4 and pREP4 are designed for high-level, constitutive expression from either the CMV or Rous sarcoma virus long-terminal repeat enhancer-promoters (Invitrogen). Both vectors are excellent for transient high-level episomal expression in primate cells. Stable expression lines can be generated by transfecting with linearized vector and appropriate drug selection.

Other vectors (e.g., pcDNA, Invitrogen) are designed for rapid cloning and constitutive expression in a variety of mammalian cells. Vectors are available that carry different selectable markers, with or without tagged fusion proteins (6xHis and MYC), and with other features, including secretion of the expressed protein or inducible expression. Many vectors employ the CMV enhancer-promoter for high-level expression in many mammalian cells. Where the CMV promoter is not optimal, vectors employing the human EF-1α or UbC promoter can be selected.

Regardless of vector choice, the fragment to be expressed should be amplified by PCR and cloned into the vector with standard recombinant DNA technology. Where possible, full-length genes should be cloned to ensure appropriate trafficking to the cell surface, post-translational modifications, and correct folding. The addition of tags to the protein should be limited to intracellular regions to avoid potential effects the tag may have on conformation, yet still providing an epitope to enable the protein to be detected. Expression vectors can be stably introduced into mammalian cells by techniques that include calcium phosphate coprecipitation,[6,7] electroporation,[8,9] and transfection with cationic lipids[10] and clones selected for vector integration with appropriate antibiotics. Cell lines should be fully characterized for appropriate expression of cell-surface protein, by Western blot analysis and flow cytometry where possible, before being used as a whole-cell immunogen. After lines are established and characterized, aliquots should be stored in liquid nitrogen and used to replace cultures that show any loss of expression.

Retroviruses provide an efficient means for delivery of single DNA expression constructs to a wide range of mammalian cell types, especially where cell lines are not easily transfected by other techniques. Retroviral vectors have the advantage that they integrate into the host genome allowing for long-term stable expression of the recombinant protein. Although retrovirus vectors provide a relative easy way to deliver genes stably to mammalian cells, their use does require special considerations and the protocols are relatively time consuming. Use of retroviral vectors necessitates adherence to standard guidelines to ensure safety; these include use of a biosafety level II tissue culture facility (USA), use of a class II laminar flow hood with HEPA filter, and performance of work in an area with limited access and with the use of double gloves, lab coats, and face protection. Guidelines for use of retrovirus may vary according to the institution. Before using any retroviral expression system, consult an institutional safety officer for specific requirements in your facility.

The laboratory of Dr. Garry Nolan at Stanford University has developed the Phoenix retroviral delivery system, based on Moloney murine leukemia virus that allows for delivery of genes to most dividing mammalian cell types. The system comes as either an *ecotropic* packaging system (capable of delivering genes to dividing murine or rat cells) or an *amphotropic* system (capable of delivering genes to dividing cells of most mammalian species, including humans). Materials, tutorials, and protocols for retroviral vector construction, viral production, and stable infection are available through the informative Nolan laboratory web site (http://www.stanford.edu/group/nolan/retroviral_systems/retsys.html). Retroviral systems are commercially available from Clontech (Mountain View, CA).

2.4.1 IMMUNIZATION SCHEDULE FOR WHOLE-CELL IMMUNOGEN

2.4.1.1 Mice: BALB/c, Female Mice, 6–8 Weeks of Age

Intact cells are usually highly immunogenic and need no adjuvant. Cells are washed three times in PBS and resuspended in PBS at an appropriate concentration. For a mouse immunization, a typical dose can be from $2–50 \times 10^6$ cells injected i.p. (0.4 ml, 10^8 cells/ml), although i.v. has also been used successfully (0.1 ml, 10^8 cells/ml). Boosts should be with a similar dose at 3- to 8-week intervals with a spleen harvested for fusion 2–4 days after the final boost.

2.5 GENETIC IMMUNIZATION

Recent advances in antibody based therapies require the development of efficient methodologies for generating antibodies against the native conformation of membrane proteins. Such proteins are usually problematic from an expression standpoint and conservation of native conformation is difficult to achieve. In these cases, or in any situation in which recombinant proteins cannot be expressed, the use of genetic vaccination techniques, such as plasmid DNA or viral vectors, can be used to generate antibodies in mice or in larger animal species.

2.5.1 PLASMID DNA

Generation of antibodies after immunization with plasmid DNA expressing a gene or cDNA of interest has obvious attractions because of the ease of generating an immunogen, without the need for expressing protein. The specificity of the induced humoral response is excellent because of lack of *E. coli* or mammalian host cell contaminants often copurified with recombinant proteins. It is also a highly flexible approach, allowing for use of truncated constructs, addition of secretion signal sequences, or coexpression with adjuvant molecules, such as cytokine genes, to enhance the humoral immune response. Furthermore, this technique lends itself to immunization of the same animal with multiple genes or even expression libraries[11] without apparent antigenic competition.

The major disadvantage associated with plasmid DNA immunization to generate polyclonal or monoclonal antibodies is the relatively low antibody titer generated, sometimes even after multiple immunizations. This observation is directly related to the low amounts of protein expressed *in vivo*, in the picogram to nanogram range, compared to tens of micrograms usually injected as recombinant protein immunogen. These low polyclonal antibody titers can in turn correlate with low yield of hybridoma generation for monoclonal antibody development. However, the benefits in some cases may outweigh this disadvantage, and there are many reports in the literature of successful use of plasmid DNA immunization for the generation of mouse and rabbit polyclonal and mouse monoclonal antibody generation to a wide variety of antigens and to both conformational and linear epitopes.[12–18]

2.5.1.1   Construction of Expression Vector

A detailed description of molecular biology procedures involved in PCR amplification and DNA cloning is beyond the scope of this review, but these standard procedures can be obtained from other resources.[3] Briefly, the gene of interest should be isolated as a cDNA sequence by reverse transcription-PCR with appropriate primers and cloned into a mammalian expression vector. There are various mammalian expression plasmids commercially available with strong mammalian promoters, such as the CMV promoter, and a poly(A) sequence, such as that of bovine growth hormone. Two such vectors suitable for immunization purposes are VAX1 (Invitrogen) and pCI (Promega Corporation, Madison, WI). After the recombinant vector has been generated, amplified by transformation into *E. coli*, and confirmed for correct orientation of the gene of interest by restriction digest, it is a good idea to confirm expression by transient transfection into a cell line, such as COS cells. However, the ability to detect the expressed protein may depend upon the availability of detection antibodies for immunocytochemistry on fixed cells or Western blot or ELISA assay on cell lysates. In the absence of such reagents, it may be necessary to use only a diagnostic restriction digest or sequencing of the insert as sufficient justification to move forward with immunization.

   The plasmid is produced in large scale by overnight culture of the transformed *E. coli*, and subsequent purification from bacterial lysates. Endotoxin-free plasmid purification kits from Qiagen are routinely used to purify plasmid of a suitable quality for *in vivo* immunization. The scale of the kit can be matched with the desired yield of plasmid DNA, from 500 μg to 10 mg, which in turn influences the volume of original bacterial overnight culture for processing. The final purified plasmid DNA should be resuspended in sterile saline for injection or in PBS to a final concentration of about 1–5 mg/ml. The material can be stored long term in aliquots at –20°C, or short term at 4°C.

2.5.1.2   Immunization with Plasmid DNA

There are many variations in the methods used to immunize with plasmid DNA. A standard protocol for immunizing mice and rabbits is provided.

2.5.1.2.1   Mice: BALB/c, Female Mice, 6–8 Weeks of Age

Dose: 100 μg per immunization.

Route and volume: Intramuscular delivery of 50-μl volumes into each tibialis anterior (shin) muscle of the hind legs with a tuberculin syringe and 28-gauge needle, for a total volume of 100 μl. Alternatively, 100 μl can be delivered into a single site intradermally at the base of the tail. Subcutaneous delivery of DNA is not effective.

Schedule: Immunize four times, no less than 10 days apart, with a more optimum frequency being every 3 weeks. Monitor serum titers before each new immunization. Collect spleens for hybridoma generation at 3–4 days after the last immunization.

2.5.1.2.2   Rabbits: New Zealand White, Female Rabbits,
            8 Weeks of Age

Dose: 1 mg per immunization.
Route and volume: Intramuscular delivery of 200-µl volumes into each quadriceps muscle of the hind legs or intradermal delivery into shaved skin on the back, distributed at 100 µl per site over numerous sites to achieve the full dose.
Schedule: Immunize at least four times, or more as necessary, at monthly intervals. Monitor serum titers at prebleed and at 3 weeks after each immunization.

2.5.1.3   Improving the Efficiency of DNA Immunization

A number of refinements of this basic approach can be employed to increase the efficiency of DNA immunization.

- Levels of protein expression from the plasmid vector can be improved substantially by the addition of the CMV immediate early gene intron A element downstream of the CMV enhancer/promoter.[19]
- Addition of a secretion signal sequence upstream of the gene of interest has been reported to increase *in vivo* antibody responses to a significant degree.[20,21]
- Uptake of the delivered plasmid by cells *in vivo* can be improved by delivery using a gene gun particle-bombardment device,[13,22] such as the Helios gene gun available from Bio-Rad Laboratories (Hercules, CA). Immunization protocols using the gene gun require 100-fold less DNA to elicit a comparable antibody response to DNA delivered by needle into muscle or skin.
- The technique of delivering DNA directly into spleen cells *in vivo*, either by needle injection or by gene gun inoculation, has been used to enhance the efficiency with which expressed protein is presented to B cells. This technique has been used to generate polyclonal and monoclonal antibodies in mice after a single immunization.[18,23] The optimal time point for harvest of spleen cells for generation of hybridomas was 5 days postintrasplenic immunization, with both IgM and IgG antibody isotypes being produced, making this an extremely rapid protocol.[18]
- Finally, the addition of adjuvants, such as coinjection with a plasmid expressing granulocyte/macrophage colony-stimulating factor,[13] or by adsorption of the plasmid DNA onto alum,[24] also improves antibody titers to expressed transgenes *in vivo*.

2.5.1.4   The Antibody Response

Antibody titers of 1/10,000 to 1/100,000 have been obtained in mice and rabbits with DNA immunization protocols.[14,15,17] The profile of antibody subclasses induced by DNA immunization in mice is generally a mixed IgG profile, with IgG2a predominating. However, both the overall titer and the balance of subclasses within the polyclonal response can be influenced by the antigen itself, the method of delivery, mouse strain, and the presence or absence of a secretion signal sequence.[25–27] Monoclonal antibodies of IgM, IgG1, and IgG2a isotypes have been derived using this method of immunization, and their affinities are generally comparable with those obtained using protein in adjuvant or viral vector immunogens.[15,18,28]

2.5.1.5   Boosting DNA with Other Immunogens

As mentioned previously, DNA immunization alone may generate insufficient polyclonal antibody titers and low efficiencies of hybridoma generation. In this case, the antibody response can be boosted by immunizing with a second immunogen, such as cells expressing the antigenic protein[29–31] or recombinant viral vectors.[32] This strategy not only boosts overall antibody titers, but also appears to synchronize B-cell activation and thus improve the numbers of stable hybridomas obtained.[29] Use of these types of boosting agents maintains the antigen in native conformation, while still avoiding the need to express recombinant protein. A particular advantage of this combination approach is the ability of DNA priming immunizations to focus the humoral response on the gene of interest, rather than on any host cell proteins or viral proteins delivered in the boosting immunization.[29,33]

2.5.2   ADENOVIRUS

As is the case with plasmid DNA, immunization with a viral vector, such as adenovirus, is particularly advantageous for generating antibodies against cell surface or other conformational proteins. Generating recombinant adenovirus is a more complex proposition than production of recombinant plasmid constructs, but in general, induced antibody titers are higher because of the *in vivo* expression of greater amounts of protein.[28,34] As stated previously, a recombinant adenovirus vector can be used as a sole immunogen or in combination with DNA priming immunizations.

2.5.2.1   Construction of Recombinant Adenovirus

There are commercial adenovirus vectors available that simplify generation of a recombinant virus. The two systems described here are based on human adenovirus serotype 5, and are replication incompetent because of the deletion of the viral E1 gene. The E3 gene is also deleted to prevent this gene's role in evasion of the host immune response.

- The AdEasy system[35] (Qbiogene, Irvine, CA) uses a transfer vector for cloning of the gene of interest. This vector is then cotransformed into *E. coli* with the viral DNA plasmid where recombination between the two occurs. Recombinants are then screened for kanamycin resistance. Viral particles are produced by transfection of the resulting DNA into the permissive HEK293A cell line supplied with the kit, which expresses the E1 gene product. The virus can be amplified, and the stock titrated by limiting dilution analysis on the same cell line, with cytopathic effect as the endpoint.
- Invitrogen supplies the ViraPower adenovirus vector system. This system uses Gateway Technology to facilitate the speed and accuracy of cloning

into the adenoviral vector. The E1 expressing HEK293A producer cell line is provided with the kit to allow amplification and titration of recombinant viral stocks.

2.5.2.2   Purification and Titration of Adenovirus

Methods for purification of recombinant adenovirus by the standard method of double cesium chloride gradient centrifugation can be found in earlier work.[36] Alternatives to the cesium chloride purification exist as chromatographic purification kits available from BD Biosciences (San Jose, CA; BD Adeno-X Virus Purification Kits) and from Vivascience AG (Hannover, Germany; Vivapure AdenoPACK 100), which do not require ultracentrifugation. The resulting viral stock should be resuspended in 10 mM Tris, pH 8, 2 mM $MgCl_2$, and 5% sucrose, aliquoted, and stored at $-80°C$. Up to three freeze-thaw cycles result in relatively little loss in titer, but beyond this, the remaining stock should be discarded. Infectious viral titers are determined by limiting dilution analysis on HEK293 cells, with cytopathic effect as an endpoint.[36]

2.5.2.3   Immunization with Adenovirus

In the absence of the E1 gene supplied in *trans*, these recombinant adenoviruses are replication incompetent and will not produce progeny particles in immunized animals. However, both in the laboratory and in the animal facility, the material should be handled under BL-2 conditions. Adenoviral stocks are quantified by total particle counts per milliliter, not all of which are viable, or as infectious units (plaque forming units or pfu) per milliliter. For the purposes of immunization, it is more relevant to use pfu/ml to calculate dose, and this will be used in the following guidelines.

2.5.2.3.1   *Mice: BALB/c, Female Mice, 6–8 Weeks of Age*

Dose: The optimal dose will vary with specific antigens, but it is reasonable to test responses induced after inoculation in the range of 10^6 to 10^8 pfu per mouse, diluted to the appropriate concentration in PBS. Thawed adenovirus should be stored on ice until inoculation.

Route and volume: Intramuscular delivery of 50 µl into each tibialis anterior (shin) muscle of the hind legs using a tuberculin syringe and 28-gauge needle, for a total volume of 100 µl. Alternatively, the 100-µl volume can be delivered into a single site intradermally at the base of the tail. Both intradermal and intramuscular routes for immunization with adenovirus induce a more robust antibody response than the s.c. route.

Schedule: Immunize two times 3 weeks apart. Further immunizations will not boost the response because of virus-neutralizing antibodies induced in the host. Collect spleens for hybridoma generation at 3–4 days after the last immunization.

2.5.2.3.2   *Rabbits: New Zealand White, Female Rabbits,*
           8 Weeks of Age

Dose: 10^7–10^9 pfu per rabbit, diluted to the appropriate concentration in PBS. Thawed adenovirus should be stored on ice until inoculation.

Route and volume: Intramuscular delivery of 200-µl volumes into each quadriceps muscle of the hind legs or intradermal delivery to shaved skin at 100 µl per site distributed over multiple sites as necessary to achieve the full dose.

Schedule: Immunize two times 3 weeks apart. Further immunizations will not boost the response because of virus-neutralizing antibodies induced in the host.

REFERENCES

1. Stoffel, K.H.W., TMbase—a database of membrane spanning proteins segments. *Biol. Chem. Hoppe-Seyler*, 374, 166, 1993.
2. Janin, J., Surface and inside volumes in globular proteins. *Nature*, 277, 491, 1979.
3. Sambrook, J. and Russell, D.W., *Molecular Cloning: A Laboratory Manual*, 3rd ed., Cold Spring Harbor Laboratory Press, Cold Spring Harbor, NY, 2001.
4. Ashkenazi, A. and Chamow, S.M., Immunoadhesins as research tools and therapeutic agents. *Curr. Opin. Immunol.*, 9, 195, 1997.
5. Sakurai, T., Roonprapunt, C., and Grumet, M., Purification of Ig-fusion proteins from medium containing Ig. *BioTechniques*, 25, 382, 1998.
6. Chen, C. and Okayama, H., High-efficiency transformation of mammalian cells by plasmid DNA. *Mol. Cell. Biol.*, 7, 2745, 1987.
7. Wigler, M., et al., Transfer of purified herpes virus thymidine kinase gene to cultured mouse cells. *Cell*. 11, 223, 1977.
8. Chu, G., Hayakawa, H., and Berg, P., Electroporation for the efficient transfection of mammalian cells with DNA. *Nucleic Acids Res.*, 15, 1311, 1987.
9. Shigekawa, K. and Dower, W.J., *Electroporation of eukaryotes and prokaryotes: a general approach to the introduction of macromolecules into cells.* Biotechniques, 1988. 6(8): p. 742–51.
10. Felgner, P.L. and Ringold, G.M., Cationic liposome-mediated transfection. *Nature*, 337, 387, 1989.
11. Yero, C.D., et al., Immunization of mice with Neisseria meningitidis serogroup B genomic expression libraries elicits functional antibodies and reduces the level of bacteremia in an infant rat infection model. *Vaccine*, 23, 932, 2005.
12. Barry, M.A., Barry M.E., and Johnston S.A., Production of monoclonal antibodies by genetic immunization. *Biotechniques*, 16, 616, 1994.
13. Chambers, R.S. and Johnston, S.A., High-level generation of polyclonal antibodies by genetic immunization. *Nat. Biotechnol.*, 21, 1088, 2003.
14. Chowdhury, P.S., Gallo, M., and Pastan I., Generation of high titer antisera in rabbits by DNA immunization. *J. Immunol. Methods*, 249, 147, 2001.
15. Leinonen, J., et al., Characterization of monoclonal antibodies against prostate specific antigen produced by genetic immunization. *J. Immunol. Methods*, 289, 157, 2004.
16. Gardsvoll, H., et al., Generation of high-affinity rabbit polyclonal antibodies to the murine urokinase receptor using DNA immunization. *J. Immunol. Methods*, 234, 107, 2000.
17. Pass, J., et al., Generation of antibodies to the urokinase receptor (uPAR) by DNA immunization of uPAR knockout mice: membrane-bound uPAR is not required for an antibody response. *Scand. J. Immunol.*, 58, 298, 2003.
18. Velikovsky, C.A., et al., Single-shot plasmid DNA intrasplenic immunization for the production of monoclonal antibodies. Persistent expression of DNA. *J. Immunol. Methods*, 244, 1, 2000.

19. Chapman, B.S., et al., Effect of intron A from human cytomegalovirus (Towne) immediate-early gene on heterologous expression in mammalian cells. *Nucleic Acids Res.,* 19, 3979, 1991.
20. Svanholm, C., et al., Enhancement of antibody responses by DNA immunization using expression vectors mediating efficient antigen secretion. *J. Immunol. Methods,* 228, 121, 1999.
21. Li, Z., et al., Immunogenicity of DNA vaccines expressing tuberculosis proteins fused to tissue plasminogen activator signal sequences. *Infect. Immun.,* 67, 4780, 1999.
22. Kilpatrick, K.E., et al., Gene gun delivered DNA-based immunizations mediate rapid production of murine monoclonal antibodies to the Flt-3 receptor. *Hybridoma*, 17, 569, 1998.
23. Moonsom, S., Khunkeawla, P., and Kasinrerk, W., Production of polyclonal and monoclonal antibodies against CD54 molecules by intrasplenic immunization of plasmid DNA encoding CD54 protein. *Immunol. Lett.*, 76, 25, 2001.
24. Kwissa, M., et al., Codelivery of a DNA vaccine and a protein vaccine with aluminum phosphate stimulates a potent and multivalent immune response. *J. Mol. Med.*, 81, 502, 2003.
25. Daly, L.M., et al., Innate IL-10 promotes the induction of Th2 responses with plasmid DNA expressing HIV gp120. *Vaccine*, 23, 963, 2005.
26. Feltquate, D.M., et al., Different T helper cell types and antibody isotypes generated by saline and gene gun DNA immunization. *J. Immunol.*, 158, 2278, 1997.
27. Haddad, D., et al., Differential induction of immunoglobulin G subclasses by immunization with DNA vectors containing or lacking a signal sequence. *Immunol. Lett.*, 61, 201, 1998.
28. Guo, J., et al., Insight into antibody responses induced by plasmid or adenoviral vectors encoding thyroid peroxidase, a major thyroid autoantigen. *Clin. Exp. Immunol.*, 132, 408, 2003.
29. Nagata, S., Salvatore G., and Pastan I., DNA immunization followed by a single boost with cells: a protein-free immunization protocol for production of monoclonal antibodies against the native form of membrane proteins. *J. Immunol. Methods*, 280, 59, 2003.
30. Tearina Chu, T.H., et al., A DNA-based immunization protocol to produce monoclonal antibodies to blood group antigens. *Br. J. Haematol.*, 113, 32, 2001.
31. Costagliola, S., et al., Genetic immunization against the human thyrotropin receptor causes thyroiditis and allows production of monoclonal antibodies recognizing the native receptor. *J. Immunol.*, 160, 1458, 1998.
32. Krasemann, S., Jurgens T., and Bodemer W., Generation of monoclonal antibodies against prion proteins with an unconventional nucleic acid-based immunization strategy. *J. Biotechnol.*, 73, 119, 1999.
33. Yang, Z.Y., et al., Overcoming immunity to a viral vaccine by DNA priming before vector boosting. *J. Virol.,* 77, 799, 2003.
34. Schwarz-Lauer, L., et al., The cysteine-rich amino terminus of the thyrotropin receptor is the immunodominant linear antibody epitope in mice immunized using naked deoxyribonucleic acid or adenovirus vectors. *Endocrinology*, 144, 1718, 2003.
35. He, T.C., et al., A simplified system for generating recombinant adenoviruses. *Proc. Natl. Acad. Sci. U. S. A.*, 95, 2509, 1998.
36. Tollefson, A., Hermiston, T.W., and Wold, W.S.M., Preparation and titration of CsCl-Banded adenovirus stock, in *Adenovirus Methods and Protocols*, 3rd ed. Humana Press: Totowa, NJ, 1999.

3 Adjuvants

Jory Baldridge, Paul Algate, and Sally Mossman

CONTENTS

3.1 INTRODUCTION

Immunologic adjuvants nonspecifically enhance or modify the immune response to coadministered antigens. Thus, antibody production is more vigorous than if antigen is administered alone. A variety of mechanisms contribute to this adjuvant-induced amplification: (1) a depot of immunogen forms, resulting in the slow release and presentation of antigen to the immune system over an extended period; (2) the immune stimulus is optimized by focusing a general immunostimulant and antigen in the same microenvironment where they can interact with antigen-presenting cells and lymphocytes simultaneously; and (3) cells of the immune system are nonspecifically activated, thus facilitating those interactions that favor the production of antibody. This chapter will deal primarily with commercially available adjuvants designed for the production of antisera in the research setting (Table 3.1).

When selecting an adjuvant for the generation of polyclonal antisera or for the induction of monoclonal antibodies, a number of factors should be considered,

TABLE 3.1
Sources for Adjuvants

Adjuvant	Supplier/Address	Web Site	Phone
Adjuprime	Pierce Chemical Co. P.O. Box 117 Rockford, IL 61105	www.piercenet.com	1-800-874-3723
Abisco (ISCOMs)	Isconova AB Uppsala Science Park SE-751 83 Uppsala, Sweden	www.isconova.se	+46 18 57 24 00
Aluminum-based –Alhydrogel (alum) –Adju-Phos	Brenntag Biosector A/S Elsenbakken 23 DK-3600 Frederikssund, Denmark	www.brenntag- biosector.com	+45 47 38 47 00
Freund's adjuvants Complete (CFA) and incomplete (IFA)	Several sources		
Gerbu adjuvant	GERBU Biochemicals GmbH Am Kirchwald 6 69251 Gaiberg Germany	www.gerbu.de	(0049)6223/951 30
Imject (alum)	Pierce Chemical Co. P.O. Box 117 Rockford, IL 61105	www.piercenet.com	1-800-874-3723
Montanide	ISA Seppic 30 Two Bridges Road, Suite 225 Fairfield, NJ 07004-1530	www.seppic.com	1 (937) 882-5597 1-201-882-5597
Quil A	Accurate Chemical & Scientific Corporation 300 Shames Drive Westbury, NY 11590	www.accurate chemical.com	1-800-645-6264 or 1-800-255-9378
Ribi adjuvant system	Corixa Corporation 553 Old Corvallis Road Hamilton, MT 59840	www.corixa.com	1-800-548-7424
TiterMax	TiterMax USA 6971 Peachtree Industrial Boulevard, Suite 103 Norcross, GA 30092	www.titermax.com	1-800-345-2987

including: (1) humane treatment of the animals, (2) institutional policies, (3) characteristics of the desired antibody, (4) ease of use, and (5) cost. Generally, the water-in-oil emulsions, such as Freund's, are among the most potent of the commercially available adjuvants, cause the most irritation or injection site reactogenicity, and are the most difficult to use. These emulsions can cause denaturization of protein antigens, thus preventing the induction of antibodies to conformational epitopes. To avoid the toxicity induced with Freund's adjuvants, oil-in-water emulsions were developed that contain minimal amounts of oil. Oil-in-water emulsions tend to be less potent, less reactogenic, easier to use, and less likely to denature protein antigens than the Freund's adjuvants. Additional immunostimulants are often added to increase the potency of such adjuvants. Aluminum salts promote predominantly Th-2 type antibody responses and are generally the least potent and least reactogenic adjuvants, but very easy to use.

There is no single adjuvant that is right for all situations, but most of the commercially available adjuvants mediate adequate antibody titers if used appropriately. For the selection of an adjuvant, readers are encouraged to discuss their needs with local Institutional Animal Care and Use Committees, experienced colleagues, and published comparisons.[1-6]

The mostly widely recognized and used adjuvants are the water-in-oil emulsions initially developed by Freund.[7,8] These emulsions are generated by mixing an aqueous solution of soluble antigen with an equal volume of oil through a process known as emulsification. During this procedure the antigen-containing water droplets become entrapped in the oil, forming particulates in a very viscous emulsion. For adjuvant activity, it is critical that these thick emulsions remain stable and that the components do not separate after mixing. After administration into the host's tissue, this viscous mixture acts as a depot of antigen.

The potency of water-in-oil emulsions can be increased by incorporating immunostimulants. The classic example is complete Freund's adjuvant (CFA), which consists of the aqueous antigen solution, mineral oil, an emulsifying agent, and heat-killed *Mycobacterium tuberculosis*, a very potent immunomodifier. Incomplete Freund's adjuvant (IFA) is identical to CFA except that the *M. tuberculosis* is omitted. Freund's adjuvants have been used and continue to be used extensively with antigens to stimulate the production of high-titer sera.

There are, however, complicating factors that need to be considered. The emulsions are rather tedious to generate and can contribute to the degradation of protein antigens.[9] Freund's adjuvants are inherently toxic, frequently resulting in granulomas, sterile abscesses, and ulcerations after administration.[10,11] To reduce the extent of these complications, most protocols recommend using CFA for the primary immunization only and IFA for all subsequent vaccinations. Furthermore, because of increasing concerns for research animals, many animal care facilities have banned or restricted the use of CFA, while encouraging the use of other adjuvants. Several alternatives are now available that are generally as effective as CFA/IFA for enhancing antibody production and have greatly reduced toxicity.[5,12,13]

TiterMax (TiterMax USA, Norcross, GA) is a water-in-oil emulsion developed specifically for the production of antibodies in research animals. TiterMax has three primary ingredients: a block copolymer CRL 89-41, squalene (a metabolizable oil),

and microparticulate silica. The nonionic block copolymer is a synthetic immunomodifier that acts in part by adhesion of protein antigens to oil, thereby enhancing
delivery of high concentrations of antigen to antigen-presenting cells. By using
squalene instead of mineral oil, TiterMax emulsions are much less toxic than
Freund's.[13] Also, emulsions of TiterMax with antigen are easier to prepare and
more stable than Freund's emulsions.[13]

The Ribi adjuvant system (RAS) was developed as an alternative to Freund's
for producing antibodies in research animals. To reduce the toxicity found in Freund's
water-in-oil emulsions, the RAS emulsions were designed as oil-in-water emulsions
with a final concentration of only 2% metabolizable squalene in oil.[14] These emulsions include one or more of the following bacterial-derived immunostimulators: (1)
synthetic trehalose dimycolate, (2) monophosphoryl lipid A, and (3) cell-wall skeleton from *Mycobacterium phlei*. The immunostimulants nonspecifically induce
immune activation resulting in the production of inflammatory cytokines that facilitate antibody production. The RAS emulsions are thin fluids that are easy to use,
compared with the thick, viscous Freund's water-in-oil emulsions. Three variations
of RAS are available for use, depending on the antigen and the animal to be
immunized.

Gerbu adjuvants were developed as a replacement for Freund's adjuvants. As
with the RAS, Gerbu adjuvants rely on an immunostimulant to boost the antibody
response to accompanying antigens. N-Acetyl-glucosaminyl-N-acetylmuramyl-L-
alanyl-D-isoglutamine or glucosaminylmuramyldipeptide is a glycopeptide isolated from the cell walls of *Lactobacillus bulgaricus*. This immunomodulator
activates cells of the immune system, resulting in the production of cytokines that
facilitate the production of antibodies. Gerbu adjuvants are aqueous in nature and
easy to use.

Aluminum salt adjuvants are the predominant adjuvants used in many human
vaccines. Proteins readily adsorb to the surface of the aluminum. When injected,
the alum-adsorbed antigens form a depot that is slowly released to the immune
system over several days. Generally, aluminum salts are considered to be weaker
adjuvants but, because of their mild reactogenicity, they provide a safe option.

ISCOMs, or immune-stimulating complexes, are micellular structures composed
of saponins, cholesterol, and phospholipids. ISCOMs have been used as experimental
adjuvants for a number of years, but Isconova has recently adapted an ISCOM
product called Abisco as ready-to-use adjuvants for antibody production in laboratory animals.

3.2 WORKING WITH ADJUVANTS

3.2.1 GENERAL GUIDELINES

a. Adjuvants designed for the production of antisera induce very strong
   inflammatory reactions on introduction into tissues. Protective eyewear
   and gloves should be worn when working with these materials.
b. Adjuvants are sold as sterile reagents and should be handled aseptically.
   Any introduction of contaminants during vaccine preparation can

adversely affect the desired antisera. The contaminants can be more immunogenic than the antigen, resulting in the production of antisera specific for the contaminant rather than the antigen. If the adjuvant becomes contaminated, it should be discarded.

c. Excessive amounts of adjuvant can be immunosuppressive, resulting in low yields of antibody. Adjuvants, therefore, should be used at or near doses recommended by the manufacturer.

d. For peptides or other poorly immunogenic antigens, it can be beneficial to covalently couple them to a helper protein before using them with adjuvants. Several protocols for attaching helper proteins to antigens can be found in Chapters 2 and 5 of this handbook and also in *Current Protocols in Immunology*.[15]

e. Higher titers of antibody are normally generated by immunizing multiple times with small amounts of antigen, rather than by immunizing one time with a large dose of antigen. This may be a consideration when antigen is scarce.

f. Some adjuvants, including RAS and TiterMax, can increase the generation of antibodies of the immunoglobulin G2a subclass in mice.[14,16,17] Murine immunoglobulin G2a antibodies may be of interest because of their ability to activate complement and to participate in antibody-dependent cellular cytotoxicity both *in vitro* and *in vivo*. This subclass of antibodies binds avidly to the FcγRI found on macrophages.

3.2.2 USING FREUND'S ADJUVANT

Some research institutions and Institutional Animal Care and Use Committees restrict the use of CFA. Check with the appropriate animal care and use committee at your institution before using CFA.

CFA should be used for primary vaccinations only. Subsequent immunizations with CFA will lead to severe lesions at the site of injection. Therefore, IFA or another adjuvant should be used for all subsequent boosts.

3.2.2.1 Emulsification Using Syringes

CFA (or IFA) can be emulsified aseptically using two syringes, subsequently requiring only the addition of the hypodermic needle.

3.2.2.1.1 Materials

CFA (or IFA)
Safety eyewear
Two identical glass syringes with Luer-locks (3, 5, or 10 ml)
Double-hub Luer-lock emulsifying needle
Antigen in saline (or other appropriate diluent)
Latex gloves
Beaker with cool tap water

3.2.2.1.2   Procedure

1. Calculate the amount of antigen required and the volume desired per injection. Allow 5 to 100 µg of antigen per mouse in a volume of ≤ 200 µl per mouse for subcutaneous (s.c.) or intramuscular (i.m.) injections. For intraperitoneal (i.p.) injections, allow 200–500 µl per mouse. *Strong detergents will weaken the emulsion and should be avoided.* In general, prepare approximately 50% more vaccine than required to ensure enough of the final product.
2. Select syringe size according to total volume of materials to be added. The total volume of adjuvant and antigen should be approximately half the volume of one syringe; for example, when preparing 2.5 ml of vaccine, use 5-ml syringes for emulsification.
3. Warm CFA (or IFA) to 37°C before use and vortex for 1–2 min or invert vial several times by hand to uniformly resuspend the *M. tuberculosis*. Aspirate the required volume of CFA (or IFA) into a syringe. Connect the syringe to a double-hub emulsion needle with Luer-lock and expel any air. Excess air will hinder the formation of a stable emulsion.
4. Aspirate the aqueous antigen into the second syringe and expel any air. Excess air will hinder the formation of a stable emulsion.
5. Attach the antigen-containing syringe to the free end of the double-hub needle. Make certain both syringes are securely fastened to the double-hub emulsion needle via the Luer-locks.
6. Firmly depress the plunger forcing all of the antigen solution through the double-hub needle and into the CFA (or IFA). In an alternating pattern, continue to force the mixture from one syringe to the other (Figure 3.1).
7. Continue to emulsify as in the preceding step, until a stable emulsion is formed. This will take several minutes. Test for a stable emulsion by expelling a small drop onto the surface of some water in a beaker. The droplet should form a stable bead on the surface of water. If the droplet disperses across the water's surface, reconnect the syringes and continue the emulsification process.
8. Force the entire volume of emulsion into one syringe. Remove the double-hub needle and add a needle for injections. Alternatively, remove the empty syringe from the dual-connector and replace it with a sterile 1-ml syringe. The emulsion can now be transferred to the 1-ml syringe for injection.

3.2.2.2   Sonication of CFA (or IFA)

Aqueous solutions of antigen can be emulsified with CFA (or IFA) by the technique of sonication.

3.2.2.2.1   Materials

CFA (or IFA)
Sterile polypropylene tube
Safety eyewear

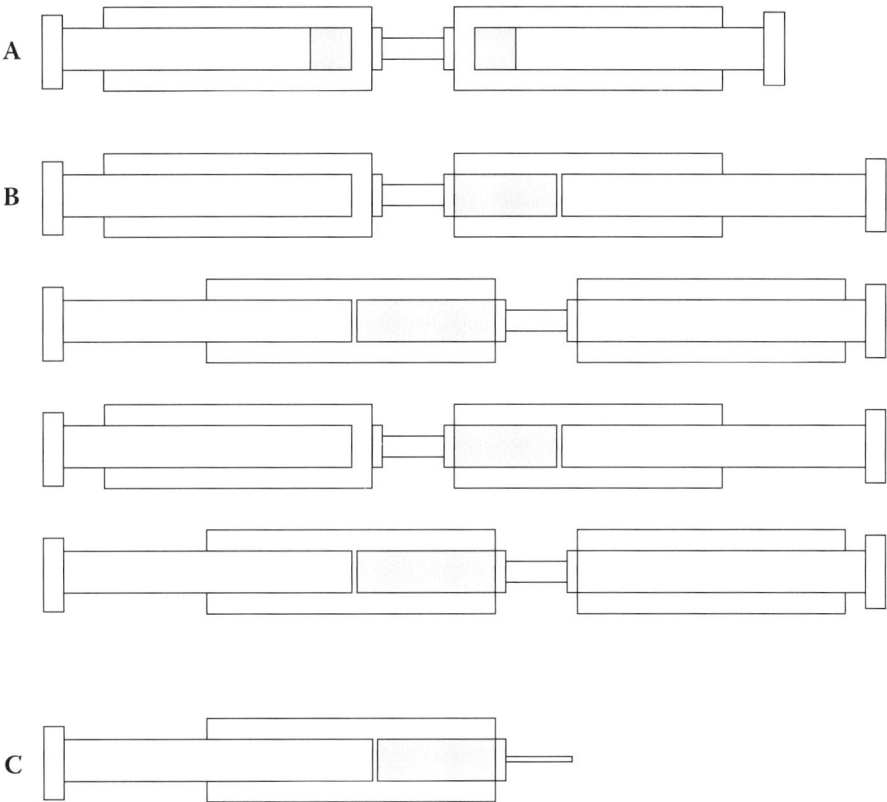

FIGURE 3.1 Emulsification of aqueous antigen and oil-based adjuvants using a dual-syringe method. (A) Attach the syringes securely to a Luer-lock, double-hulled connector. (B) First, push the aqueous antigen solution into the oil, then continue to force the mixture through the connector in an alternating pattern. (C) After a stable emulsion has formed, remove the connector and add a needle for injection or transfer the emulsion to a 1-ml syringe for injection.

Syringe for injection
Antigen in phosphate-buffered saline (PBS)
Probe sonicator
Latex gloves
Ice bath
Beaker with cool tap water

3.2.2.2.2 Procedure

1. Calculate the amount of antigen required and the volume desired per injection. Allow 5–100 µg of antigen per mouse, in a volume of ≤200 µl per mouse for i.m. or s.c. injections. For i.p. injections, allow 200–500 µl volume per mouse. *Detergents will weaken the emulsion and should be avoided.* In general, prepare approximately 50% more vaccine than required to ensure enough of the final product.

2. Warm CFA (or IFA) to 37°C before use and vortex vigorously for 1–2 min or invert vial several times by hand to uniformly resuspend the *M. tuberculosis*. Transfer CFA (or IFA) and an equal volume of aqueous antigen into a sterile polypropylene tube.

3. Sonication can generate a great deal of heat, which may lead to denaturation of protein antigens. To avoid overheating the mixture, place the tube on ice during sonication or between pulses. Begin emulsifying the mixture by pulsing for 5–20 sec at a time with the probe sonicator. Move the tube up or down as needed to ensure that entire volume becomes emulsified. Be sure to wear ear protection.

4. Invert tube slowly to determine the thickness of the emulsion. Continue sonicating with 10-sec pulses until a stable emulsion is formed. The process is complete when the emulsion remains immobile in the inverted tube or when a drop forms a stable bead on the surface of tap water.

5. Transfer to a 1-ml syringe for injection by aspirating the emulsion into the syringe. Alternatively, the transfer technique of Spack and Toaves[18] can be used. Obtain a sterile plunger from a syringe whose barrel diameter is the same as the tube in which the emulsification was performed. Swab the bottom of the emulsification tube with alcohol to sterilize, then pierce a hole in the bottom of the tube using a sterile 18-gauge needle. Hold the top of a sterile 1-ml syringe (without its plunger) directly beneath the hole in the emulsification tube. Insert a snug-fitting syringe plunger into the emulsification tube and push the emulsion into the 1-ml syringe.

3.2.3 PROTOCOL FOR USING TITERMAX

3.2.3.1 Materials

TiterMax
Safety eyewear
Syringe for injection
Antigen in PBS
Latex gloves
One three-way stopcock or double-hub emulsion needle
Two 3-ml all-plastic syringes (rubber plungers will stick to syringe barrels) or siliconized all-glass Luer-lock syringes.

3.2.3.2 Procedure

1. Warm TiterMax to room temperature and vortex for 30 sec or longer to ensure that TiterMax is a homogenous suspension.

2. For 1.0 ml of emulsion, load one syringe with 0.5 ml of TiterMax. Load second syringe with 0.25 ml of aqueous antigen. *Detergents will weaken the emulsion and should be avoided.* Retain 0.25 ml of aqueous antigen for further use.

3. Connect the two syringes to the three-way stopcock. It is important to begin the emulsification process by forcing the aqueous antigen into the

TiterMax. Force the materials back and forth through the stopcock for about 2 min (see Figure 3.1).

4. Force all of the emulsion into one syringe and disconnect the empty syringe. Load the empty syringe with the remaining 0.25 ml of antigen previously withheld.
5. Reconnect the syringes. Push the aqueous antigen into the emulsion first, then repeat the emulsification process for an additional 60 sec.
6. Push all the emulsion into one syringe, and disconnect the empty syringe.
7. Test for a stable emulsion by expelling a small drop of the mixture onto the surface of some water in a beaker. The droplet should form a stable bead on the surface of water. If the droplet flattens or disperses across the water's surface, reconnect the syringes to the three-way stopcock and continue the emulsification process until a stable emulsion is formed, which may take several minutes.

3.2.4 USING THE RIBI ADJUVANT SYSTEM

3.2.4.1 Materials

Ribi adjuvant
Antigen in saline
Safety eyewear
Vortex mixture
Latex gloves
Syringes

3.2.4.2 Procedure

1. To use the RAS, it is recommended that the antigen be solubilized at a concentration of 50–500 µg/ml in saline (or PBS). Some antigens require detergents to ensure solubility. In these cases, try to make the final detergent concentration 0.2% or less.
2. RAS is supplied in vials as an oil concentrate containing 500 µg of each immunostimulant. Each vial is intended for reconstitution to a final volume of 2 ml with an antigen-containing saline solution. Before use, warm the vial for approximately 10 min at 37°–45°C in a water bath or incubator. This will facilitate in the mixing of the oil into the antigen solution during the next step.
3. Inject 2 ml of antigen solution directly into the stoppered vial (leave the cap seal in place). Vortex vigorously for 2–3 min. Monophosphoryl lipid A, trehalose dimycolate, and *Mycobacterial* cell wall skeletons immunostimulants will be at a final concentration of 250 µg/ml in the antigen–adjuvant emulsions.

OR

Alternatively, if all the adjuvant will not be used initially, reconstitute the vial with 1 ml of PBS or saline alone (without antigen) and vortex vigorously. With a sterile syringe and 22-gauge needle, withdraw the required volume of adjuvant and mix with an equal volume of antigen-containing saline, then vortex vigorously. Any unused portion of adjuvant will be stable for several months when stored at 4°C (warm to 37°C before use).

Note: The antigen–adjuvant emulsions formed with RAS are not thick, viscous emulsions such as those formed with Freund's or TiterMax. They will readily disperse if applied to the surface of water.

1. Withdraw the emulsion into a sterile 1-ml syringe for injection and remove any air bubbles.
2. Vaccinate the animals by injecting the antigen–adjuvant emulsion accordingly:
   a. Mice: a total dose of 0.2 ml administered at two sites s.c. (0.1 ml/site) or the entire 0.2 ml i.p.
   b. Rats or guinea pigs: a total dose of 0.5 ml administered at two sites s.c. (0.2 ml/site) and 0.1 ml i.p.
   c. Rabbits: a total dose of 1 ml administered as 0.3 ml i.m. in each thigh, 0.1 ml s.c. at the back of the neck, and the remaining 0.3 ml ID at several other sites.
   d. Goats: a total dose of 1 ml administered in each hind leg i.m. (0.5 ml/site).
3. It is recommended that the animals be given vaccine boosts at monthly intervals until the desired titers are obtained.

3.2.5 USING GERBU ADJUVANTS

3.2.5.1 Materials

Gerbu adjuvant
Antigen
Safety eyewear
Latex gloves
All-plastic syringes
Sterile test tube or Eppendorf tube
Syringe for injection

3.2.5.2 Procedure

1. Determine the amount of antigen required and aseptically transfer it to a sterile Eppendorf tube (or test tube).
2. Use a sterile plastic syringe to transfer the appropriate volume of Gerbu adjuvant to the antigen-containing tube and mix briefly. Generally a 1:1 (volume/volume) ratio of antigen to adjuvant is recommended.
3. Withdraw the antigen–adjuvant solution into a syringe and vaccinate the animals by injecting the antigen–adjuvant emulsion accordingly (the volumes cited are for the adjuvant only).
   a. Mice: 50 µl of adjuvant/mouse.
   b. Rats: 100 µl of adjuvant/rat.

    c. Hamsters: 200 µl of adjuvant/hamster.

    d. Guinea pigs: 250 µl of adjuvant/guinea pig.

    e. Rabbits: 1 ml of adjuvant/rabbit.

    f. Goats: 4 ml of adjuvant/goat.

4. It is recommended that the animals receive booster vaccinations every 14–28 days as needed.

3.2.6 USING IMJECT ALUM

3.2.6.1 Materials

Imject alum
Antigen
Safety eyewear
Latex gloves
Sterile beaker
Sterile stir-bar
Sterile pipets
Stir plate
Syringe for injection

3.2.6.2 Procedure

1. Determine the amount of antigen required and aseptically transfer it to a sterile beaker containing a stir-bar. Place the beaker and its contents on a stir plate.
2. Add Imject Alum a drop at a time to the antigen solution with constant stirring. Add 0.3–1 ml alum per 1 ml of antigen.
3. Cover the beaker and mix the combined solutions for 30 min at room temperature to ensure that the antigen has effectively adsorbed to the alum.
4. Draw the suspension into a sterile 1-ml syringe for injection and remove any air bubbles.
5. Vaccinate the animals by injecting the antigen–adjuvant emulsion accordingly:
    a. Mice: a total dose of 0.05–0.2 ml administered at two sites s.c.
    b. Rats or guinea pigs: a total dose of 0.2–0.5 ml administered at two sites s.c.
    c. Rabbits: a total dose of 0.5–1 ml administered as 0.25 ml s.c./site at separate sites at the back of the neck.

Booster vaccinations should be administered at 2- to 4-week intervals for best results.

ACKNOWLEDGMENTS

We thank Kathy Price and Susan Mackay for assistance in the preparation of this chapter.

REFERENCES

1. Jennings, V.J., Review of selected adjuvants used in antibody production, *ILAR J.,* 37, 119, 1995.
2. Hanley, W.C., Artwhol, J.E., and Bennett, B.T., Review of polyclonal antibody production procedures in mammals and poultry, *ILAR J.,* 37, 93, 1995.
3. Robuccio, J.A., Griffith, J.W. Chroscinski, E.A., Cross, P.J., Light, T.E. and Lang, C.M., Comparison of effects of five adjuvants on the antibody response to influenza virus antigen in guinea pigs, *Lab. Animal Sci.,* 45, 420, 1995.
4. Deeb, B.J., DiGiacomo, R.F. Kunz, L.L., and Stewart, J.L., Comparison of Freund's and Ribi adjuvants for inducing antibodies to synthetic antigen (TG)-AL in rabbits, *J. Immunol. Methods,* 152, 105, 1992.
5. Lipman, N.S., Trudel, L.J. Murphy, J.C., and Sahali, Y., Comparison of immune response potentiation and invivo inflammatory effects of Freund's and Ribi adjuvants in mice, *Lab. Animal Sci.* 43, 193, 1992.
6. Leenaars, P.P.A.M., Hendriksen, C.F.M., Koedam, M.A., Claassen, I., and Claassen, E., Comparison of adjuvants for immune potentiating properties and side effects in mice, *Vet. Immunol. Immunopathol.,* 48, 123, 1995
7. Freund, J., Casals, J., and Hosmer, E. P., Sensitization and antibody formation after injection of tubercle bacilli and paraffin oil, *Proc. Soc. Exp. Biol. Med.,* 37, 509, 1937.
8. Freund, J., The effect of paraffin oil and mycobacteria on antibody formulation and sensitization, *Am. J. Clin. Pathol.,* 21, 645, 1951.
9. Kenney, J. S., Hughes, B. W., Masada, M. P., and Allison, A. C., Influence of adjuvants on the quality, affinity, isotype and epitope specificity of murine antibodies, *J. Immunol. Methods,* 121, 157, 1989.
10. Broderson, J. R., A retrospective review of lesions associated with the use of Freund's adjuvant, *Lab. Anim. Sci.,* 39, 400, 1989.
11. Claassen, E., de Leeuw, W., de Greeve, P., Hendriksen, C., and Boersma, W., Freund's complete adjuvant: an effective but disagreeable formula, *Res. Immunol.,* 143, 478, 1992.
12. Mallon, F. M., Graichen, M. E., Conway, B. R., Landi, M. S., and Hughes, H. C., Comparison of antibody response by use of synthetic adjuvant system and Freund complete adjuvants in rabbits, *Am. J. Vet. Res.,* 52, 1503, 1991.
13. Hunter, R. L., Olsen, M. R., and Bennett, B., Copolymer adjuvants and TiterMax, in *The Theory and Practical Applications of Adjuvants,* Stewart-Tull, D. E. S., Ed., Wiley, New York, 1995, 51.
14. Rudbach, J. A., Cantrell, J. L., and Ulrich, J. T., Methods of immunization to enhance the immune response to specific antigens *in vivo* in preparation for fusions yielding monoclonal antibodies, *Methods Mol. Biol.,* 45, 1–8, 1995.
15. Maloy, W. L. Coligan, J. E., and Paterson, Y., Production of antipeptide antisera, in *Current Protocols in Immunology,* Coligan, J. E., Kruisbeek, A. D., Margulies, D. H., Shevach, E.M., and Strober, W., Eds., John Wiley & Sons, New York, 1998, pp. 9.4.1–9.4.11.
16. van de Wijgert, J. H. Verheul, A.F., Snippe, H., Check, I. J., and Hunter, R. L., Immunogenicity of *Streptococcus pneumoniae* type 14 capsular polysaccharide: influence of carriers and adjuvants on isotype distribution, *Infect. Immunol.,* 59, 2750, 1991.

17. Glenn, G. M., Rao, M., Richards, R. L., Matyas, G. R., and Alving, C. R., Murine IgG subclass antibodies to antigens incorporated in liposomes containing lipid A. *Immunol. Lett.*, 47, 73, 1995.

18. Spack, E. G. and Toavs, D., Transfer techniques for minimizing waste of sonified adjuvant emulsions, *Biotechniques*, 20, 28, 1996.

4 Production of Polyclonal Antibodies

Lon V. Kendall

CONTENTS

4.1 INTRODUCTION

Polyclonal antibodies (pAbs) remain a widely utilized tool in research and diagnostics. They identify specific antigens, principally protein antigens, and are used in Western blots, radioimmune assays (RIAs), enzyme-linked immunoabsorbent assays (ELISAs), indirect and direct fluorescent antibody test, hemaglutination tests, immunohistochemistry (IHC), immunoprecipitation (IP) assays, immunodiffusion, affinity chromatography, enzymology, and isolation of gene products. Their use is so widespread that there are well over 100 biotech companies that specialize in the production of polyclonal antibodies.

Polyclonal antibodies are derived from multiple B-cell clones that have differentiated into antibody-producing plasma cells in response to an immunogen. Immunogens or antigens (Ags) are any substance such as protein, lipid, or carbohydrate, which can elicit a humoral immune response. This is in contrast to monoclonal antibodies, which are derived from a single B-cell clone and produce antibody to a very specific antigen.

To generate pAb, immunogens are intentionally inoculated into the host, similar to the process of vaccination. When immunogens are introduced to a host under controlled conditions, the immune response processes them, which ultimately results in proliferation and differentiation of B cells into antibody-secreting plasma cells. Blood serum can be collected and antibodies isolated for use. To enhance the antibody response and maximize the antibody titer, adjuvants are included with the immunogen. Adjuvants are substances that enhance the immune response, generally by creating a depot for slow release of the antigen. A more complete discussion is presented in Chapter 3 of this volume. To understand how polyclonal antibodies are synthesized, the reader must understand how the immune system generates antibodies in response to immunogens and antigens. This information is provided in more detail in Chapter 4 of this volume and there are several excellent texts that review this process.[1,2] Nonetheless, a brief review of antigen processing and antibody production, in the context of intentional inoculation, is provided.

4.2 ANTIGEN-PRESENTING CELLS

Antigens (Ags) are typically thought of as a molecule that is capable of generating an antibody response, also known as humoral or acquired immunity. The process of antibody production begins when the Ag is processed by antigen-presenting cells (APCs). The principal APCs are dendritic cells, macrophages, and B cells (Figure 4.1). Dendritic cells (DCs), often referred to as professional antigen-presenting cells, are located within lymphoid organs, in the skin (where they are termed Langerhans cells), heart, intestinal tract, genital tract, lung, and eye. These cells are most important in the primary immune response. Particulate antigens and soluble antigens are taken up by immature DCs in the periphery and are moved by that cell into the lymphatic system and subsequently into the local lymph node where they become mature DCs.[3] Mature DCs are potent stimulators of T cells; however, they are unable to sample antigens by endocytosis. Antigens are also sampled by immature DCs located at tissue sites by nonspecific micropinocytosis, then migrate to the lymphatic system where they present antigen and upregulate costimulatory molecules for T-cell activation. Bacterial DNA with cytosine phosphate guanine (CpG) motifs induce a rapid activation of DCs that results in the production of proinflammatory cytokines interleukin (IL-)6, IL-12, IL-18, interferon (IFN-) α, and IFN-γ, that upregulate costimulatory molecule expression on the surface of the DCs. Heat-shock proteins from bacteria have a similar mechanism.[2] The high expression of costimulatory molecules is critical to T-cell activation and subsequent antibody production in B cells.

The other major type of APC is macrophages. These cells are located throughout the body and activate and regulate adaptive immune responses, functioning as part of the innate immune response.[4] Immune responses are regulated by the macrophage phagocytizing particulate antigens and presenting them on the cell surface. The humoral immune response is initiated when lymphocytes recognize antigen processed by the APC. To be recognized properly, they must be presented in the context of the major histocompatability complex (MHC) that is present on the cell surface (Figure 4.2). There are two classes of MHC molecules, class I (cI MHC) and class

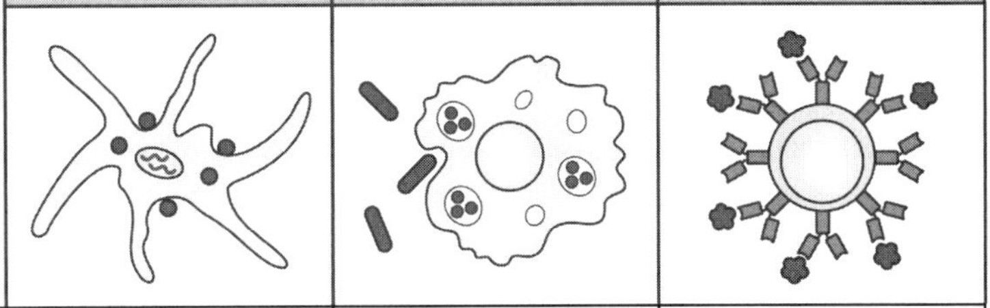

	Dendritic cells	Macrophages	B cells
Antigen Uptake	Macropinocytosis and phagocytosis by tissue dendritic cells	Phagocytosis	Antigen-specific receptor
MHC Expression	Low on tissue dendritic cells High on dendritic cells in lymphoid tissue	Inducible by bacteria and cytokines	Constitutive Increases on activation
Co-stimulatory delivery	Constitutive by mature dendritic cells	Inducible	Inducible
Antigen Presented	Peptides, viral antigens, allergens	Particulate antigens	Soluble antigens

FIGURE 4.1 Properties of the various antigen-presenting cells. Dendritic cells, macrophages, and B cells are the main cell types involved in the initial presentation of foreign antigens. These cells vary in their means of antigen uptake, major histocompatibility complex expression, costimulatory expression, the type of antigen they present, and their location in the body. Copyright 2005. From *Immunobiology: The Immune System in Health and Disease*, 6th ed., Janeway, C.A., Jr., Travers, P., Walport, M., and Shlomchik, M. Reproduced by permission of Garland Science/Taylor & Francis LLC.

FIGURE 4.2 Antigen processing by antigen-presenting cells (APC). Antigen is taken up by the APC by phagocytosis or receptor mediated endocytosis. In this example, the antigen is taken up by cell surface receptors and internalized. The antigen is broken down by proteosomes to small peptide fragments in the endosome. The endosome fuses with the major histocompatibility complex (MHC) vesicle and binds a specific peptide. The peptide:MHC complex is presented on the cell surface for presentation to T cell receptors. Copyright 2005. From *Immunobiology: The Immune System in Health and Disease*, 6th ed., Janeway, C.A., Jr., Travers, P., Walport, M., and Shlomchik, M. Reproduced by permission of Garland Science/Taylor & Francis LLC.

II (cII MHC), which express antigen from different cellular compartments and the two classes are recognized by two distinct populations of T cells. DCs and macrophages are able to present antigens in the context of either cI MHC or cII MHC complexes.

Class I MHC present intracellular cytosolic antigens that are typically derived from viral proteins within the cytosol. The cytosolic antigenic proteins are degraded into small peptides of 8–10 amino acids within the DC or macrophage's proteosome. The peptide is then processed within the endoplasmic reticulum, where it binds to a groove in the cI MHC molecule. Once bound, the antigen:cI MHC complex is transported to the cell surface through the Golgi apparatus. The antigen:cI MHC complex is recognized by the T-cell receptor (TCR) of CD8 T cells and potentiates the cellular immune response and cytotoxicity of the T cell. Although this is an important aspect of the immune response in vaccine research, it is not the principal mechanism of antibody production.

Antigens that are localized in the vesicular compartment of a cell are processed differently. Some pathogens, including *Mycobacterium* sp., reside within the intracellular vesicle, where they escape surveillance by the cytosolic proteosome. Other antigens are internalized by endocytosis and contained in an endosome. In either case, the intracellular vesicle or the endosome become increasingly acidic as it progresses to the interior of the cell. The acidic environment activates proteases and degrades the antigenic proteins. Class II MHC molecules, which are synthesized in the endoplasmic reticulum, are transported to the cell surface in a separate cII MHC vesicle. The endosomes fuse with the cII MHC vesicle in the cytoplasm, the cII MHC proteins bind the peptide fragments, and the vesicle transports them to the cell surface. The antigen:cII MHC complex is then recognized by the TCR of CD4 T cells.

Costimulator molecules are required for T-cell activation. In the absence of costimulation, T cells can become nonresponsive. As described previously, several bacterial antigens stimulate the expression of costimulator molecules to activate T cells. By mixing bacterial antigens with a protein antigen under experimental conditions, the costimulatory molecules are expressed thereby enhancing the immunogenicity of the protein. Hence the bacteria can serve as an adjuvant to the protein.[2]

4.2.1 ANTIGEN RECOGNITION BY B CELLS

B cells also function as APCs, typically for soluble proteins. Before becoming an antibody-secreting plasma cell, B cells express their immunoglobulin or B-cell receptor (BCR) on the cell surface. B cells have 200–500,000 identical BCRs on the cell surface that recognize a specific soluble antigen. The antigen binds to the receptor (similar to antigen:antibody binding) and it is internalized by endocytosis. As with macrophage or DC processing, the antigen is processed in vesicles, and the B cell presents antigen:cII MHC complexes on the cell surface for engagement by CD4 T cells. Expression of costimulatory molecules by B cells is also required for T-cell help. This again can be enhanced with bacterial antigens serving as an adjuvant. In addition to recognizing and processing antigen, BCR engagement by a foreign antigen can lead directly to antibody production.

4.2.2 Antigen Recognition by T cells

T cells recognize antigen displayed by APC in the context of antigen:MHC complexes via the TCR. The TCR has a similar structure and diversity as the BCR of B cells, with approximately 30,000 identical antigen receptor molecules on the surface of each T cell.[2] Unlike the BCR, which recognizes only the antigenic fragment, antigens must be presented in the context of an MHC molecule to be recognized by T cells. T cells also require coreceptors, known as CD4 and CD8, for appropriate activation. CD8 coreceptors (CD8 T cells) recognize antigen:cI MHC complexes, whereas CD4 coreceptors (CD4 T cells) recognize antigen:cII MHC complexes. CD4 T cells are also known as helper T cells and help stimulate cell mediated immunity and humoral immunity. CD8 T cells are known as cytotoxic T cells and contribute to eliminating cells that express viral antigens.

Before B cells are induced by the CD4 T cells to secret antibody, the CD4 T cells must be activated by APCs. Activation of T cells requires two signals. The first signal is initiated through the interactions of the TCR and CD4 coreceptor of naive T cells with the antigen:cII MHC complex on APCs. The second signal is a costimulatory signal from the APCs. The best characterized costimulatory molecules of APCs are the B7 molecules (CD80 and CD86).[5] B7 molecules are constitutively expressed in high levels on mature DCs, which is why they are such potent stimulators of immune responses. The B7 molecules activate the CD28 receptor on T cells, resulting in differentiation and proliferation of T cells (Figure 4.3). Additional molecules are expressed to sustain the

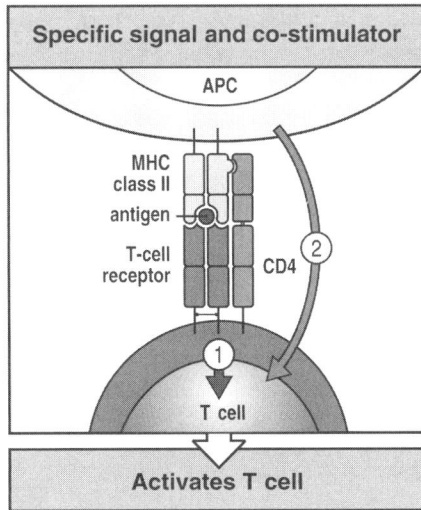

FIGURE 4.3 Activation of naive T cells requires two signals. Binding of the peptide:major histocompatibility complex by the T-cell receptor and, in this example, the CD4 coreceptor, transmits a signal (arrow 1) to the T cell that antigen has been encountered. Activation of naive T cells requires a second signal (arrow 2), the costimulatory signal, to be delivered by the same antigen presenting cell. Copyright 2005. From *Immunobiology: The Immune System in Health and Disease*, 6th ed., Janeway, C.A., Jr., Travers, P., Walport, M., and Shlomchik, M. Reproduced by permission of Garland Science/Taylor & Francis LLC.

costimulation including CD40 ligand. CD40 ligand binds the CD40 of APCs, activating B7 molecule expression and potentiating the T-cell proliferation. Without co-stimulation, the T cells are rendered anergic or tolerant. Once activated by both signals, T cells produce IL-2, resulting in T-cell proliferation and differentiation.

There are two subsets of CD4 T-helper cells described based on their cytokine secreting profile.[6] Type 1 T helper (Th1) cells are characterized by IL-2 and IFN-γ production, which principally stimulate cell-mediated immune functions such as macrophage activation and opsonizing antibody production. Type 2 T helper (Th2) cells are characterized by IL-4, IL-5, and IL-10, which stimulate humoral immune responses and antibody secretion.[6] Armed effector CD4 T cells engage the antigen:cII MHC complex of B cells and activate proliferation and differentiation into antibody-secreting plasma cells. Cytokines secreted by activated CD4 T cells drive the antibody response and determine the immunoglobulin isotype secreted. Th1 cytokines enhance opsonizing antibody production such as IgG2a in mice, whereas Th2 cytokines enhance neutralizing antibody responses such as IgG1 in mice from B cells. This antibody response requires that the TCR recognizes the same antigen as the BCR. These events (antigen processing, antigen recognition, costimulation, and cytokine production) result in clonal expansion of the B cells to plasma cells with each plasma cell secreting antibody of a single specificity (Figure 4.4).

4.2.3 B-CELL RECEPTOR AND ANTIBODY PRODUCTION

Certain antigens are capable of inducing B-cell proliferation and differentiation without T-cell help. Examples include the lipopolysaccharide of gram-negative bacteria and polysaccharide capsules of certain Gram-positive bacteria. Each of these is a highly repetitive structure that binds to the BCR and cross-link adjacent BCRs. This cross-linking causes the B cells to produce IgM. Helper T cells that similarly recognize a component of the cross-linked antigen can augment the B-cell response and induce isotype switching.[2] Activation in this way typically results in proliferation of multiple B cells, regardless of their antigen specificity and is true polyclonal activation.

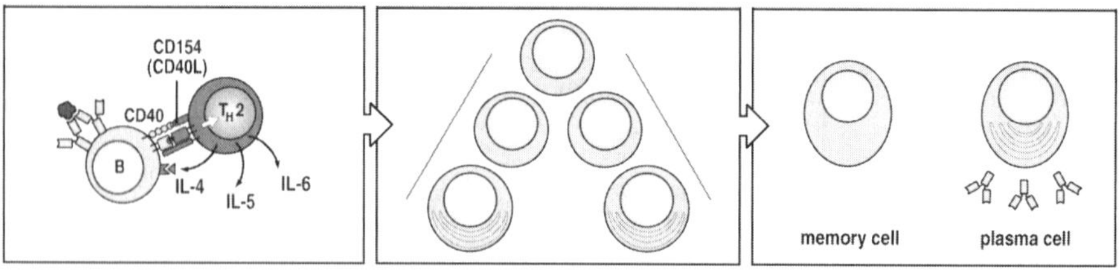

FIGURE 4.4 Helper T cells stimulate the proliferation and differentiation of antigen binding B cells. The interaction of an antigen-binding B cells with an armed helper T cells leads to the expression of costimulatory molecules (CD40 and CD40 ligand) and the secretion of B-cell stimulatory cytokines (interleukin-4, interleukin-5, and interleukin-6), leading to proliferation and differentiation to an antibody secreting plasma cell. It can alternatively become a memory B cell. Copyright 2005. From *Immunobiology: The Immune System in Health and Disease*, 6th ed., Janeway, C.A., Jr., Travers, P., Walport, M., and Shlomchik, M. Reproduced by permission of Garland Science/Taylor & Francis LLC.

4.2.4 IMMUNOLOGIC MEMORY

The events described explain the primary immune response to antigens. It is characterized primarily by a low affinity antibody, primarily of the IgM isotype. After a primary immune response, the immune system establishes memory T and B cells.[7] When the immune system encounters the same antigen on a different occasion, immunologic memory allows for a more rapid and robust immune response with antibodies of a greater affinity, primarily of the IgG isotype. Antibody affinity is enhanced on reexposure through somatic hypermutation and selection of B-cell clones, with the highest affinity for the antigen.[2] PAb production takes advantage of immunologic memory to booster the immune response.

4.2.5 ANTIBODIES

Antibodies secreted from B cells in response to antigenic stimulation are very specific in their recognition of protein antigens. If a mammal is immunized with a protein antigen, that protein is degraded into smaller different peptide fragments (epitopes) by APCs. This results in several different antigens presented by APCs to T cells for T-cell activation and the consequent clonal expansion of multiple B-cell lineages. Each B cell thereby produces an antibody to a specific epitope of the original protein antigen, hence the term *polyclonal antibody.*

The repertoire of antigens that antibodies are capable of recognizing is incredibly diverse. This diversity is made possible by genetic recombination resulting in recognition of more than 10^5 different antigens. Antibody molecules are typically drawn as a Y-shaped molecule with a variable region at the N-terminus and a constant region at the C-terminus. The N-terminus is the portion of the antibody that recognizes antigens, referred to as the *variable region*. The C-terminus determines the effector function of the antibody (i.e., how it behaves in response to antigen). The variable and constant regions of the antibody are further composed of four polypeptide chains termed the *heavy chain* and the *light chain*. The antigen recognition site is composed of the variable region of both the heavy and light chains (Figure 4.5). Variation and rearrangement of the genes responsible for forming the variable region determine antigen specificity. Variability and rearrangement of the genes responsible for forming the constant region determine effector function and identify of different isotypes or classes of antibody. There are four classes of immunoglobulin: IgM, IgG, IgE, and IgA, principally found in substantial amounts in response to an antigen with defined effector functions. A fifth class, IgD, is coexpressed with IgM on the B-cell surface and is secreted in small amounts. The function of IgD is not known. IgM and IgG are the most important immunoglobulins when referring to production of pAbs; the antibody repertoire is further influenced in that different mammalian species produce several subclasses of IgG. Table 4.1 provides a list of the IgG subclasses of various species.

4.2.6 ROLE OF ADJUVANTS

Although an antigen presented to the immune system will elicit an antibody response, this response may be inadequate to develop sufficient amounts of antibody for research

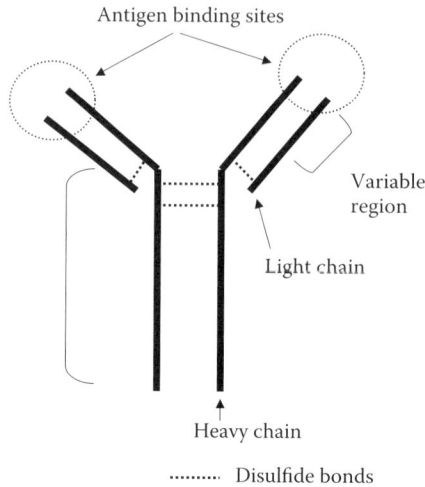

Antigen binding sites

Variable region

Light chain

Heavy chain

.... Disulfide bonds

FIGURE 4.5 Structure of an antibody molecule. Antibody molecules are composed of two heavy chains and two light chains. The N-terminus region of the antibody is antigen recognition site and is composed of the variable regions of the light and heavy chains. The C-terminus is the constant region that determines the effector function of the antibody. Copyright 2005. From *Immunobiology: The Immune System in Health and Disease*, 6th ed., Janeway, C.A., Jr., Travers, P., Walport, M., and Shlomchik, M. Reproduced by permission of Garland Science/Taylor & Francis LLC.

TABLE 4.1
Immunoglobulin (Ig) Subclasses of Different Species[10,135]

Species	Immunoglobulins Subclasses
Human	IgG1, IgG2, IgG3, IgG4
Mouse	IgG1, IgG2a, IgG2b, IgG2c, IgG3
Rat	IgG1, IgG2a, IgG2b, IgG2c
Guinea pig	IgG1, IgG2
Rabbit	Only 1 IgG isotype
Sheep	IgG1, IgG2, IgG3
Chicken	IgY (comparable to IgG)
Pig	IgG1, IgG2a, IgG2b, IgG3, IgG4
Horse	IgGa, IgGb, IgGc, IgG(B), IgG(Ta), IgG(Tb)

purposes (or in vaccinology to provide a protective immune response). Adjuvants are substances that increase the immune response to an antigen and are widely used in the immunization process to enhance antibody production. Adjuvants can be used to enhance the immune response to antigen, reduce the amount of antigen or number of immunizations needed, or act as an antigen delivery system for antigen uptake.[8] There are several classification schemes for adjuvants. For the purpose of pAb production, they can be considered either antigen delivery systems or immunostimulants.[9]

Adjuvants that act as an antigen delivery system provide sustained release of antigen to APC by forming a depot. Examples include emulsions and liposomes. Immuno-stimulant adjuvants promote proliferation and differentiation of naive immune cells.[10] Examples include bacterial derived adjuvants such as lipopolysaccharide of Gram-negative bacteria, muramyl dipeptide (a peptidoglycan derived from bacterial cell walls), cytokines, or bacterial DNA.[8,9] The best adjuvants simultaneously deliver antigen to the immune system and modulate the immune system to enhance its response to that antigen.[11] It is important to note there is no universal adjuvant available to deliver the variety of antigens and the optimal adjuvant should be deter-mined for each antigen.[12] More than 100 adjuvant preparations have been described to date.[12] Selection of the adjuvant is important as they each enhance different qualities of the antibody response such as quantity, affinity, isotype, and epitope specificity.[13] Nonetheless, there are adjuvants that consistently provide the desired response and are widely used in the production of polyclonal antibodies. A brief review of the most commonly used adjuvants and their mechanism of action is provided here. These are reviewed in greater detail in Chapter 3 as well as review articles.[8,9,14–17]

Commonly used adjuvants primarily function by forming an antigen depot, a mixture of adjuvant and antigen. This allows for slow and sustained release of antigen to APCs; that results in prolonging B-cell stimulation and antibody synthesis. This also promotes immunologic memory for prolonged antibody responses without repeated injections. The emulsion based adjuvants (such as Freund's complete adju-vant), aluminum hydroxide gel, solid phase adsorbents, and encapsulating materials such as liposomes form antigen depots.[10]

Emulsions are the most commonly used adjuvants for polyclonal antibody produc-tion with a variety of water-in-oil and oil-in-water emulsions available. The "gold standard" is Freund's complete adjuvant (FCA), which is a water-in-oil emulsion containing mannide monooleate (a surfactant) and heat-killed *Mycobacterium tuber-culosis* bacteia.[18] Although the traditional FCA contains whole *M. tuberculosis* organ-isms, other preparations may contain components of the *M. tuberculosis* or other mycobacterium.[12] Freund's incomplete adjuvant (FIA) is identical to FCA, except that it lacks the *Mycobacterium* components. Both FCA and FIA maintain the antigen in an aqueous phase of the emulsion resulting in sequestering, allowing slow release of antigen for exposure to the immune system. The major disadvantages to the use of FCA are its associated toxicities. The mineral oil is not metabolizable and the myco-bacterial components can incite an inflammatory reaction, severe granulomatous reac-tion (Figure 4.6) and ulceration at the injection site.[8] This granulomatous reaction can lead to large, sterile abscesses that exert pressure on the skin creating discomfort and may rupture. The side effects of FCA can be reduced if the concentration of adjuvant is reduced at the injection site. For these reasons, the use of FCA is limited to exper-imental uses and even then there is pressure to discontinue or greatly limit its use.[19,20]

TiterMax (TiterMax USA Inc., and Sigma-Aldrich, St. Louis, MO) is another water-in-oil emulsion that contains squalene, instead of mineral oil that is readily metabolized, an emulsifier (sorbitan monooleate), and a surface-acting copolymer (CRL8941).[11] The copolymers are less toxic than other surface-acting agents and have a pronounced adjuvant activity through antigen presentation, complement acti-vation, chemotactic properties, and macrophage activation.[21,22] The copolymers act

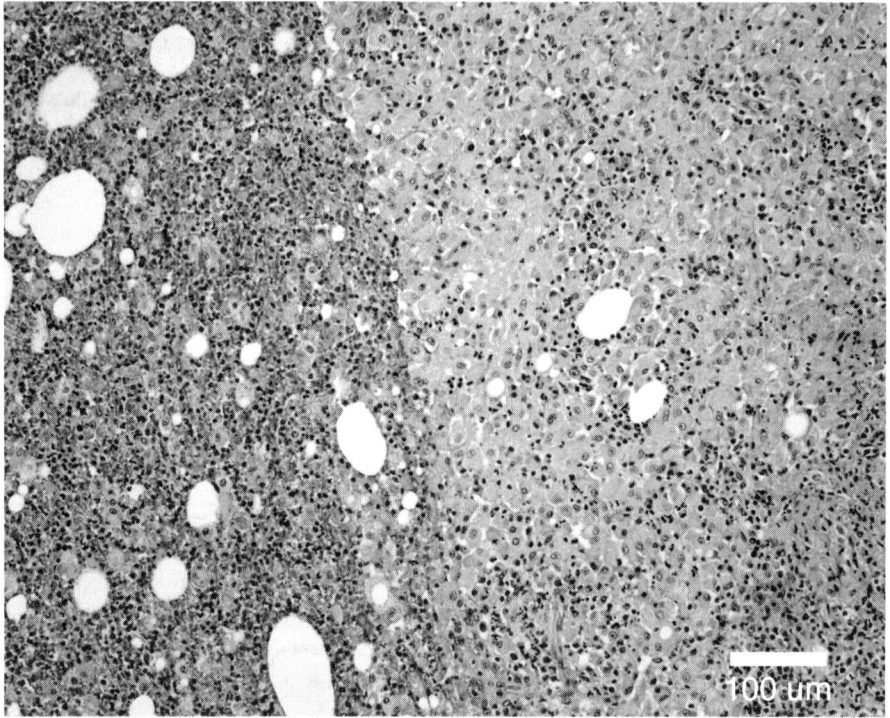

FIGURE 4.6 This photomicrograph depicts the granulomatous response to Freund's complete adjuvant in a rabbit. The area to the right is composed of epithelioid macrophages admix a heterophilic infiltrate. The large clear vacuoles represent the mineral oil. The area to the left has become necrotic with diffuse heterophilic infiltrate.

by forming adhesions with antigens, thereby displaying a high concentration of antigen on the surface compared with the amount of antigen found in an aqueous solution.[21] Such copolymers have been used to sequester proteins, peptides, polysaccharides, whole viruses, and recombinant proteins for pAb production.[14] Depending on the antigen and species used, there have been mixed reports on the antibody response using TiterMax adjuvant. Several studies have compared the antibody response generated in response with several antigens using different adjuvants. TiterMax had a delayed antibody response and lower titer compared with FCA.[23–25] However, TiterMax had minimal reactions in rabbits and did not appear to cause prolonged pain and distress.[26] In another study, mice were immunized with a hapten conjugated to a carrier protein (hen egg albumin) in one of several adjuvants. The antibody response with the TiterMax adjuvant was longer lasting with fewer injections than the other adjuvants, including FCA. The antibody isotypes IgG1, IgG2a, and IgG2b were at substantial titers. The same study also compared the effectiveness of different adjuvants in promoting an antibody response with a steroid hormone conjugated to a carrier protein (bovine serum albumin) in rabbits and goats. The antibody response using TiterMax was equivalent to FCA.[27] This exemplifies the need to evaluate antigen/adjuvant combinations on an individual basis for effectiveness.

Unlike FCA and TiterMax, the Ribi adjuvant system (RAS) (Corixa Corp. Seattle, WA, and Sigma-Aldrich, St. Louis, MO) is an oil-in-water emulsion. An antigen is incorporated in a small amount of metabolizable squalene oil and the mixture is then emulsified in a saline solution containing a surfactant.[28] This alone is a weak adjuvant and so bacterial derived immunostimulants from mycobacteria or gram-negative bacteria are added.[29] Several components of mycobacteria have been demonstrated to be immunostimulants such as trehalose 6,6'-dimycolate, also known as cord factor, which enhances the immune response when combined with other immunostimulants.[16,29]

There are three commercially available formulations of the RAS that contain the oil-in-water emulsion and trehalose, plus one or more combination of the following bacterial derived immunostimulants. In addition to trehalose 6,6'-dimy-colate, mycobacterial cell wall skeleton is used as an immunostimulant in RAS. The cell wall skeleton is a mycobacterial cell wall extract containing the muramyl dipeptide that is largely responsible for the adjuvant activity.[30] Another component of Ribi adjuvants is monophosphoryl lipid A, which is a chemically modified form of lipid A derived from bacterial endotoxin (lipopolysaccharide).[31,32] The lipid A component of lipopolysaccharide has been demonstrated to increase humoral and cell-mediated immunity.[33,34] Although the use of RAS potentially has less toxicity associated with it compared to FCA,[26,35] most studies have demonstrated a lower antibody titer compared with the use of FCA/FIA as an adjuvant.[23,24,36–38]

Although this is not an exhaustive list, these emulsions are the most commonly used in pAb production. Other emulsions used as adjuvants include Specol, Montanide incomplete Seppic adjuvants, and Syntex adjuvant formulation.[16,39] Aluminum salt adjuvants are the most commonly used adjuvant in human and veterinary vaccines. They increase the antigenicity by acting as an antigen depot,[40] but are cleared within 3–4 weeks after injection, but are then followed by a rapid decline in antibody titers.[16] Each of these emulsions and the aluminum salt adjuvants resulted in fewer adjuvant-induced lesions compared with FCA, but typically realize a lower antibody titer, hence their limited use in pAb production.[13,16,41,42] Standard protocols for producing pAbs using FCA, TiterMax, and RAS are given in Table 4.2.

Other adjuvant preparations have been used to induce pAb production in the research setting but have mixed results. For example, liposome adjuvants incorporate antigen in a phospholipid bilayer. This protects the antigen from rapid decay and has a depot effect.[43] Liposomes induce a humoral immune response and generate most IgG subclasses during the primary immunization as a result of the depot effect and subsequent processing by macrophages.[44] There is also evidence that liposomes transport the antigen to draining lymph nodes, where they can interact with APCs.[43] In vaccine research, liposomes are able to induce cell-mediated immunity. It is thought that during antigen processing by macrophages, some antigen escapes the lysosome and enters the cytoplasm where it is processed in the context of cI MHC, promoting a cell-mediated immune response.[45] The ability to simultaneously induce humoral and cell-mediated immune responses makes liposomes favorable adjuvant in vaccine studies. Immune responses can be enhanced by adding other immunostimulants to the liposome mixture, such as lipid polysaccharide or bacterial-derived muramyl dipeptide.[14,46] Although they may have fewer side effects than emulsion

TABLE 4.2
Standard Immunization Protocols Using Emulsion Adjuvants for Producing Polyclonal Antibodies[134]

Adjuvant	Protein Concentration	Adjuvant Amount	Final Antigen: Adjuvant Ratio	Comments
Freund's complete adjuvant (1)	2 ml of 0.25 to 0.5 mg/ml purified protein	2 ml (each ml contains 1 mg *M. tuberculosis*, 0.85 ml paraffin oil, and 0.15 ml mannide monooleate)	1:1	Sufficient to immunize 4 rabbits or 80 mice
Freund's incomplete adjuvant (1)	2 ml of 0.25 to 0.5 mg/ml purified protein	2 ml (each ml contains 0.85 ml paraffin oil and 0.15 ml mannide monooleate)	1:1	Sufficient to immunize 4 rabbits or 80 mice
TiterMax (1)	0.5 ml of 0.25 to 0.5 mg/ml purified protein	0.5 ml (contains block polymer [CRL-8941, CRL-8300], squalene and monooleate)	1:1	Sufficient to immunize 10 rabbits or 20 mice
RAS	2 ml of 0.25 to 0.5 mg/ml purified protein	Mix with 1 vial RAS (contains 0.5 mg monophosphoryl lipid A, trehalose 6,6′-dimycolate, CSW, in 44 μl squalene and Tween 80	1:1	Sufficient to immunize 2 rabbits or 10 mice

(1) Available from Sigma-Aldrich, Inc.   (2) Available from Corixa, Inc.

adjuvants, liposomes can be difficult to properly prepare and may not be effective with most antigens.[14] Hence their use in polyclonal antibody production is limited.

Immune-stimulating complexes (ISCOMs) are composed of antigen, cholesterol, phospholipids, and saponin (Quil A) that form mixed micelles.[47] Although the mechanism of the adjuvant activity has not been fully determined,[48] these preparations are potent adjuvants of both soluble and particulate antigens. ISCOMs stimulate an antibody response comparable to using FCA, with high titer and broad isotype production.[48] Similar to liposomes, they have fewer reported side effects than emulsion adjuvants,[49] but may be limited in the types of antigens they present and may be challenging to prepare. Therefore they are not commonly used for polyclonal antibody production. A variation of ISCOMs is the ISCOMATRIX adjuvant (ZLB Behring/CSL Ltd., King of Prussia, PA).[50] The ISCOMATRIX adjuvant is structurally similar to the ISCOM except it does not contain the antigen. It is easily prepared compared with the ISCOM and mixed with antigen before immunization. ISCOM and ISCOMATRIX formulations have been shown to be safe and effective at producing humoral and cell-mediated immunity in a number of animal models including mice, rabbits, and sheep. They have maintained an effective antibody response using 10-fold to 100-fold less antigen than other antigen/adjuvant combinations.[48,50]

Synthetic oligodeoxynucleotides are a recent adjuvant used in vaccine development.[51–53] The oligodeoxynucleotide sequences containing unmethylated CpG motifs act as an immune adjuvant that resemble bacterial DNA sequences. These CpG motifs are recognized by the innate immune system by pathogen-associated molecular patterns that are recognized by a limited repertoire of pattern recognition receptors.[54] This allows the immune response to recognize and respond to conserved microbial structures. CpG motifs are recognized by cell-surface receptors, rapidly internalized and interact with the toll-like receptor 9 in the endocystic vesicle with a resultant cellular activation and upregulation of proinflammatory cytokine expression.[51,55] This enhances antigen processing, upregulation of costimulatory molecules, and secretion of cytokines. B-cell activation and proliferation and immunoglobulin secretion are also influenced by CpG motifs.[51] The cytokine milieu so created consists of IL-6 that promotes B-cell activation[52] and IFN-γ and IL-12 that promote a Th1 cell–mediated response. This cytokine milieu influences antibody isotype.[56] Although typically used with DNA vaccines, immunostimulation can be enhanced by incorporating DNA encoding cytokines such as IL-4 to increase antibody responses.[56] Other mechanisms for enhancing the adjuvant activity of CpG include altering the backbone chemistry, altering the delivery system, or combining them with other adjuvants.[53] Furthermore, CpG motifs can be used as an adjuvant when coadministered with protein antigens. However, the effects on the immune system vary with the antigen, the CpG motif length, and the host species. For example, CpG motifs constructed with a phosphorothioate backbone are better at producing a Th2 response with a broader spectrum of antibody isotypes secreted.[56] Too many CpG sequences may lessen an immune response[52]; synthetic CpG motifs also alter the response in different species. For example, the flanking nucleotide sequence GACGTT is optimal for activating the mouse or rabbit immune response, but not that of man.[56] CpG has been demonstrated to elicit B-cell proliferation in several species, including mice, sheep, horses, and chickens.[57] Its use as an adjuvant for pAb production is now being taken advantage of with genetic immunization.[58,59] Genetic immunization involves inoculation of a plasmid containing a DNA sequence to the protein of interest. The plasmid is delivered intradermally with a gene gun, the macrophages or APCs express the protein that then incites an immune response.[60] The immune response can be enhanced by incorporating gene sequences of immunostimulants such as cytokines into the plasmid or can be enhanced with traditional adjuvants.[58]

Other adjuvants used in vaccine research may prove beneficial in pAb production. These include synthetic lipoproteins,[61,62] other synthetic bacterial proteins,[63] parasitic proteins that enhance antibody responses,[64] mucosal adjuvants such as cholera toxin or labile toxin of *Escherichia coli*,[65] or the manipulation of costimulatory molecules such as CD40 and CD28.[66,67]

4.2.7 ANTIGEN CHARACTERISTICS

Antibody production is influenced by the antigen characteristics including the size of the antigen, the state of aggregation, and the state of nativity. Antigens greater than 5 kDa readily stimulate antibody responses; however, smaller polypeptide and non-protein antigens need to be conjugated to larger immunogenic carrier proteins, such as keyhole limpet hemocyanin or bovine serum albumin.[68] This is especially true if the antigen is a polypeptide consisting of 15–20 amino acids (approximately 2 kDa).[10]

The antigen state must also be considered when generating pAb. Antigens presented to the immune system in their native state will produce antibodies to native proteins. Antigens presented in a denatured state will elicit an antibody response to denatured proteins. Therefore one must consider how the antibodies will be used before generating polyclonal antisera. For example, if one is using antibodies to detect bacterial proteins on a denaturing gel, then a denatured protein should be used in antibody production. But if antibodies are being used to detect virus in a capture ELISA, then native viral proteins may be more appropriate for antibody production. The use of water-in-oil adjuvants require vigorous mixing that can alter the antigen state,[69] so one must also consider the adjuvant being used and how it may influence the antigen and consequently the antibody response. Some antigen preparations may lead to tolerance rather than an antibody response. For example, soluble deaggregated proteins given intravenously have been demonstrated to result in tolerance rather than antibody production.[70]

As a general rule, the larger the antigen, the better the pAb response. Larger proteins have greater opportunities to be recognized by a TCR because there is more antigen available for processing by APCs. This results in multiple smaller antigen fragments to the same protein. These engage the TCR and, in turn, lead to a greater diversity of B cell–producing antibody. If smaller peptides are used, the antibody response may not recognize the parent protein because the small peptide may be internalized in the native state. As such, terminal peptides of the antigen tend to make better antibody response to native proteins.[10]

Before immunization, the antigen needs to be properly prepared to avoid impurities and unwanted antibody response. Some contaminants may be toxic to the host, such as chemicals or bacterial endotoxin. Endotoxin can elicit a pyogenic or inflammatory response that could alter the antibody response. Most antigens can be prefiltered through a 0.22-micron filter that minimally disrupts the antigen conformation; however, this does not remove endotoxin.[10]

Each antigen has a broad "window of immunogenicity." However, the amount of antigen used to elicit an antibody response must be carefully considered because too much or too little may cause immunosuppression, tolerance, or skewing of the immune response to promote cell-mediated immunity.[71–73] The optimal quantity of antigen is dependent on the antigen properties, adjuvant used, the route of immunization and the species used, and, ideally, is determined for each antigen.[74] In general, nanogram to microgram quantities of antigen plus adjuvant are needed to elicit a high titer antibody response.[75] For example, rabbits immunized with 500–1000 μg, mice 10–200 μg, or goats or sheep with 250–5000 μg of soluble antigen plus FCA, develop high titer antibody responses.[75,76] Smaller animals may require a smaller concentration of antigen/adjuvant combination, but the dose of antigen required to elicit an antibody response does not increase or decrease relative to body size.[76] A 4-kg rabbit would require about the same dose of antigen as a 25-g mouse.[77] Antigen dose is important because high antigen doses can also lead to low affinity B-cell activation, whereas low antigen dose can result in high-affinity B-cell activation. The use of adjuvants therefore allow a smaller amount of antigen to be used, elicit a high antibody titer, and minimize the chance of developing tolerance.[10] For a more comprehensive review of antigens, see Chapter 2.

4.2.8 ROUTE OF IMMUNIZATION

The route of immunization is influenced by species, antigen characteristics, adjuvant mixture, quantity of antigen, and volume, and also determines which lymphoid organs are activated and the type of antibody response that will be induced.[74] The route of immunization should also consider the welfare of the animal because adjuvants may cause pain and distress.[42,78,79] Given all these considerations, there is not a single immunization protocol that is universal for all the antigen, adjuvant, and species combinations. The typical routes of injection are intravenous (i.v.), intramuscular (i.m.), subcutaneous (s.c.), intraperitoneal (i.p.), or intradermal (i.d.). Less commonly used routes of immunization designed to deliver antigens more directly to lymphoid tissues are intranodal (directly in the lymph node) or intrasplenic. Foot-pad and intra-articular routes have also been used. These latter routes, although successful, are less than desirable because of the adverse effects they have on the welfare of the animals, particularly when given with an adjuvant.[10] This includes severe swelling of the foot pad or joint preventing the animal from ambulating normally.

The effectiveness of the antibody response is determined by the efficiency of antigen delivery to the lymphoid tissues[76] and the antigen/adjuvant combination. To achieve an optimal antibody response, the antigen should be broadly distributed to engage an increased number of immune cells.[27,80] Multiple sites of injection not only have the potential to enhance the antibody response, but also may minimize the side effects associated with adjuvants as it is given in smaller amounts at each site.[81]

The i.v. route will deliver antigen primarily to the spleen and secondary lymph nodes. Some workers consider this the route of choice for small particulate antigens; however, there is a risk of embolism when the i.v. route is used with emulsion adjuvants or large particulate adjuvants.[82] When using soluble antigens without adjuvant, the i.v. route is not effective at establishing an antigen depot site outside of lymphoid tissues. This may hinder the ability to sustain a high-titer response when used for primary immunization, and may increase tolerance.[10] Secondary immunization with an aqueous antigen by the i.v. route may be ineffective because antigen is aggregated *in vivo* as antigen:antibody complexes are formed.[83] However, there is a greater potential for anaphylaxis if the secondary immunization is given i.v. than if it is given i.m. or s.c. This should be avoided but the administration of antihistamines before inoculation may be helpful.[84] Not all adjuvants can be given i.v. and some may cause a systemic inflammatory response in systemic organs such as the liver and lung.[10] Vein access is limited in some species making i.v. injection difficult, and the volume injected is limited based on body size. These are particularly true in rodent species.

Intraperitoneal injection of antigen provides an efficient means of delivery of antigen to the lymphatics[85] and broadly distributes the antigen. Large volumes can be injected into the peritoneal cavity and it is easy to access. The i.p. route accommodates many types of adjuvants including liposomes and emulsion adjuvants. The i.p. route is frequently used in mice and occasionally in other species. Because the rapid absorption of antigen is similar to the i.v. route, booster injection of aqueous antigen given the i.p. route may cause anaphylaxis.[10] The main disadvantage of the i.p. route is the potential to cause severe, acute, and painful peritonitis for the first few days postimmunization.[26,86]

Skeletal muscle rapidly absorbs small molecules (<2 kDa) through vascular capillaries; therefore, the i.m. route is a good site of immunization for small molecular weight antigens. Immunization by the i.m. route also results in compartmentalization of the antigen for processing.[87] Large molecules and particulate antigens are unlikely to penetrate the vasculature unless there has been capillary damage or severe inflammation,[10] but are likely absorbed by the lymphatics and delivered to the draining lymph nodes. However, muscle lymphatic flow is limited, so antigen delivery is not rapid.[85] Large volumes of antigen can be injected into larger muscle masses. This again limits the use of this technique in rodents. Large volumes injected into the muscle often disperse along the fascial planes between muscles.[88] If given with an adjuvant, the inflammatory response can extend to adjacent nerve bundles, resulting in temporary or permanent nerve damage.[89]

The s.c. route is the most readily accessible and easiest to inject. Antigen may be dispersed to multiple sites even when given at a single injection site, depending on the tautness of the animal's skin. For example, rabbits have very loose skin, whereas guinea pigs do not; therefore, it may be easier for the injection medium to migrate to multiple sites. Inflammatory reactions from emulsion adjuvants may also increase dispersion of antigen.[10] Even though large amounts can be given at a single site with effective antibody responses, it is recommended to give immunizations in multiple sites to minimize the adverse reactions associated with the adjuvant, and minimize the potential for inducing tolerance.[73,74] Antigen is slowly absorbed after s.c. injection primarily through the lymphatics; however, this can also be effected by blood flow and activity of underlying musculature, antigen characteristics, species and injection site.[10] The s.c. route is often used for secondary immunizations for aqueous antigens to minimize the potential for anaphylaxis.

Subcutaneous chambers have been implanted in a number of species as a variation on s.c. injections with variable success.[90–95] This technique requires a surgical procedure in which a perforated golf ball is implanted in the subcutaneous space. The surgical site heals over a 4- to 6-week period, during which a granuloma forms around the implant. The chamber is then injected with antigen. Adjuvant may be required, or the granulomas associated with healing may act as an adjuvant.[91] There are no data describing the quality of the antibody obtained using s.c. implants; however, it is unlikely that affinity maturation occurs within in the granulomas. The use of s.c. implants has been described in rabbits, rats, sheep, and chickens.

Another variation of the s.c. route to raise pAb using microgram quantities of antigen involves immunizing with nitrocellulose impregnated strips.[96] The protein band of interest is recovered from a nitrocellulose blot, destained, and surgically implanted s.c. An excellent antibody response was demonstrated within 4 weeks without the use of an adjuvant, and the inflammatory responses were markedly reduced compared with FCA.

The high concentration of DCs in the skin and the consequent rapid delivery of antigen to the lymph nodes make the intradermal (i.d.) route favored by many investigators. However, the use of emulsion adjuvants leads to severe ulcers. This can be minimized by injecting at multiple sites, which may also bolster the immune response. For example, a one-time injection of 2 ml of an antigen/adjuvant emulsion was injected i.d. in 30–50 sites with as little as 20 μg of antigen, resulting in peak titers 2–4 months after primary immunization in rabbits.[80,97,98] Immunizations in this manner may not require booster injections.[68] The i.d. route has also been successfully used in rats and sheep.[99,100] The i.d. route is frequently used to deliver DNA immunization onto gold beads or using gene gun.[60]

A variation on the i.d. route is injection of the hind footpad. This is a popular method because it increases antigen retention and drains to the easily identified popliteal lymph node. Because the footpad is weight bearing, it is very sensitive when inflammation is present; therefore, careful consideration should be used in determining if this is the optimal method for polyclonal antibody production. If performed it should be avoided on the front feet and limited to only one back foot. It should not be performed in rabbits.[86] The inflammatory response is much more severe when given using an emulsion adjuvant. The foot pad injection is more useful for isolating high number of B cells from a local lymph node than for polyclonal antibody production.[74] However, similar results can be achieved with less pain and distress by injecting at the base of the tail or in the popliteal region.[20]

When small doses of antigen or antigens that are poorly immunogenic are available for immunization, the intranodal or intrasplenic routes may be considered for direct delivery of antigen to lymphoid tissues. This can be performed by transcutaneous inoculation or using a surgical preparation to directly visualize the organs. There has been variable success in developing a polyclonal antibody response using either of these routes, particularly when adjuvants elicit an inflammatory response. Some suggest the inflammatory response disrupts the architecture of the tissue altering its functions.[68] There are many reports that other routes of inoculation are as effective as intranodal including footpad injection and s.c. injection with antigen coupled to Sepharose or nitrocellulose.[101,102]

Table 4.3 provides a summary of the advantages and disadvantages of the injection routes. Table 4.4 provides maximum volume of injection used with either oil and viscous gel adjuvants or aqueous antigen/adjuvant mixtures.

As with antigen/adjuvant combination, there is no best route for immunization to produce pAb and it is dependent on the antigen, adjuvant, and the species. In general, aqueous soluble antigens given with adjuvants produce an antibody response most favorably when given i.d. followed by i.p., s.c., i.m., and i.v.[76] Other routes of injection yield variable responses in different studies and may be interchangeable.[10]

Several studies have compared different antigens and different routes on antibody titers. Some have found the i.p. route to be superior to the s.c. route in mice,[103] whereas others have found the s.c. route to be superior to the i.p. route.[13] Antibody responses in sheep have found that s.c. was as effective as i.d. when using antigen emulsified in FCA.[100] Rabbits immunized with antigen emulsified in FCA by the i.d. and i.m. route found higher titers with i.d. immunization, but antibody avidity

TABLE 4.3
Injection Routes Advantages and Disadvantages and Species Recommendations

Route	Details	Advantages	Disadvantages	Species
Subcutaneous (s.c.)	Most frequently used route Preferred route Do not inject material in part of animal used for restraint	Relatively large volumes can be administered Inflammatory process can be easily monitored	Slow absorption	Common route used in all species
Intramuscular (i.m.)	Skeletal muscles are well vascularized	Rapid adsorption, in particular, with muscle activity In large animals, relatively large volumes can be administered	Injection into closed space, which is painful Antigen and adjuvant can spread along interfacial plans and nerve bundles and may damage nerves and have other serious side effects Local reactions can be easily over looked	Not recommended for injection of oil adjuvants in rodents
Intraperitoneal (i.p.)	An efficient route for antigen delivery	Relatively large volumes of inoculum can be accommodated	Relative high percentage of injection failures Oil adjuvants produce peritonitis Risk for anaphylactic shocks at booster injection	Used primarily in rodents
Intravenous (i.v.)	Route of choice for particulate antigen Not recommended for insoluble antigens Not recommended for injection of oil adjuvant Antigen is delivery primarily to spleen and secondary lymph nodes	Rapid distribution of antigen	No oil or viscous gel adjuvant can be used High risk of anaphylactic shock at booster injection	Can be used in all species, but only for soluble antigens
Intradermal (i.d.)	High concentration of dendritic cells in the skin and rapid delivery of antigen to the lymph nodes Given at multiple locations	Small quantities of antigen already effective	Use of oil adjuvants leads to ulcerative process	Not recommended as a route for rodents

Modified from Hendriksen, C. F. and Hau, J., Production of Polyclonal and Monoclonal Antibodies, in *Handbook of Laboratory Animal Science*, Hau, J. and Van Hoosier, G.L. CRC Press, Boca Raton, 2003, pp. 391–408. Copyright 2003. Reproduced with permission of Routledge/Taylor & Francis LLC.

TABLE 4.4

Recommended Injection Volumes per Species for Oil and Aqueous Adjuvants

Using Oils and Viscous Gel Adjuvants

Species	s.c.	i.d.	i.m.	i.p.	i.v.
Mice	0.1	0.05 (1)	NR	NR	NA
Rats	0.1–0.2	0.05 (1)	NR	NR	NA
Guinea pigs	0.2	0.05 (1)	NR	NR	NA
Rabbits	0.1–0.25	0.025–0.05	NR	NR	NA
Sheep/goats	0.5	0.05	NR	NR	NA
Chickens	0.25	0.05	NR	NR	NA

Using Aqueous Antigen/Adjuvant Preparations

Species	s.c.	i.d.	i.m.	i.p.	i.v.
Mice	0.5	0.05 (1), 0.025 (2)	0.05	1.0	0.2
Rats	0.5–1.0	0.05 (1)	0.1 (1)	5.0	0.5
Guinea pigs	1.0	0.05 (1)	0.2 (1)	5.0–10.0 (1)	0.5–1.0
Rabbits	1.5	0.05	0.2–0.5	10.0–20.0 (1)	1.0–5.0
Sheep/goats	2.0	0.05	2.0	NA	30
Chickens	0.5	0.05	1.0	NA	0.5

Adapted from Leenaars, M. and Hendriksen, C.F., Critical steps in the production of polyclonal and monoclonal antibodies: evaluation and recommendations, *Ilar J* 46 (3), 269–79, 2005; and Hendriksen, C.F. and Hau, J., Production of Polyclonal and Monoclonal Antibodies, in *Handbook of Laboratory Animal Science*, Hau, J. and Van Hoosier, G.L. CRC Press, Boca Raton, 2003, pp. 391–408. Copyright 2003.

Volumes are in ml. NR: not recommended. NA: not available. (1) Not recommended. (2) Use smallest volume possible for foot pad insections.

was similar.[104] The most common route for immunizing animals to induce polyclonal antibodies is provided in Table 4.5.

4.2.9 IMMUNIZATION PROTOCOLS

Before initiating an immunization protocol to generate pAb, one must consider all the variables involved in the process, and the correct species should be selected. This may be based on the phylogenetic relationship of the antibodies for the intended use and selecting an animal that is as phylogenetically distant as possible. The amount of antibody desired should be considered, because large volumes of antibody are challenging to generate in smaller rodent species. The inoculation schedule should consider the dose, the optimal antigen concentration, the volume of injection, and the site of immunization. The antigen preparation should be adequate to immunize with a pure antigen to avoid undesired, contaminating antibodies that will need

TABLE 4.5
Recommended Age of Animals to Initiate Immunizations and Sites

Species	Age	Sites
Mice and rats	6 weeks	s.c., i.p.
Chickens	18–20 weeks	i.m.
Rabbits and guinea pigs	3 months	i.d., s.c., i.m., i.v.
Goats	6–7 months	i.d., s.c., i.m., i.v.
Sheep	7–9 months	i.d., s.c., i.m., i.v.

Adapted from The Report and Recommendations of ECVAM Workshop 35 (altweb.jhsph.edu/publications/ECVAM.ecvam35.htm).

to be removed. Booster schedules should be considered including the frequency, the route and the purpose of the study.[87]

Although there is no single best immunization protocol for inducing antibody responses, there are some generalities that can be applied. It is typically insufficient to immunize animals only once and expect a high titer antibody response, even with the use of adjuvants. Booster immunizations are frequently used to enhance and prolong the antibody response, because they activate memory cells. Memory cells have a shorter latency period before producing antibodies, produce more IgG than IgM, and the antibodies have a greater affinity and avidity than primary responses.[2,87] Ideally, one should follow the antibody titers and booster when the titers plateau or shortly after they begin to decline, because antigen in the presence of antibody does not enhance antibody responses.[105] Primary antibody response wane 3–6 weeks after the primary immunization indicating boosters can occur at this time, typically 3–4 weeks.[10,74] Increasing the frequency of the booster immunization does not enhance the antibody titer or avidity and may actually decrease it.[104,106,107] Although prolonging the booster immunization may enhance the avidity of the antibody response, boostering with too high a dose of antigen could result in tolerance. Booster immunizations typically result in a maximal antibody response 10 days after each booster.[87]

Booster immunization need not be given in the same site or by the same route. Intradermal immunizations have localized antigen distribution, and boosters could be given by an alternative route to deliver the antigen more broadly. For example, booster immunizations can readily be given by the s.c. route after primary immunization by alterative routes.[84] Soluble antigens should not be boostered by the i.v. route because the potential for anaphylaxis is high; therefore, boosters should occur by the s.c. or i.m. routes. The most frequently used protocol for producing antibody responses to soluble antigens involves the use of FCA for the primary immunization followed by antigen in FIA for booster immunizations.[10] Immunizations without depot forming adjuvants typically reach peak antibody responses in 1–2 weeks, and subsequent booster may need to occur monthly to maintain a high antibody titer.[10] When a depot-forming adjuvant is used, antibody titers may remain increased for several weeks to months. If an emulsion is used for the primary immunization, then booster immunizations may be done without the use of the adjuvant as the emulsion maintains a depot effect that may last for several weeks to months.[74] Certainly one does not want to give booster immunization into a subcutaneous granuloma. Some suggest that the number of booster immunizations be limited to two or three, and, if there is no antibody response, the experiment is terminated. However, low-molecular-weight antigens may require additional boosters.[74] There is a recent report of obtaining an effective pAb response in 1 month with a modified immunization protocol.[108] In this report the rabbit was immunized with the antigen with FCA on Day 1 and Day 3, followed by a booster on Day 28, and serum collected on Day 35.

The frequency of intermittent collection of sera does not appear to adversely affect antibody response, and in most instances promotes the antibody response. It is thought that this is due to the release of antigen from circulating antigen:antibody complexes exposing memory B cells to freed antigen.[10]

4.2.10 SPECIES SELECTION

Selection of the most appropriate animal model for producing polyclonal antibodies is dependent on several factors: the phylogenic relationship of the antibody to the antigen, the use of the antibody collected, the characteristics of the antibody response, the amount of sera required for the amount of antibody, the availability of the animal model, the age and sex of the animal model, and the ease of blood collection.[10,74,109]

The phylogenic relationship between the antigen source and the immunized animal can influence the immune response to the desired antigen. For example, the antibody response to a murine antigen may not be as pronounced in mice or rats as it might be in a rabbit. This is particularly true for intracellular antigens that would ordinarily be recognized by the host during thymic development and generate tolerance. On the other hand, highly specific antibody response to a limited number of epitopes may require immunization with closely related species or the same species with genetic divergence.[10] Mammals hyperimmunized with an antigen from a phylogenically distant species can make up to 10 mg/ml of antibody, whereas those hyperimmunized with an antigen from an allotypic species may make less than 1 mg/ml of antibody.[10] Chickens used for polyclonal antibody production produce IgY (equivalent to mammalian IgG); however, it does not cross-react with mammalian IgG.[110,111] It is particularly useful for highly conserved mammalian antigens such as intracellular proteins because there is a substantial phylogenic distance between the donor and the recipient.[74]

The proposed use of the antibody will influence the animal selected. For example if the antibodies are going to be used in an ELISA, the primary antibody should not be the same species as the conjugate, or there will be a high level of cross-reactivity. This may also take a greater amount of sera than if one were to use a different technology such as multiplex microbead analysis.[113] This may influence the decision to use rabbits instead of mice for producing antibodies, because more can be collected from the larger sized rabbit. Antibody subclass characteristics such as opsonization ability and complement fixation may influence the immunization protocol and the animal model selected. Although mice may produce five different subclasses of antibody, one may be desired over the others. For example, IgG2a and IgG2b have strong complement fixation and opsonization compared with IgG1.

Younger animals are preferred models for polyclonal antibody production because as animals mature they develop strong IgG responses with increased immunologic memory.[87] In addition, they have not had a significant immune challenge to potential pathogens and the environment in general. However, if the animal is too young, then one may see a nonspecific IgM response, or if it is too old, the immune responses decline in intensity and diversity.[10,107] In some species, such as mice, in which the immune system is not completely developed in utero, antigen exposure too early in the course of development could lead to tolerance. Table 4.5 provides the recommended age of animals for each species to begin immunizations.

Animal housing conditions can also influence the immune response. Normal flora and infectious agents can modulate the immune response and may develop antibodies that cross react to other antigens.[114,115] Animals housed in specific pathogen free environments are preferred because they will not encounter the immune challenges in conventional housing.

Gender is also a factor to consider. Female animals have typically been chosen over males. They are typically less aggressive than males, easier to handle, and can be group housed. Females have been reported to be more sensitive to lower antigen doses and have an increased primary and secondary response to antigens of a longer duration compared with males.[10] Females have higher circulating antibodies with stronger immune responses as estrogens increase B-cell responses, whereas testosterone decreases B-cell response.[116] However, there is controversy over the effects of hormones in the generation of polyclonal antibodies.[109] Stress hormones may adversely affect the immune response as cortisol suppresses lymphocyte proliferation. This has been demonstrated in pigs in which the immune response of a dominant pig was greater than the subordinates when housed as a group.[117] Therefore social or single housing of animals, particularly rodents and rabbits, may influence the antibody response. The nutritional status of the animals can influence the immune response as protein deficiencies can impair immune responses,[118] and those with supplements such as vitamins A, C, and E can enhance response.[119]

When considering an animal model for antibody production, it is important to consider the amount of antibody needed because it relates to the size of the animal and the ease of collecting blood. Usually more than one animal is immunized. This creates a more diverse antibody pool and protects against animals that are nonresponders. It typically takes 4–8 weeks to generate polyclonal antibodies after immunization.[109] The most frequently used species for polyclonal antibody production are the rabbit, mouse, rat, guinea pig, goat, sheep, and chicken.

Rabbits are the most common mammalian species used. They are readily available from commercial vendors and their size and lifespan facilitate easy housing, handling, immunizing, and collection. They produce antibodies of a high titer with high affinity, precipitating antisera,[89] which is easily collected from the central ear artery. Rabbits are lagomorphs and phylogenically distant to rodents.

Several rodent species are used for polyclonal antibody production, but mice are the most common. Mice are available in many different strains from commercial vendors, allowing investigators to select the best responder to their antigen. For example, BALB/c mice have strong humoral antibody response to antigens with a predominant IgG1 response, whereas C57BL/6 mice have a comparably diminished antibody response with a predominant IgG2a response.[120,121] The use of inbred strains may be more appropriate because the immune response is likely to have little individual variation.[122] However, not all inbred strains are the same. Those originating from different vendors may have undergone substrain divergence that could alter the immune response seen in an inbred strain obtained from different sources.[87] Generating a large amount of antibody using mice requires a substantial number of animals as only 200 µl can be safely acquired from a 25-g mouse at regular bleeding intervals and injection volumes are limited due to the small tissue mass (see Table 4.4). Some investigators have overcome this limitation by producing polyclonal antibodies using an ascites method.[123–126] This method is a variation on the production of monoclonal antibodies by the ascites cells that involves tumor production in the peritoneal cavity of a rodent, usually the mouse. It has been shown that inducing ascites causes a significant degree of pain and distress and therefore alternatives to antibody production by the ascites method is suggested.[127–129] If the ascites method for polyclonal antibody production must be used, then it should be performed with one terminal tap on development of ascites to minimize pain and distress to the animal.

Other rodents have been used for polyclonal antibody production, but they do not offer any advantages over rabbits for routine polyclonal antibody production. For specific responses, rats may be the species of choice for IgE and guinea pig antibodies are excellent at complement fixation.[10]

Farm animals, such as sheep, goats, and horses, offer some advantages over smaller mammalian species. Blood collection is easily achieved via the jugular vein and can be performed on a routine basis without many complications. Their longer life span allows collection over a longer period. One can typically achieve a greater amount of antisera with fewer farm animals than with other species. However, they are expensive to purchase and require special housing that may not be available at most facilities.

Chickens are a particularly useful animal used in polyclonal antibody production. Immunized chickens produce antibody IgY in the egg yolk at high concentrations.[110–112] A single egg can yield more than 10 times the amount of antibody than weekly serum collection from rabbits, with equivalent avidity, yet it does not require handling or needle injection into the animal to collect blood.[130] Simply collect the egg. Chickens are phylogenically distant from mammalian species, thereby minimizing cross-reactivity, and inbred strains are available, thereby minimizing individual responses. Furthermore, IgY does not activate mammalian complement and it does not react with bacterial protein A or G, mammalian Fc receptors, or rheumatoid factors. IgY is stable for months within the egg, but the stability can be prolonged if the immunoglobulin is purified from the yolk and stored appropriately. The disadvantage of using chickens is the limited number of commercially available conjugated antibodies directed against IgY.[112]

After immunization, the animals should be monitored for anaphylaxis, particularly when soluble antigens are given i.v. or i.p. Then animals should be monitored daily for adverse reactions, such as granulomas formation, and to assess the welfare of the animal. If the adverse reactions are too severe, then the animal may need to be humanely euthanized.

Frequency and amount of blood collected are limited in each species. As a general rule, no more than 10% of the blood volume should be removed every 10 days. Removal of more than 10% of an animal's blood volume can result in hypovolemic shock and anemia.[131] Recovery periods can vary with the amount withdrawn. Because single samplings are usually performed for the collection of antisera, the recommended recovery period is 1, 2, and 4 weeks if removing 7.5, 10, 15% of the blood volume, respectively.[132] The recommended blood volumes and sites of collections are given in Table 4.6.

Terminal exsanguinations can be performed to maximize the one-time collection of antisera. For maximum collection, this procedure is performed by cardiocentesis under general anesthesia for smaller species such as rabbits and rodents. Larger species may be exsanguinated via catheterization of the jugular vein under general anesthesia. Consult your institutional veterinarian for advice on how to appropriately

TABLE 4.6
Blood Volumes and Recommended Safe Bleeding Volumes and Sites for Each Species[74,132,136]

Species (Average Weight)	Blood Volume in ml (Average ml/kg)	Maximum Volume Collected	Recommended Sites
Mice (25 g)	1.8 (79)	0.3	Tail vein, lateral saphenous, submandibular vein, retroorbital sinus, tail tip amputation
Rats (250 g)	16 (64)	2.0	Tail vein, retroorbital plexus, jugular vein (2)
Guinea pigs (900 g)	68 (75)	5.0	Metatarsal vein, cranial vena cava (2)
Rabbits (4 kg)	224 (56)	25 (1)	Marginal ear vein
Sheep/goats (90 kg)	6.0 liters (68)	200–600 (1)	Jugular vein, cephalic vein

(1) Depends on body weight.

(2) Requires anesthesia.

Adapted from Diehl, K.H., Hull, R., Morton, D., Pfister, R., Rabemampianina, Y., Smith, D., Vidal, J.M., and van de Vorstenbosch, C., A good practice guide to the administration of substances and removal of blood, including routes and volumes, *J Appl Toxicol* 21, 15, 2001; Hawk, C.T. and Leary, S.L., *Formulary for Laboratory Animals,* Iowa State University Press, Ames, 1995; and Hendriksen, C.F. and Hau, J., Production of polyclonal and monoclonal antibodies, in *Handbook of Laboratory Animal Science*, Hau, J. and Van Hoosier, G.L., eds., CRC Press, Boca Raton, 2003, pp. 391–408.

TABLE 4.7
Suggested Anesthetics for Exsanguinations for Each Species

Species	Anesthesia (Dose and Route)	Expected Volume (ml)
Mice	1. Ketamine (100 mg/kg) and xylazine (10 mg/kg), i.p. 2. Pentobarbital (45 mg/kg), i.p. 3. Isoflurane inhalation	1.0–1.5
Rats	1. Ketamine (75 mg/kg) and xylazine (10 mg/kg), i.p. 2. Pentobarbital (45 mg/kg), i.p. 3. Isoflurane inhalation	6.0–8.0
Guinea pigs	1. Ketamine (40 mg/kg) and xylazine (5 mg/kg), i.p. 2. Pentobarbital (40 mg/kg), i.p. 3. Isoflurane inhalation	20–30
Rabbits	Acepromazine (1 mg/kg), i.m. given 15–20 minutes before anesthesia 1. Ketamine (10 mg/kg) and xylazine (3 mg/kg), i.v. 2. Pentobarbital (35 mg/kg), i.v. 3. Isoflurane inhalation	90–120
Sheep/goats	1. Ketamine (4 mg/kg) and xylazine (0.2 mg/kg, sheep; 0.05 mg/kg, goats), i.v. Pentobarbital (30 mg/kg), i.v. 2. Telazol (3 mg/kg), i.m. 3. Isoflurane inhalation (requires intubation)	2000–3000

Adapted from Flecknell, P., *Laboratory Animal Anaesthesia: A Practical Introduction for Research Workers and Technicians,* Academic Press, London, 1996; and Diehl, K.H., Hull, R., Morton, D., Pfister, R., Rabemampianina, Y., Smith, D., Vidal, J.M., and van de Vorstenbosch, C., A good practice guide to the administration of substances and removal of blood, including routes and volumes, *J Appl Toxicol* 21, 15, 2001.

deliver anesthesia to the different species. After the exsanguinations, the animal is humanely euthanized in accordance with the most current American Veterinary Medical Association Panel on Euthanasia or local equivalent organization.[133] Table 4.7 provides suggested anesthesia protocols for exsanguinations and expected volumes.

4.3 CONCLUSION

Use of pAbs in research continues to be of great importance. Although the production of polyclonal antibodies is a relatively straightforward process, there are several aspects to consider before embarking on production. One needs to consider the type of antibody response desired (i.e., isotype), the adjuvant to be used, the route of immunization and the schedule, and the species to be used. There is no single best combination for maximal production. This chapter, as well as several other excellent reference texts, will serve to guide investigators producing polyclonal antibodies.

ACKNOWLEDGMENTS

I would like to thank Mary Wood (University of California Center for Animal Alternatives) for assistance with literature search, Michelle Swan for administrative assistance, and Drs. Laurel Gershwin and Kate Wasson for their review of the chapter.

REFERENCES

1. Paul, W.E., *Fundamental Immunology*, 5th ed., Lippincott Williams & Wilkins, Philadelphia, 2003.
2. Janeway, C.A., Jr., et al., *Immunobiology: The Immune System in Health and Disease*, 6th ed., Garland Science, New York, 2005.
3. Moser, M., Dendritic cells, in *Fundamental Immunology*, 5th ed., Paul, W.E., Lippincott Williams & Wilkins, Philadelphia, 2003.
4. Gordon, S., Macrophages and the immune response, in *Fundamental Immunology*, 5th ed., Paul, W.E., Lippincott Williams & Wilkins, Philadelphia, 2003.
5. Lenschow, D.J., et al., CD28/B7 system of T cell costimulation, *Annu. Rev. Immunol.*, 14, 233, 1996.
6. Mosmann, T.R. and Coffman, R.L., TH1 and TH2 cells: different patterns of lymphokine secretion lead to different functional properties, *Annu. Rev. Immunol.*, 7, 145, 1989.
7. Tough, D.F. and Sprent, J., Immunologic memory, in *Fundamental Immunology*, 5th ed., Paul, W.E., Lippincott Williams & Wilkins, Philadelphia, 2003.
8. Petrovsky, N. and Aguilar, J.C., Vaccine adjuvants: current state and future trends, *Immunol. Cell. Biol.*, 82, 488, 2004.
9. Singh, M. and O'Hagan, D.T., Recent advances in veterinary vaccine adjuvants, *Int. J. Parasitol.*, 33, 469, 2003.
10. Hanly, W.C., Artwohl, J.E., and Bennett, B.T., Review of polyclonal antibody production procedures in mammals and poultry, *Ilar J.*, 37, 93, 1995.
11. Hunter, R.L., Olsen, M.R., and Bennett, B., Copolymer adjuvants and titermax, in *The Theory and Application of Adjuvants*, Stewart-Tull, D.E.S., John Wiley & Sons Ltd, West Sussex, England, 1995.
12. Stewart-Tull, D.E.S., Freund-type mineral oil adjuvant emulsions, in *The Theory and Application of Adjuvants*, Stewart-Tull, D.E.S., John Wiley & Sons Ltd, West Sussex, England, 1995.
13. Kenney, J.S., et al., Influence of adjuvants on the quantity, affinity, isotype and epitope specificity of murine antibodies, *J. Immunol. Methods*, 121, 157, 1989.
14. Jennings, V.M., Review of selected adjuvants used in antibody production, *Ilar J.*, 37, 119, 1995.
15. *Topics in Vaccine Adjuvant Research*, CRC Press, Boca Raton, FL, 1991. eds. Spriggs, Dale R., and Koff, Wayne C.
16. Stills, H.F., Jr., Adjuvants and antibody production: dispelling the myths associated with Freund's complete and other adjuvants, *Ilar J.*, 46, 280, 2005.

17. Stewart-Tull, D.E.S., *The Theory and Application of Adjuvants,* John Wiley & Sons Ltd, West Sussex, England, 1995.
18. Freund, J., The effect of paraffin oil and mycobacteria on antibody formation and sensitization; a review, *Am. J. Clin. Pathol.,* 21, 645, 1951.
19. NIH Intramural Recommendation for the use of Complete Freund's Adjuvant, *ILAR News,* 30, 1988.
20. Canadian Council on Animal Care, *Guidelines on: Antibody Production*, Ottawa, Ontario, Canada, 2002.
21. Hunter, R.L. and Bennett, B., The adjuvant activity of nonionic block polymer surfactants. II. Antibody formation and inflammation related to the structure of triblock and octablock copolymers, *J. Immunol.,* 133, 3167, 1984.
22. Howerton, D.A., et al., Induction of macrophage Ia expression *in vivo* by a synthetic block copolymer, L81, *J. Immunol.,* 144, 1578, 1990.
23. Leenaars, P. P., et al., Evaluation of several adjuvants as alternatives to the use of Freund's adjuvant in rabbits, *Vet. Immunol, Immunopathol.,* 40, 225, 1994.
24. Smith, D.E., et al., The selection of an adjuvant emulsion for polyclonal antibody production using a low-molecular-weight antigen in rabbits, *Lab. Anim. Sci.,* 42, 599, 1992.
25. Tejada-Simon, M.V. and Pestka, J.J., Production of polyclonal antibody against ergosterol hemisuccinate using Freund's and Titermax adjuvants, *J. Food Prot.,* 61, 1060, 1998.
26. Leenaars, P.P., et al., Assessment of side effects induced by injection of different adjuvant/antigen combinations in rabbits and mice, *Lab. Anim.,* 32, 387, 1998.
27. Bennett, B., et al., A comparison of commercially available adjuvants for use in research, *J. Immunol. Methods,* 153, 31, 1992.
28. Ribi, E., et al., Biologically active components from mycobacterial cell walls. IV. Protection of mice against aerosol infection with virulent mycobacterium tuberculosis, *Cell Immunol.,* 16, 1, 1975.
29. Altman, A. and Dixon, F.J., Immunomodifiers in vaccines, *Adv. Vet. Sci. Comp. Med.,* 33, 301, 1989.
30. Ellouz, F., et al., Minimal structural requirements for adjuvant activity of bacterial peptidoglycan derivatives, *Biochem. Biophys. Res. Commun.,* 59, 1317, 1974.
31. Rudbach, J.A., Johnson, D.A., and Ulrich, J.T., Ribi adjuvants: chemistry, biology and utility in vaccines for human and veterinary medicine, in *The Theory and Application of Adjuvants*, Stewart-Tull, D.E.S., John Wiley & Sons Ltd, West Sussex, England, 1995.
32. Ribi, E., et al., Lipid A and immunotherapy, *Rev. Infect. Dis.,* 6, 567, 1984.
33. Chiller, J.M., et al., Relationship of the structure of bacterial lipopolysaccharides to its function in mitogenesis and adjuvanticity, *Proc. Natl. Acad. Sc. U. S. A.,* 70, 2129, 1973.
34. Kotani, S., et al., Immunobiological activities of synthetic lipid A analogs with low endotoxicity, *Infect., Immun.,* 54, 673, 1986.
35. Deeb, B.J., et al., Comparison of Freund's and Ribi adjuvants for inducing antibodies to the synthetic antigen (TG)-AL in rabbits, *J. Immunol. Methods,* 152, 105, 1992.
36. Johnston, B.A., et al., An evaluation of several adjuvant emulsion regimens for the production of polyclonal antisera in rabbits, *Lab. Anim. Sci.,* 41, 15, 1991.
37. Lipman, N.S., et al., Comparison of immune response potentiation and *in vivo* inflammatory effects of Freund's and RIBI adjuvants in mice, *Lab., Anim. Sci.,* 42, 193, 1992.
38. Mallon, F.M., et al., Comparison of antibody response by use of synthetic adjuvant system and Freund complete adjuvant in rabbits, *Am. J. Vet. Res.,* 52, 1503, 1991.

39. Bokhout, B., et al., A selected water-in-oil emulsion: composition and usefulness as an immunological adjuvant, *Vet. Immunol. Immunopathol.*, 2, 491, 1981.

40. Lindblad, E.B., Aluminium adjuvants, in *The Theory and Application of Adjuvants*, Stewart-Tull, D.E.S., John Wiley & Sons Ltd, West Sussex, England, 1995.

41. Allison, A.C. and Byars, N.E., Immunological adjuvants: desirable properties and side-effects, *Mol. Immunol.*, 28, 279, 1991.

42. Leenaars, P.P., et al., Comparison of adjuvants for immune potentiating properties and side effects in mice, *Vet. Immunol. Immunopathol.*, 48, 123, 1995.

43. Allison, A.C. and Gregoriadis, G., Liposomes as immunological adjuvants, *Recent Results Cancer Res.*, 56, 58, 1976.

44. Gregoriadis, G., Liposomes as immunological adjuvants, in *The Theory and Application of Adjuvants*, Stewart-Tull, D.E.S., John Wiley & Sons Ltd, West Sussex, England, 1995.

45. Reddy, R., et al., Liposomes as antigen delivery systems in viral immunity, *Semin. Immunol.*, 4, 91, 1992.

46. Takada, H. and Shozo, K., Muramy dipeptide and derivatives, in *The Theory and Application of Adjuvants*, Stewart-Tull, D.E.S., John Wiley & Sons Ltd, West Sussex, England, 1995.

47. Dalsgaard, K., et al., Immune stimulating complexes with Quil A, in *The Theory and Application of Adjuvants*, Stewart-Tull, D.E.S., John Wiley & Sons Ltd, West Sussex, England, 1995.

48. Sanders, M.T., et al., ISCOM-based vaccines: the second decade, *Immunol. Cell. Biol.*, 83, 119, 2005.

49. Speijers, G.J., et al., Local reactions of the saponin Quil A and a Quil A containing iscom measles vaccine after intramuscular injection of rats: a comparison with the effect of DPT-polio vaccine, *Fundam. Appl. Toxicol.*, 10, 425, 1988.

50. Pearse, M.J. and Drane, D., ISCOMATRIX adjuvant for antigen delivery, *Adv. Drug Deliv. Rev.*, 57, 465, 2005.

51. Dalpke, A.H. and Heeg, K., CpG-DNA as immune response modifier, *Int. J. Med. Microbiol.*, 294, 345, 2004.

52. Klinman, D.M., CpG DNA as a vaccine adjuvant, *Expert Rev. Vaccines,* 2, 305, 2003.

53. Mutwiri, G.K., et al., Strategies for enhancing the immunostimulatory effects of CpG oligodeoxynucleotides, *J. Control Release,* 97, 1, 2004.

54. Medzhitov, R. and Janeway, C.A., Jr., Innate immunity: the virtues of a nonclonal system of recognition, *Cell,* 91, 295, 1997.

55. Klinman, D.M., et al., Use of CpG oligodeoxynucleotides as immune adjuvants, *Immunol. Rev.,* 199, 201, 2004.

56. Ada, G. and Ramshaw, I., DNA vaccination, *Expert Opin. Emerg. Drugs,* 8, 27, 2003.

57. Mutwiri, G., et al., Biological activity of immunostimulatory CpG DNA motifs in domestic animals, *Vet. Immunol. Immunopathol.*, 91, 89, 2003.

58. Sasaki, S., et al., Adjuvant formulations and delivery systems for DNA vaccines, *Methods,* 31, 243, 2003.

59. Chambers, R.S. and Johnston, S.A., High-level generation of polyclonal antibodies by genetic immunization, *Nat. Biotechnol.*, 21, 1088, 2003.

60. Johnston, S.A. and Tang, D.C., Gene gun transfection of animal cells and genetic immunization, *Methods Cell. Biol.*, 43, 353, 1994.

61. Esche, U.v.d., et al., Immunostimulation by bacterial components: I. Activation Of macrophages and enhancement of genetic immunization by the lipopeptide P3CSK4, *Int. J. Immunopharmacol.*, 22, 1093, 2000.

62. Kellner, J., et al., The influence of various adjuvants on antibody synthesis following immunization with an hapten, *Biol. Chem. Hoppe Seyler,* 373, 51, 1992.

63. Wu, J.Y., et al., Evaluation of cholera vaccines formulated with toxin-coregulated pilin peptide plus polymer adjuvant in mice, *Infect. Immunol.,* 69, 7695, 2001.

64. Holland, M.J., et al., Proteins secreted by the parasitic nematode Nippostrongylus brasiliensis act as adjuvants for Th2 responses, *Eur. J. Immunol.,* 30, 1977, 2000.

65. Stevceva, L. and Ferrari, M.G., Mucosal adjuvants, *Curr. Pharm. Des.,* 11, 801, 2005.

66. Barr, T.A., et al., A potent adjuvant effect of CD40 antibody attached to antigen, *Immunology,* 109, 87, 2003.

67. Carlring, J., et al., Anti-CD28 has a potent adjuvant effect on the antibody response to soluble antigens mediated through CTLA-4 by-pass, *Eur. J. Immunol.,* 33, 135, 2003.

68. Hurn, B., Practical problems in raising antisera, *Br. Med. Bull.,* 30, 26, 1974.

69. Byars, N.E. and Allison, A.C., Syntex adjuvant formulation, in *The Theory and Application of Adjuvants,* Stewart-Tull, D.E.S., John Wiley & Sons Ltd, West Sussex, England, 1995.

70. McCoy, K.L., et al., Tolerance defects in New Zealand Black and New Zealand Black X New Zealand White F1 mice, *J. Immunol.,* 136, 1217, 1986.

71. Maurer, P.H. and Callahan, H.J., Proteins and polypeptides as antigens, *Methods Enzymol.,* 70, 49, 1980.

72. Hu, J.G. and Kitagawa, T., Studies on the optimal immunization schedule of experimental animals. VI. Antigen dose-response of aluminum hydroxide-aided immunization and booster effect under low antigen dose, *Chem. Pharm. Bull. (Tokyo),* 38, 2775, 1990.

73. Zinkernagel, R.M., Localization dose and time of antigens determine immune reactivity, *Semin. Immunol.,* 12, 163, 2000.

74. Hendriksen, C.F. and Hau, J., *Production of Polyclonal and Monoclonal Antibodies,* CRC Press, Boca Raton, FL, 2003.

75. Harlow, E. and Lane, D., *Antibodies: A Laboratory Manual,* Cold Spring Harbor Laboratory, Cold Spring Harbor, NY, 1988.

76. Hurn, B.A. and Chantler, S.M., Production of reagent antibodies, *Methods Enzymol.,* 70, 104, 1980.

77. Muller, S., Immunization with peptides, in *Synthetic Peptides as Antigens,* Muller, S., Elsevier, Amsterdam, 1999.

78. Halliday, L.C., et al., Physiologic and behavioral assessment of rabbits immunized with Freund's complete adjuvant, *Contemp. Top. Lab. Anim. Sci.,* 39, 8-, 2000.

79. Halliday, L.C., et al., Effects of Freund's complete adjuvant on the physiology, histology, and activity of New Zealand White Rabbits, *Contemp. Top. Lab. Anim. Sci.,* 43, 2004.

80. Vaitukaitis, J.L., Production of antisera with small doses of immunogen: multiple intradermal injections, *Methods Enzymol.,* 73, 46, 1981.

81. Stills, H.F., Jr. and Bailey, M., The use of Freund's complete adjuvant, *Lab Anim.,* 20, 25, 1991.

82. Herbert, W.J., Mineral oil adjuvants and the immunization of laboratory animals, in *Handbook of Experimental Immunology,* 3rd ed., Weir, D.M., Blackwell Scientific Publications, Oxford, 1978.

83. Leskowitz, S. and Waksman, B.H., Studies on immunization. 1. The effect of route of injection of bovine serum albumin in Freund adjuvant on production of circulating antibody and delayed hypersensitivity, *J. Immunol.,* 84, 58, 1960.

84. Herbert, W.J., Laboratory animal techniques for immunology, in *Handbook of Experimental Immunology*, 3rd ed., Weir, D.M., Blackwell Scientific Publications, Oxford, 1978.

85. O'Driscoll, C.M., Anatomy and physiology of the lymphatics, in *Lymphatic Transport of Drugs*, Charman, W.N. and Stella, V.J., CRC Press, Boca Raton, FL, 1992.

86. Amyx, H.L., Control of animal pain and distress in antibody production and infectious disease studies, *J, Am. Vet. Med. Assoc.*, 191, 1287, 1987.

87. Schunk, M.K. and Macallum, G.E., Applications and optimization of immunization procedures, *Ilar J.*, 46, 241., 2005.

88. Droual, R., et al., Investigation of problems associated with intramuscular breast injection of oil-adjuvanted killed vaccines in chickens, *Avian Dis.*, 3, 473, 1990.

89. Stills, H.F., Polyclonal antibody production, in *The Biology of the Laboratory Rabbits*, 2nd ed., Manning, P.J., et al., Academic Press, San Diego, CA, 1994.

90. Ermeling, B.L., et al., Evaluation of subcutaneous chambers as an alternative to conventional methods of antibody production in chickens, *Lab. Anim. Sci.*, 42, 402, 1992.

91. Clemons, D.J., et al., Evaluation of a subcutaneously implanted chamber for antibody production in rabbits, *Lab. Anim. Sci.*, 42, 307, 1992.

92. Ried, J.L., et al., Production of polyclonal antibodies in rabbits is simplified using perforated plastic golf balls, *Biotechniques*, 12, 660, 1992.

93. Wolff, K.L., et al., Production of antibody in induced granulomas, *J. Clin. Microbiol.*, 4, 384, 1976.

94. Hillam, R. P., et al., Local antibody production against the murine toxin of Yersinia pestis in a golf ball-induced granuloma, *Infect. Immunol.*, 10, 458, 1974.

95. Hajer, I., et al., Immunoglobulin response to bluetongue virus soluble antigen in subcutaneous chambers, *Am. J. Vet. Res.*, 38, 815, 1977.

96. Coghlan, L.G. and Hanausek, M., Subcutaneous immunization of rabbits with nitrocellulose paper strips impregnated with microgram quantities of protein, *J. Immunol. Methods*, 129, 135, 1990.

97. Herbert, W.J., The mode of action of mineral-oil emulsion adjuvants on antibody production in mice, *Immunology*, 14, 301, 1968.

98. Vaitukaitis, J., et al., A method for producing specific antisera with small doses of immunogen, *J. Clin. Endocrinol. Metab.*, 33, 988, 1971.

99. Hillier, S.G., et al., The active immunisation of intact adult rats against steroid-protein conjugates: effects on circulating hormone levels and related physiological processes, in *Steroid Immunoassay*, Cameron, E.H.D., et al., Alpha Omega Publishing, Ltd., Cardiff, Wales, 1975.

100. Scaramuzzi, R.J., et al., Production of antisera to steroid hormones in sheep, in *Steroid Immunoassay*, Cameron, E.H.D., et al., Alpha Omega Publishing, Ltd., Cardiff, Wales, 1975.

101. Horne, C.H. and White, R.G., Evaluation of the direct injection of antigen into a peripheral lymph node for the production of humoral and cell-mediated immunity in the guinea-pig, *Immunology*, 15, 65, 1968.

102. Nilsson, B.O., et al., Immunization of mice and rabbits by intrasplenic deposition of nanogram quantities of protein attached to Sepharose beads or nitrocellulose paper strips, *J. Immunol. Methods*, 99, 67, 1987.

103. Hu, J.G., et al., Studies on the optimal immunization schedule of experimental animals. V. The effects of the route of injection, the content of Mycobacteria in Freund's adjuvant and the emulsifying antigen, *Chem. Pharm. Bull. (Tokyo)*, 38, 1961, 1990.

104. Lader, S., et al., A comparative assessment of immunisation procedures for radioimmunoassays, in *Radioimmunoassay and Related Procedures in Medicine,* International Atomic Energy Agency, Vienna, 1974.

105. Chande, C., et al., Sequential study of IgG antibody response in immunized rabbit and development of immunization protocol for raising monospecific antibody, *Indian J. Pathol. Microbiol.,* 39, 27, 1996.

106. Serody, J.S., et al., T cell activity after dendritic cell vaccination is dependent on both the type of antigen and the mode of delivery, *J. Immunol.,* 164, 4961, 2000.

107. Hu, J.G., et al., Studies on the optimal immunization schedule of experimental animals. IV. The optimal age and sex of mice, and the influence of booster injections, *Chem. Pharm. Bull. (Tokyo),* 38, 448, 1990.

108. Hu, Y.X., et al., Get effective polyclonal antisera in one month, *Cell Res.,* 12, 157, 2002.

109. Leenaars, M. and Hendriksen, C.F., Critical steps in the production of polyclonal and monoclonal antibodies: evaluation and recommendations, *Ilar J.,* 46, 269, 2005.

110. Larsson, A., et al., Chicken antibodies: taking advantage of evolution—a review, *Poult. Sci.,* 72, 1807, 1993.

111. Tini, M.J., et al., Generation and application of chicken egg-yolk antibodies, *Comp. Biochem. Physiol. A Mol. Integr. Physiol.,* 131, 569, 2002.

112. Schade, R., et al., The Production of Avian (Egg Yolk) Antibodies: IgY. The Report and Recommendations of ECVAM Workshop 21, 1994.

113. Khan, I.H., et al., Simultaneous serodetection of 10 highly prevalent mouse infectious pathogens in a single reaction by multiplex analysis, *Clin. Diagn. Lab. Immunol.,* 12, 513, 2005.

114. Klaasen, H.L., et al., Apathogenic, intestinal, segmented, filamentous bacteria stimulate the mucosal immune system of mice, *Infect. Immun.,* 61, 303, 1993.

115. O'Rourke, J., et al., Differences in the gastrointestinal microbiota of specific pathogen free mice: an often unknown variable in biomedical research, *Lab. Anim.,* 22, 297, 1988.

116. Da Silva, J.A., Sex hormones and glucocorticoids: interactions with the immune system, *Ann. N. Y. Acad. Sci.,* 876, 102, 1999.

117. de Groot, J., et al., Long-term effects of social stress on antiviral immunity in pigs, *Physiol. Behav.,* 73, 145, 2001.

118. Konno, A., et al., Effects of a protein-free diet or food restriction on the immune system of Wistar and Buffalo rats at different ages, *Mech. Ageing Dev.,* 72, 183, 1993.

119. Lopez-Varela, S., et al., Functional foods and the immune system: a review, *Eur. J. Clin. Nutr.,* 56(Suppl 3), S29, 2002.

120. Kendall, L.V., et al., Antibody and cytokine responses to the Cilia-Associated Respiratory bacillus infected BALB/c and C57BL/6 mice, *Infect. Immun.,* 68, 4961, 2000.

121. Snapper, C.M. and Finkelman, F.D., Immunoglobulin class switching, in *Fundamental Immunology,* 4th ed., Paul, W.E., Lippincott-Raven, Philadelphia, 1999.

122. Melo, M.E., et al., Strain-dependent effect of nasal instillation of antigen on the immune response in mice, *Isr. Med. Assoc. J.,* 4(11 Suppl), 902, 2002.

123. Cartledge, C., et al., Production of polyclonal antibodies in ascitic fluid of mice: time and dose relationships, *J. Immunoassay,* 13, 339, 1992.

124. Kurpisz, M., et al., Production of large amounts of mouse polyclonal antisera, *J. Immunol. Methods,* 115, 195, 1988.

125. Lacy, M.J. and Voss, E.W., Jr., A modified method to induce immune polyclonal ascites fluid in BALB/c mice using Sp2/0-Ag14 cells, *J. Immunol. Methods,* 87, 169, 1986.

126. Mahana, W. and Paraf, A., Mice ascites as a source of polyclonal and monoclonal antibodies, *J. Immunol. Methods,* 161, 187, 1993.
127. Peterson, N.C., Behavioral, clinical, and physiologic analysis of mice used for ascites monoclonal antibody production, *Comp. Med.,* 50, 516, 2000.
128. Jackson, L.R., et al., Monoclonal antibody production in murine ascites. I. Clinical and pathologic features, *Lab. Anim. Sci.* 49, 70, 1999.
129. Toth, L.A., et al., An evaluation of distress following intraperitoneal immunization with Freund's adjuvant in mice, *Lab. Anim. Sci.,* 39, 122, 1989.
130. Svendsen Bollen, L., et al., Antibody production in rabbits and chickens immunized with human IgG. A comparison of titre and avidity development in rabbit serum, chicken serum and egg yolk using three different adjuvants, *J. Immunol. Methods,* 191, 113, 1996.
131. Wagner, A.E. and Dunlop, C.I., Anesthetic and medical management of acute hemorrhage during surgery, *J. Am. Vet. Med. Assoc.,* 203, 40, 1993.
132. Diehl, K.H., et al., A good practice guide to the administration of substances and removal of blood, including routes and volumes, *J. Appl. Toxicol.,* 21, 15, 2001.
133. 2000 Report of the AVMA Panel on Euthanasia, *J. Am. Vet. Med. Assoc.,* 218, 669, 2001.
134. Cooper, H.M. and Paterson, Y., Production of antibodies, in *Current Protocols in Immunology,* Coligan, J.E., et al., John Wiley & Sons, Inc., 1995.
135. Tizard, I.R., *Veterinary Immunology: An Introduction,* 6th ed., W.B. Saunders Company, Philadelphia, 2000.
136. Hawk, C.T. and Leary, S.L., *Formulary for Laboratory Animals,* Iowa State University Press, Ames, 1995.
137. Flecknell, P., *Laboratory Animal Anaesthesia: A Practical Introduction for Research Workers and Technicians,* Academic Press, London, 1996.

5 Production of Monoclonal Antibodies

Kathleen C. F. Sheehan

CONTENTS

5.1 INTRODUCTION

Three decades after the advent of hybridoma technology, the development of mono-clonal antibodies (mAb) continues to uphold the hopes of Kohler and Milstein as reagents "valuable for medical and industrial use."[1] Hybridoma technology and the generation of mAbs have revolutionized research, providing limitless quantities of unique reagents capable of identifying, isolating, eliminating, activating, or targeting specific antigens. Improved immunization strategies now allow for more rapid and efficient generation of humoral responses to a wide variety of antigenic stimuli using a broad collection of hosts. Technical advances have also provided better and more rapid means of hybridoma screening and antibody production. In addition, advances in genetic engineering permit the modification of mAbs for clinical use with increased efficacy and decreased toxicity. Therefore, although we continue to modify and improve on many of the originally described technical aspects, hybridoma technology remains a vital tool in scientific discovery and clinical therapy.

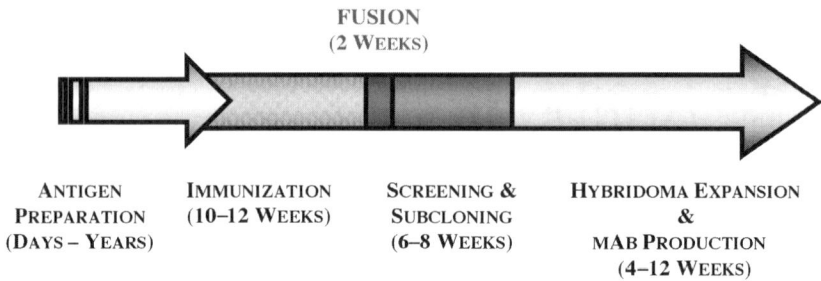

FIGURE 5.1 Timeline for monoclonal antibody development.

At the outset of any antibody development venture, it is important to first determine whether the task at hand can best be accomplished using polyvalent or polyclonal antisera or mAbs. Polyvalent antisera consist of a heterogeneous mixture of antibodies that each bind to unique sites on each antigen, display different binding affinities, and represent different classes or subclasses of immunoglobulin. Reactivity with multiple epitopes results in high-affinity binding; however, polyclonal sera may also contain irrelevant antibodies. Although polyvalent antisera are relatively easy to generate, inconsistencies may exist between bleeds and quantities will be limited. mAbs, the product of a single, clonal hybridoma, are highly specific and homogeneous in their binding characteristics. Individual mAbs can be selected for unique epitope binding or functional activities and generally display nanomolar binding affinities. Unlike polyvalent mixtures, mAbs may more easily lose reactivity if the antigenic epitope is denatured, and heteroclitic antibodies can exhibit unexpected cross reactivity. Although unlimited quantities can be generated, the commitment to develop these reagents is lengthy and costly. Thus it is important to discern whether the advantages of mAbs warrant the expenditure of effort above and beyond that needed to generate polyclonal antisera (Figure 5.1).

5.2 MATERIALS

5.2.1 THE IMMUNOGEN

After the decision is made to develop mAbs the first set of considerations involves the nature of the immunogen (see Chapter 2). The immunogen (or antigen) can be expressed in many forms: whole cells, cellular extracts, purified protein, fusion proteins, peptide, carbohydrate, lipid, or DNA. Furthermore, the immunogen need not be pure or monospecific for the generation of mAbs, as the specificity of each clone will be individually addressed during antibody screening. Whole cells or cellular lysates (10^7 cells/injection) can serve as a source of intracellular or extracellular molecules in their native form; however, the abundance or scarcity of a particular agent may affect the efficiency of immunization. Cell lines engineered to overexpress a particular protein are often excellent sources of antigen. Purified recombinant proteins or fusion proteins are also commonly used (10–100 μg/injection) and may result in reactivity to a number of different epitopes. *If fusion proteins are used for immunization, it is important to also generate either the immunogen alone or as part of a different fusion protein to identify the proper antigenetic specificity when screening for antibody production.* Similarly, peptide antigens (>10 amino acids) must be administered as conjugates with carrier protein. Because these carrier proteins are highly immunogenic, the peptides should also be linked to more than one carrier to enable differential screening for peptide versus carrier protein (i.e., keyhole limpet hemocyanin and bovine serum albumin). Antipeptide antibodies define small linear sequences and have been particularly effective at recognizing modified sequences, such as differentiating between the presence or absence of a phosphate group on a particular peptide.

Recently, new methods to induce antigenic protein expression *in vivo* have been developed using plasmid-based techniques. Genetic immunization is quite effective after either intrasplenic or intramuscular injection,[2] intradermal administration with a device, such as the Gene Gun apparatus,[3,4] or by intravenous injection by hydrodynamic immunization.[5-7] These methods differ in the mechanism by which the plasmid DNA is delivered to host cells. However, each method relies on the administration of a unique coding sequence as part of a mammalian expression vector, with subsequent host cell uptake of the injected DNA and *in vivo* production of the protein. Moreover, genetic immunization can be used successfully when the antigen is difficult to generate and can also be used for gene products, for which little is known. Plasmid-based immunization generates significant quantities of the target protein *in vivo*, resulting in the generation of a robust humoral response. This technique can also be utilized with gene-targeted mice to develop antibodies specific for the targeted antigen.[5,5a]

Carbohydrate determinants, which are often expressed on microbial pathogens, can elicit moderate humoral responses, but rarely lead to secondary priming. Hence, carbohydrate determinants often generate only immunoglobulin (Ig)M responses, such as those seen in response to blood group antigens. The response to carbohydrate antigens may also be improved by conjugation to carrier proteins. Lipids, alone or expressed as lipoproteins, can be used to generate antibodies reactive with hapten-like fragments of the molecule after antigen presentation by CD1 molecules. Even more difficult is the development of humoral responses to nucleic acids.

In general, the types of antigenic protein determinants recognized by antibody molecules can be classified into three categories: conformational determinants, linear determinants, and neoantigenic determinants. Conformational determinants result from three-dimensional structures that may be constructed from separate segments of the protein (nonlinear sequences). Conformational determinants represent antigen expressed in its native form and may be lost when the molecule is denatured. Linear determinants (a sequence of adjacent amino acids) are not destroyed by protein denaturation but may or may not be exposed when a protein is in its native, folded state. Antipeptide antibodies that bind to linear determinants are an example of linear determinants. Neoantigenic determinants denote newly expressed epitopes created by proteolysis of native antigen. Antibodies that recognize neoantigens will generally not react with native protein.

Clearly the form of the immunogen will affect the spectrum of the host humoral response. Immunization with whole cells, cellular lysates, or *in vivo*–generated protein will result in the development of a broad spectrum of antibodies that identify an array of antigenic determinants present on the native protein (conformational, linear, and neoantigens). These antibodies may be more suited for a broad range of functional activities, including flow cytometry studies or functional neutralization. Peptide immunization will generally result in exposure of a single epitope. Antibodies reactive with these linear sequences do not always bind to native globular proteins; however, these epitopes are often maintained even if the native protein becomes denatured. Such peptide-reactive antibodies often function well in Western blot analyses or immunohistochemistry. Thus the end use of the antibodies in development, should influence the choice of immunogen such that it favors the isolation of antibodies with the desired function. Although there are no guarantees that any given form of an immunogen will produce antibodies with a specific functional activity, it can help to bias the development of unique specificities. Roughly 2 mg of protein antigen will be needed for immunization and preliminary screening assays.

Before immunization, it is imperative that the antigen be screened for the presence of any potential pathogens. This applies to every cell line, cell-derived product, or antigen produced in the presence serum. The screening assay, referred to as mouse antibody production screening (MAPS) consists of inoculation of quarantined, specific pathogen-free animals with each unique antigen (2×10^7 cells or 10–100 µg protein), followed by incubation for a period of 4 weeks. Thereafter, sera from inoculated animals are collected and compared with preinjection sera from the same animals for reactivity to a panel of known pathogens. Positive results signify that the material is contaminated and should be discarded; negative results indicate that the antigen is safe for *in vivo* injection.

5.2.2 THE SPECIES IMMUNIZED

Hybridoma technology has been perfected using lymphocytes isolated from the mouse, rat, Armenian hamster, and rabbit. The mouse is an excellent candidate for immunization. Multiple inbred strains are readily available at reasonable cost, mice are easy to manipulate, a wealth of reagents are available to detect mouse Ig, and mice respond well to foreign proteins. In addition, mouse lymphocytes fuse productively with several murine myeloma lines (Table 5.1). Moreover, gene-targeted mice can be used to develop antibodies reactive with the targeted protein, provided that the animal retains a functional immune system. Rats similarly provide a ready host particularly for the generation of antibodies reactive with murine proteins. Both murine and rat myelomas are available that are suitable for rat B-cell fusions. Armenian hamsters present a unique model because they are phylogenetically far enough removed from mice to respond well to murine (as well as rat or human) proteins and yet are able to fuse productively with murine myeloma lines.[8,9,9a] More importantly, Armenian hamster antibodies are nonimmunogenic in mice, making them ideal for use as *in vivo* models of disease.[9–11,11a] This model system has been used extensively to develop high-affinity neutralizing reagents to murine cytokines, receptors, and transcription factors.[9–16] Rabbit hybridomas can be generated from the fusion of immune splenocytes with a rabbit myeloma line.[17] Thus, depending

TABLE 5.1
Recommended Myelomas

Myeloma	Species	Derivation	Reference
P3X63Ag8.653	Mouse	MOPC21→P3K→	25
		P3X63Ag8	
Sp2/0-AG14	Mouse	MOPC21→P3K→	1
		P3X63Ag8→Fuse with splenocytes	26
NSO/1	Mouse	MOPC21→P3K→	27
		NSI/1-Ag4–1	
YB2/0	Rat	S210→Y3-Ag 1.2.3 fused to AO	28
		splenocytes→YB2/3HL	

on the derivation of the immunizing agent and the end use of the antibody, a number of animal models are available for the development of unique monoclonal reagents of desired specificity and function.

One alternative to cell fusion for the development of mAbs is the use of the Immortomouse.[18] These transgenic mice carry a temperature-sensitive mutant of SV40 T antigen under the control of the H-2K^b promoter. After immunization of these animals, splenocytes cultured at the permissive temperature (33°C) will express the thermolabile SV40 T antigen resulting in B-cell transformation without the need for cell fusion with an immortalized cell line and avoiding the instability often generated by the production of heterokaryons. These continuously growing B cells can then be screened for specific antibody production.

Efforts to more efficiently develop fully human antibodies with well-established hybridoma technology has led to the development of the "Abgenix XenoMouse."[19,20] These animals are completely devoid of murine J_H and C_κ genes and render the animal deficient in all mouse immunoglobulins. These murine Ig$^{-/-}$ mice are also transgenic for human Ig_H and Ig_κ chains. Thus, XenoMouse strains have the genetic components to generate diverse primary immune responses similar to that seen in an adult human, with development of a robust secondary response after antigen exposure. Conventional immunization of these animals and fusion of the immune splenocyte with murine myeloma cell lines provide a novel strategy to develop large panels of high-affinity fully human antibodies reactive with any number of proteins. Such antibodies can display neutralizing activity, and many are now approved for human therapy.

5.2.3 IMMUNIZATION STRATEGIES

Immunization strategies vary from laboratory to laboratory, with many strategies equally effective. The goal is to expand within the host population antigen reactive B cells. After each exposure to protein antigen *in vivo*, the frequency of reactive B cells increases, the amount of antibody produced rises during the secondary response to antigen, the overall immunoglobulin response switches from a predominance of IgM to IgG and the binding affinity of the specific antibody is enhanced. In general, the stronger the *in vivo* response and the more B cells reactive with the given antigen, the greater the chances of isolating a hybridoma expressing antibody with the desired binding capacity and functional activity. *In vitro* methods of immune activation and B-cell expansion have not been shown to be highly reliable. Indeed, the best results have occurred using *in vitro* activation of splenocytes obtained from immunized animals. Such methods have also resulted in the generation of predominantly IgM secreting hybridomas.

The immunogenicity of a particular antigen is based on a number of factors, but may be enhanced by the use of adjuvants (see Chapter 3). Adjuvants (from the Latin word *adjuvare* meaning to help) act as a vehicle to prevent antigen clearance, extend the length of time antigen is exposed to the immune system, recruit cells to the site of the antigen, and stimulate lymphocyte proliferation because of the presence of molecules that enhance costimulatory signals and activate the immune system. Several adjuvant systems have been described, including Freund's adjuvant, the mainstay of immunologic adjuvants for decades. However, complete Freund's adjuvant (CFA) can also induce unwanted side effects. Hence, alternative synthetic water-in-oil emulsion adjuvants have been developed with reduced toxicity (such as Ribi or TiterMax). In addition, alum, which forms a coprecipitate with antigen, has low toxicity and has even been used in some vaccines. Alum has been found to be effective in circumstances in which Freund's adjuvant has been ineffective in stimulating humoral responses to specific epitopes. Finally, bacterial-derived genetic sequences, immunostimulatory CpG oligonucleotides (containing unmethylated cytosine and guanine dinucleotides), have recently been used to enhance innate immunity, induce dendritic cell, monocyte and macrophage maturation, elicit Th1 cellular responses, and stimulate B-cell proliferation and immunoglobulin synthesis, leading to *in vivo* protective immunity.[21,22]

Typical immunization strategies rely on repeated exposure to antigen to enhance and mature the specific immune response. Table 5.2 outlines one type of immunization schedule. In this example, antigen (10–100 µg protein or 10^7 cells) is emulsified with CFA. Equal volumes of antigen and CFA are combined to produce a

TABLE 5.2
Typical Immunization Schedule

Day	Manipulation	Adjuvant	Site
0	Primary immunization	Complete Freund's adjuvant	Subcutaneous
14	Boost #1	Incomplete Freund's adjuvant	Subcutaneous
28	Boost #2	Incomplete Freund's adjuvant	Subcutaneous
36	Serum collection and titer	—	
42	Rest before fusion	Incomplete Freund's adjuvant	
	OR		
	Boost #3		Subcutaneous
52	Prefusion boost	None	Intravenous
55	Harvest splenocytes and fuse	—	

thick emulsion that is injected (in mice, rats, or hamsters) at multiple subcutaneous sites (0.5–1.0 ml total volume). Heat-killed mycobacterium in the CFA will cause an inflammatory response and recruit lymphocytes to the site of antigen deposition. Over 10–14 days, the primary response generates a weak antibody response (predominantly low-affinity IgM).

The first boost (second exposure to antigen) is performed with incomplete Freund's adjuvant (IFA). (IFA does not contain the heat-killed bacteria present in CFA; repeated exposure to CFA can lead to granuloma formation and is contraindicated.) The antigen-IFA emulsion is again administered at multiple subcutaneous sites. The secondary response to protein antigens will produce a larger humoral response with an increase in IgG isotypes as compared with IgM. Although a significant response may be detected, a second boost is recommended to mature the response, further expanding the repertoire of reactive B cells and increasing serum titers. Eight to ten days after the second boost, serum should be collected from the immunized animal and screened for reactivity compared with sera collected before immunization. Similar injection schedules may be applied to genetic immunization using plasmid DNA as well. Numerous immunization schedules have been published that vary in the choice of adjuvants, injection volumes, antigen concentrations, dosage times, and the number of injections. The method described previously delineates one schedule that repeatedly performs well in a number of model systems.

The reactivity of serum derived from immunized animals should be assessed at multiple dilutions and in several assay systems to determine the strength and nature of the humoral response. Because of the complexity of serum components, sera samples should always be tested at dilutions greater than 1:50 dilution to reduce nonspecific binding. Moreover, titrations should include dilutions greater than 1:1,000, even surpassing a 1:100,000 dilution, to provide the best view of the developing *in vivo* immune response. The serum titer is defined as the reciprocal of the greatest dilution for which clearly detectable reactivity with antigen can be measured for any given screening assay.

Several screening methods may be available (discussed in detail in the following section) depending on the nature of the immunogen including enzyme-linked immunosorbent assay (ELISA), flow cytometry, immunoprecipitation, or functional neutralization. Serum titers are best evaluated using an assay system designed to monitor the desired function. That is, if the end goal is to develop mAbs capable of binding recombinant protein, then an ELISA may provide the most efficient screening tool. However, if mAb reactive with native cell surface receptors are needed, then serum should be screened by flow cytometry or other assays in which the antigen is expressed in its native form. It is important to realize that antibodies that function in one system may fail to bind in other situations. Antibodies reactive with conformational determinants may not react with denatured proteins in Western blots; likewise, antibodies raised to peptide sequences buried within a globular protein may not be capable of blocking functional activity. Therefore, it is important to have a clear understanding of the nature of the immunogen and consider the downstream antibody requirements in designing the immunization and screening strategy. In general, a serum titer of greater than 1:1,000 is recommended to continue on to hybridoma development. If the serum titer is less than 1:1,000, additional boosts should be administered to enhance *in vivo* reactivity to antigen.

Occasionally, only minimal reactivity may be measured even after repeated antigenic boosts. Although it is possible to recover antigen specific hybridomas from animals that display low serum titers, the frequency of these clones is generally low and often results in the generation of IgM-secreting hybridomas. In these situations, it may be advisable to revisit the initial immunization strategy, altering the form of the immunogen or the adjuvant system used. Antigens that do not elicit a vigorous response in the presence of CFA/IFA may be highly immunogenic when administered with alum.

After animals have been identified that display significant serum titers (>1:1,000 dilution), the animals should be "rested" for 3–6 weeks. This resting time allows for the heightened secondary response to decay so that a final exposure to antigen will result in new B-cell activation and expansion. Rested animals should receive a final prefusion boost of antigen diluted in either pyrogen-free saline or phosphate-buffered saline (~200 μl total volume) in the absence of any adjuvant. It is most effective to administer the final boost via an intravenous injection to deliver the antigen directly to the spleen, where newly activated blasted B cells will be harvested for cell fusion. If the immunogen is not compatible with intravenous (i.v.) injection (the antigen is particulate or contains buffer components that may be toxic), the antigen may be administered intraperitoneally, again in the absence of adjuvant. Components that are incompatible with i.v. injection include Freund's adjuvant, detergent concentrations greater than 0.1%, urea greater than 1 M, endotoxins, Tris, sodium azide, or buffer and salts above physiologic concentrations. Three days after the i.v. boost (4 days after an intraperitoneal boost), immune splenocytes or lymph node cells are harvested for cell fusion.

5.3 PROCEDURES

5.3.1 MEDIA AND MYELOMAS

The development of antibody-secreting B-cell hybridomas relies on the productive cell fusion of antigen-reactive B cells with immortalized B-cell tumors, myelomas, and the subsequent identification and isolation of continuously growing cell lines capable of producing antibody with the desired specificity. The outgrowth of selected hybridomas requires fastidious tissue culture skills and extra care of fragile cultures.[23] Although several different media formulations and culture conditions will support the development and growth of hybridomas, herein is described one methodology that reliably produces outstanding results (Table 5.3). It is best to select a rich medium for cell growth, because the cell-fusion process, as well as downstream expansion and low-density subcloning, is stressful to cells. One such formulation begins with Dulbecco's Modified Eagle's Medium (DMEM) with high glucose (4.5 g/l) (RPMI-1640 or Iscoves' Modified Eagle's Medium may also be used). The DMEM is supplemented with 10–20% fetal bovine serum (FBS), 4-mM L-glutamine, 1-mM sodium pyruvate, 50-U/ml penicillin, 50-μg/ml streptomycin, and 50-μM 2-mercaptoethanol (2-ME) (hereon referred to as complete DMEM medium).

TABLE 5.3
Reagents for Hybridoma Production and Growth

	Reagent	Final Concentration
Media components	DMEM (4.5 g/l glucose)	—
	Fetal bovine serum	10–20%
	L-glutamine	4 mM
	Sodium pyruvate	1 mM
Antibiotics	Penicillin	50 U/ml
	Streptomycin	50 µg/ml
	2-Mercaptoethanol	50 µM
Supplements	HAT selection mixture	1×
	Hypoxanthine	100 µM
	Aminopterin	0.4 µM
	Thymidine	16 µM
	HT supplement	1×
	Hybridoma cloning factor	1–5%
Fusion reagents	PEG 1500	50%

Because of the lengthy cell-culture process, all media components should be sterile-filtered using 0.22-micron filters and stored at 4°C for no more than 2 weeks. FBS is preferable to other serum sources because it contains low concentrations of bovine Ig that may interfere with screening assays or antibody purification. Unique lots of FBS (or serum replacements) should be screened before use for hybridoma development because there are significant lot-to-lot differences in their ability to support hybridoma growth. All reagents must be free of mycoplasma contamination and should be low in endotoxin (<0.25 EU/ml). Both L-glutamine and sodium pyruvate are added to DMEM to support vigorous cell growth. Antibiotics are optional; however, because the hybridoma cultures are subject to frequent manipulation over a period of weeks, it is judicious to include antibiotics to cover both gram-positive and gram-negative organisms. (In the event of mycoplasma contamination, penicillin and streptomycin in the medium should be replaced with ciprofloxacin (10 µg/ml) for a minimum of five passages.) Inclusion of 2-ME can be controversial, because its mechanism of action is unknown. Although hybridomas can certainly be generated in its absence, many cultures may benefit from its inclusion. After a hybridoma is grown in the presence of 2-ME, the 2-ME should not be removed from the culture medium, as doing so may result in a significant drop in Ig secretion or viability.

Other more recent modifications to hybridoma growth procedures include the development of mediums that require no serum or are devoid of all animal proteins. These are good options for the downstream large-scale expansion and production of mAb. Many provide for high-density cell culture and simplify mAb purification. It remains to be seen how efficient these new growth media will be in the initial generation of antigen-specific hybridomas. Hybridomas may need to be slowly adapted into these new growth mediums to maintain antibody production. In many cases Ig production increases with reduction in the serum concentration.

After cell fusion, the cultures are subject to drug selection to eliminate the growth of myeloma cells that fail to yield productive hybridomas. A number of agents can be used, based on the specific mutations in individual myeloma cell lines. HAT selection is most common and is the drug of choice for the myelomas described in Table 5.2. HAT supplement can be purchased from a number of commercial sources and consists of hypoxanthine (100 μM), aminopterin (0.4 μM), and thymidine (16 μM). Aminopterin will block de novo purine and pyrimidine synthesis in all cells. Hypoxanthine and thymidine are included to support synthesis via the salvage pathways (Figure 5.2). During the original growth and expansion of hybridomas, additional growth factors or supplements may be added. Although some laboratories will use "conditioned media" generated from "feeder layers" of splenocytes or peritoneal macrophages, commercial supplements that provide a source of interleu-kin-6 can also be used (Origen HCF [hybridoma cloning factor], IGEN Inc., Gaith-ersburg, MD). If feeder cells are used, the cells (4×10^3 peritoneal lavage cells or 1×10^4 splenocytes per well in a volume of 75 μl) should be plated 1–2 days in advance of use. This ensures the production of factors to help support hybridoma growth and allows researchers to verify that the feeder plates are free of any contamination that may compromise hybridoma growth. Alternatively, commercial supplements should be used at concentrations recommended by the manufacturer, typically 1–5% (volume/volume) and may be added directly into the culture medium at the time of use. Both feeder cell layers and commercial supplements are equally effective at supporting the growth of newly forming hybridomas.

A number of murine and rat myeloma lines have been described that contribute cellular immortality and drug selection (see Table 5.1). Among the murine myelomas listed, each was originally derived from the same Balb/c mineral oil–induced plas-macytoma (MOPC21), *in vitro* adapted (P3K) and selected with 8-azaguanine to obtain clones deficient in enzymes of the salvage pathway of purine nucleotide synthesis (HGPRT⁻), as well as nonsecretion of immunoglobulin heavy or light chains. These myeloma cells grow continuously in culture. However, in the presence of selection agents capable of blocking de novo nucleotide synthesis (aminopterin,

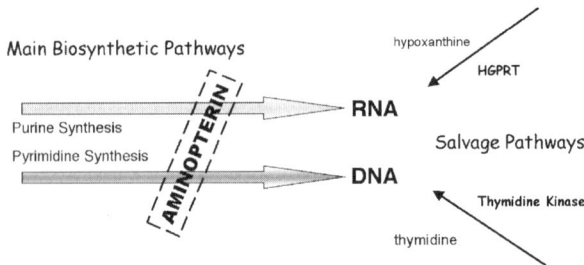

FIGURE 5.2 Mechanism of HAT selection.

methotrexate or azaserine) these HGPRT-negative myelomas cannot survive as the salvage pathway is disrupted by enzyme deficiency (see Figure 5.2). Thus, these myelomas provide the perfect model for hybridoma development by selective agents.

Myeloma cells should be banked in liquid nitrogen with a new vial expanded for each cell fusion. It is imperative that myeloma stocks are free of mycoplasma contamination and should be tested for the presence of any infectious agents, using the identical procedure described for the screening of immunogens (mouse antibody production screening). Myelomas should be cultured for no longer than 2 weeks before fusion and continually maintained in log phase growth at densities less than 10^6 per milliliter. Typically, myelomas are grown in supplemented DMEM containing 10% FBS, as described previously, and passaged three times per week at plating densities of $2.5–10.0 \times 10^4$/ml, depending on the rate of growth of the particular myeloma line. Routinely, a sample of the myeloma line should be plated in medium containing the selection agent (HAT) to ensure that all cells die and that the myeloma line has not been contaminated, as well as to confirm the efficacy of the selecting agent. Before fusion with antigen-reactive B cells, the myeloma cells are washed three times in serum-free medium at 4°C, because serum proteins will interfere with membrane fusion. The myelomas should be counted and resuspended at 1×10^7 cells/ml in DMEM without serum and held at 4°C. Viability should be greater than 95%.

5.3.2 IMMUNOREACTIVE B-CELL PREPARATION

Animals with significant serum titers (>1:1000) provide the antigen reactive B cells for cell fusion. These cells may be derived from either the spleen or lymph nodes. Animals are sacrificed by carbon dioxide inhalation, and the tissue is removed using aseptic technique and stored in serum-free medium at 4°C for up to 1 h. Single-cell suspensions are prepared by gentle disruption of the capsule with forceps or needles, never with ground glass slides that can impair membrane integrity. As with the myeloma preparation, splenocytes or lymph node–derived cells are washed (centrifugation at 1000 rpm for 10 min at 4°C) three times in medium lacking serum, counted and resuspended in 20 ml of serum-free medium, and placed in a 50-ml polypropylene conical tube on ice. Typically, 1×10^8 murine splenocytes or 0.6×10^8 Armenian hamster splenocytes are isolated from immunized animals.

5.3.3 FUSION PROTOCOL AND PLATING

Before cell fusion, all materials and equipment should be assembled and components warmed to the appropriate temperature (Figure 5.3). The biosafety cabinet should be thoroughly cleaned and swabbed with 70% ethanol. All media components should be sterile filtered. Electronic pipette-aids should have new in-line filters installed, and multichannel pipeters should be swabbed to reduce any opportunities for contamination. If used, plates containing feeder cell layers should be examined microscopically for any evidence of contamination. Other supplies needed include a timer and 2-, 10-, and 25-ml pipettes. Immunoreactive B cells and myeloma cells are resuspended in serum-free medium and held at 4°C. The following components are warmed to 37°C: insulated beaker to serve as a water bath, 50% polyethylene

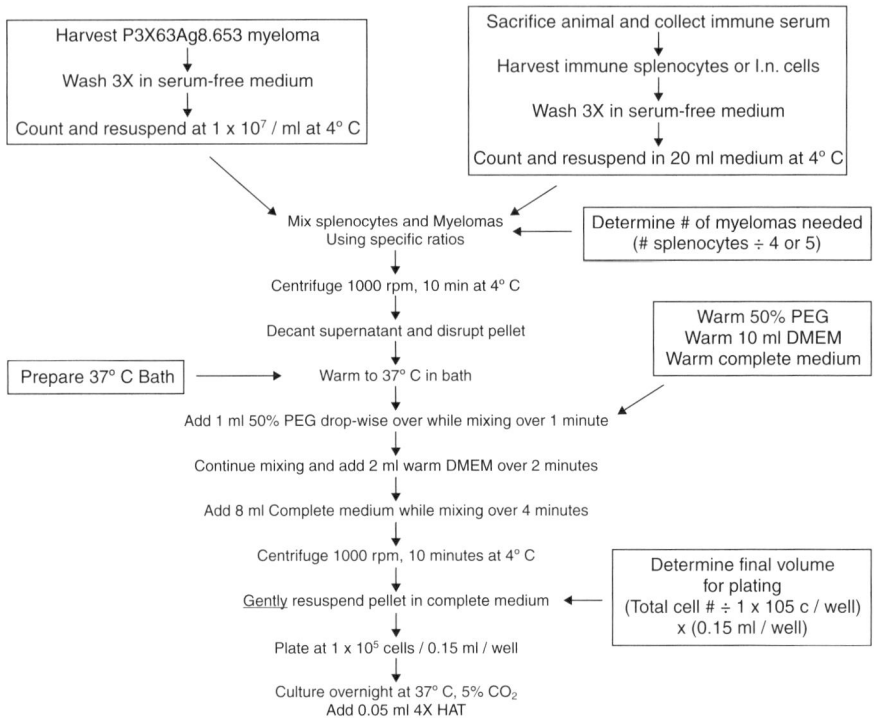

FIGURE 5.3 Protocol for B-cell fusion.

glycol solution (1 ml per spleen), 10 ml supplemented DMEM containing 10% FBS, and supplemented DMEM containing 20% FBS for final plating. Cell counts, including the percent viable cells, should be recorded. Calculations to determine the number of myeloma cells needed (see the following section) and the final volume of medium needed for plating (1×10^5 cells / well) should be completed.

Immune lymphocytes are mixed with myeloma cells at defined ratios before cell fusion. The ratio of lymphocytes to myeloma cells varies greatly between species, myeloma lines, and laboratories. In our laboratory, murine or rat splenocytes are mixed at a ratio of five splenocytes per one myeloma cell (P3X63Ag8.653). Armenian hamster splenocytes are mixed at a ratio of four splenocytes per myeloma. To generate B-cell hybridomas, myeloma cells are added at the appropriate ratio to the immune splenocytes or lymph-node cells. For example, 2×10^7 myeloma cells should be mixed with 1×10^8 murine splenocytes. The cell populations are mixed and centrifuged together at 1000 rpm for 10 min at 4°C in a 50-ml conical tube. Thereafter, the supernatant is decanted completely, and the cell pellet is disrupted by agitation (by running the tube along a grated surface) so that the cell pellet is evenly coated over the bottom apex of the tube. The tubes containing the mixed myelomas and B cells are then placed in an insulated beaker containing 37°C water.

The following procedures are performed at 37°C. At 37°C, cell membranes are more fluid, which favors the cell-fusion process mediated by PEG 1500. The warming cell pellet, the prewarmed 50% PEG, and the warmed DMEM–10% FBS are placed in the tissue-culture hood. To fuse the immunoreactive B cells with the myeloma cells, the 50% PEG mixture is added dropwise to the cell pellet (1 ml over 1 min) while continually swirling the mixture in the 37°C water bath. (The conical tube containing the cell mixture should be angled so that each drop of PEG added lands gently along the upper rim of the cell-suspension coating the base of the tube to evenly mix with the single-cell suspension.) The fused cells are gently rotated for an additional 30 sec at 37°C. Next, the PEG is slowly diluted by a series of timed additions of warmed DMEM supplemented with 10% FBS. In succession, 2 ml of warm DMEM –10% FBS are added over 2 min while gently rotating the mixture in the 37°C bath, followed by an additional 8 ml of medium over 2 min with gentle mixing. This diluted mixture of splenocytes, myeloma cells, and newly formed hybridomas are centrifuged at 1000 rpm for 10 min at 4°C. The supernatant is then decanted, and the cell pellet is gently resuspended in DMEM complete medium supplemented with 20% FBS with a large-bore (25 ml) pipette. The newly formed conjugates are fragile and should not be vigorously agitated or vortexed.

To plate the newly generated hybridomas, the cells are resuspended in an appropriate volume to dispense approximately 1×10^5 cells per well in 96-well microtiter plates with low evaporation lids. If the hybridomas are to be plated into wells already containing feeder layers (75 µl/well PEC or splenocytes), the newly fused cells are resuspended at a concentration of 1.3×10^6/ml so that the addition of 75 µl/well of hybridomas results in a plating density of 1×10^5 cells per well. Each well will contain a total volume of 150 µl. If feeder cells are not used, the complete DMEM medium containing 20% FBS should be supplemented with growth factors, such as hybridoma cloning factor (Origen HCF, IGEN Inc.) at 1–5%. Under these conditions, the newly generated hybridomas are resuspended at a concentration of 6.7×10^5 cells/ml so that the addition of 150 µl/well results in a plating density of 1×10^5 cells per well, identical to that described previously. Fusion plates containing 1×10^5 fused cells in a total volume of 150 µl are cultured overnight at 37°C in an atmosphere of 5% CO_2 and 95% humidity. One day after fusion, the cells are placed under selection by the addition of HAT-containing medium (complete DMEM medium containing 20% FBS plus 4× HAT supplement, 50 µl/well). At this time, individual wells will contain large numbers of viable cells with little or no cellular debris evident. Addition of HAT (50 µl of medium containing 4× HAT) 24 h after cell fusion (instead of at the time of fusion) allows cells to recover from the initial stress of the fusion process and results in higher yields of growth positive wells.

After the cells are placed under selection (with the major biosynthetic pathways for purine and pyrimidine synthesis blocked by aminopterin), myeloma cells (deficient in HGPRT) that did not fuse with splenocytes or that did not fuse productively so that enzymes needed for the salvage pathway were acquired, will begin to die off. Likewise, splenocytes that did not fuse with myeloma cells will only survive a few days in culture. This is visibly evident in culture wells that show reduced viable cell numbers and increased amounts of cellular debris over the first 3–5 days of *in vitro* culture. Beginning on Day 5 after fusion, the fusion plates are "fed" three times per week. Half the volume of culture supernatant (~100 µl) is removed by aspiration or using a multichannel pipeter. A sterile pipette tip is inserted half way into the culture wells and supernatant gently removed without disturbing the cells settled on the bottom of the well. The supernatant is then replaced by addition of 100–125 µl of fresh medium (complete DMEM containing 20% FBS) that contains 1× HAT (including growth factors if feeder cells are not initially used). Routine feeding replenishes nutrients to newly developing hybridomas, maintains healthy, vigorously growing cell cultures, removes wastes and cellular debris from dying cells and also reduces the presence of any immunoglobulin secreted into the culture medium by B cells that failed to fuse or by unstable hybridomas that cannot proliferate. The removal of transiently expressed antibody by frequent media replacement reduces the likelihood of identification of false positive wells during the screening process. This is most apparent in fusions derived from animals with high serum titers (>1:50,000). Fusion plates are maintained in HAT containing medium for a period of approximately 2 weeks. At this point the HAT supplement can be replaced with HT supplement, as any unfused myeloma cells have ceased to exist and there is no longer a need for aminopterin selection.

Approximately 14 days after fusion, the majority of the culture wells will show vigorous cell growth. Upon inspection of the culture wells, small grape-like clusters of hybridomas will become evident and expand. Under the plating conditions described (1×10^5 cells/well), multiple clusters of cells may be evident in each well. These cultures are not necessarily clonal. In general, at the plating densities recommended, greater than 80% of the wells plated should contain actively dividing cells. Decreased percentages may reflect the health status of the myeloma at the time of fusion (i.e., not in log phase growth), insufficient activation of B-cell blast after the final i.v. boost, inefficient cell fusion, improper culture conditions, or mycoplasma contamination. After the majority of the growth-positive wells reach 50% confluence, sufficient levels of antibody should be present in the culture supernatant (up to 1 µg/ml) to identify antigen reactive cultures. To begin screening, half the volume of each individual well is harvested. Supernatants harvested for screening should represent 2 days of growth such that maximal levels of antibody are present in the cultures. Individual pipette tips must be used for each well to prevent cross-contamination. After the supernatant is transferred to replicate plates, the hybridomas are fed as previously described. As these hybridoma cultures are in log phase growth, it is imperative to finalize the screening results within 48 h so that positive cultures can be monitored and expanded. Hybridomas are still maintained in complete DMEM medium supplemented with 20% FBS and HT.

5.3.4 SCREENING

Many screening strategies are available. Common methodologies include ELISA, immunoprecipitation, binding inhibition, flow cytometry, Western blot analysis, immunohistochemistry, and functional neutralization. The single most important consideration when choosing a screening methodology is "you get what you screen for." Screening strategies must reflect the downstream function desired of the antibody under development. For example, antibodies selected based on ELISA reactivity may not bind native molecules expressed on cell surfaces by flow cytometry.

Likewise, hybridomas selected on the basis of functional neutralization may be useless for Western blot analysis or immunohistochemistry.

When selecting a screening strategy, the following parameters must be addressed: the ability to process up to 1,000 individual samples (~120 µl per sample), the ability to identify positives within 48 h, the availability of necessary reagents/equipment, the specificity of the reaction, the sensitivity of the assay system, and most important, the applicability of the data. Assays used for hybridoma screening should always be evaluated using immune serum well in advance of the actual fusion. This will ensure that all technical aspects, reagents, and equipment are optimized. It is also important to evaluate the screening assay using the medium used to generate the hybridomas. This rich medium contains 20% FBS along with growth factors that could interfere with some assays. For example, the HAT/HT selection medium contains high concentrations of thymidine that can prohibit efficient incorporation of ^{3}H-thymidine in proliferation assays. Often multiple, tiered screening assays are necessary to identify clones with the desired characteristics. Culture supernatants may be screened at 48-h intervals in a number of assay systems, and the compiled data used to identify the appropriate antibody-producing lines.

The nature of the immunogen may also help to dictate the utility of particular screening assays. Antibodies raised against peptide antigens can readily be assayed by ELISA with plate-bound peptide and then further screened for additional functions. Flow cytometry is readily used to detect antibodies reactive with surface globular proteins before screening for functional neutralization or inhibition of ligand binding. Another consideration is the isotype of the selected antibody. Secondary reagents restricted in their ability to bind specific immunoglobulin isotypes may be used to detect antigen-reactive antibodies that express IgG heavy chains as opposed to IgM. Importantly, the identification of all positive cultures (and the disposal of all negative cultures) must be based on two independent screening assays, even if the identical protocol is used. Because individual fusion wells are screened, repeated analysis will serve to assess reproducibility of the assay system, reduce the risk of false positives or false negatives, facilitate identification of the most strongly reactive lines, and provide multiple parameters for selection of antibody-secreting hybridomas. Positive culture wells are expanded for further analysis, whereas negative cultures are discarded. It is important to remember that the original culture wells screened are *not* clonal and may display a range of binding properties.

5.3.5 Subcloning and Cryopreservation

On average, 1–3% of the growth-positive wells in a typical fusion will contain hybridomas that secrete antigen-reactive antibodies. Once a positive well is identified (using two independent screening assays), the cells are immediately subcloned to isolate individual clonal antibody-producing hybridomas and, are simultaneously expanded to generate a frozen cell stock from this positive culture. Positive culture wells will often contain more than one stable hybridoma and therefore must be cloned to isolate cultures that contain a single hybridoma secreting one unique antigen-specific antibody of one defined isotype. Nonsecreting hybridomas and clones with an unstable assortment of chromosomes may also be present in the original well that may be capable of overgrowing the desired antibody-producing clone. Single-cell cloning can be a lengthy (weeks to months) and tedious process but is essential for the isolation of stable hybridomas that produce high levels of the desired antibody. Several equivalent methods of subcloning may be used, including limiting dilution and soft-agar cloning.

Here we describe a limiting dilution subcloning technique that is straightforward and simple to perform. Regardless of the method chosen, hybridomas should always to be subject to a minimum of two rounds of subcloning to reduce the possibility of colonies arising from two cells that were stuck together. This also provides the opportunity to not only isolate individual clonal populations, but also select for those cultures with optimum growth characteristics and that secrete the highest levels of antibody.

Low-density cultures of hybridomas require the presence of feeder cells or conditioned medium to supply necessary growth factors. As described, feeder cell layers can be prepared from syngeneic splenocytes, thymocytes, or macrophages. Primary cells are an excellent source of growth factors and, because of their limited viability *in vitro*, will not contaminate long-term cultures of hybridomas. Single-cell suspensions are prepared from tissue obtained from naive animals. These cells are plated at a density of 2×10^4 cells per well (note the higher density of feeder cells needed during subcloning process as compared to the number specified for initial plating at the time of cell fusion) in a volume of 0.1 ml per well (96-well plates). Again, feeder cells should be plated at least 1 day before use to allow for growth factors to accumulate in the supernatant and to verify that the primary cultures are free of contamination. Alternatively, a number of growth factor supplements are commercially available that can replace the use of feeder layers; many contain interleukin-6. These products can be simply added to the medium at the time of cloning according to manufacturer's recommendations.

Single-cell cloning can be achieved using the following protocol. Hybridomas are diluted such that cells are plated at three concentrations: 100 cells per well, 10 cells per well, and 1 cell per well. An individual antibody-positive culture well is allowed to grow to nearly 80% confluence and then is gently resuspended and an aliquot counted with the number of viable cells determined by trypan blue exclusion. A total of 5000 viable cells are transferred into a polypropylene conical tube containing 5 ml of complete medium, including HT supplement (it is best not to alter the medium components during the cloning process) to produce a cell suspension of 1000 cells/ml. Using this stock solution of cells, serial 10-fold dilutions are performed resulting in cell suspensions of 100 cells/ml and 10 cells/ml, respectively. In the top row of two 96-well plates (Row A) 100 μl of the 1000 cell/ml suspension is added to wells containing feeder cells or supplemented medium (100 μl/well) resulting in the plating of 100 viable hybridoma cells per well. This is a density of cells that will grow to confluence within approximately 10 days. In Row B of each plate, 100 μl of the 100 cell/ml suspension is added, resulting in the plating of 10 hybridoma cells per well. One hundred microliters of the 10 cell/ml suspension is added to the remainder of the plate (Rows C–H) resulting in the dispersion of 1 cell/well (Figure 5.4).

Ideally, clones will be harvested from wells plated at 1 cell/well; however, occasionally the positive clone is rare or slow growing and can only be isolated from

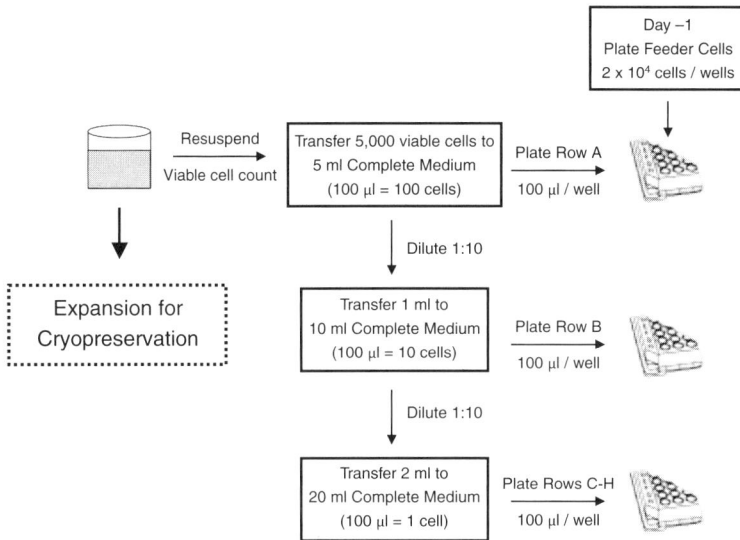

FIGURE 5.4 Procedure for hybridoma expansion and limiting dilution subcloning.

wells plated at higher densities, hence, plating some wells at 10 or 100 cells per well, provides a backup for the isolation of unique hybridomas. Such isolates, if needed, are not necessarily clonal and should be subject to multiple rounds of cloning. Optimally, clones are chosen from the single cell per well plating. Individual wells should be inspected visually for the presence of isolated colony formation. In addition, statistical analysis suggests that of wells plated at 1 cell/well, positivity of greater than 63% of the tested growth positive wells suggest clonality.[24] A minimum of two rounds of limiting dilution cloning should be performed for each original positive. It is prudent to identify the top three subclones from each original positive well and expand and bank each clone for further subcloning or analysis. If there is difficulty obtaining a clonal population, the hybridomas may be sorted by flow cytometry before subcloning. Hybridomas can be stained for the expression of surface Ig, and the top 1% of positive cells isolated and then plated at limiting dilution. This method may facilitate the more rapid isolation and identification of clones producing the highest levels of antibody. Failure to identify any positive clones may be due to technical difficulties, miscalculation in the plating density, or may reflect severe instability of the originally identified culture. Growing cultures of the original positive well should be maintained throughout the cloning process and monitored for stability of antibody production. If these passaged cultures remain antibody positive, they can serve as the source of repeated limiting dilution subcloning. If these maintained cultures are negative for antibody production, the line may be unstable. A frozen vial of the original culture can then be thawed, expanded *in vitro* for 2–3 days, and then resubcloned into six or more 96-well plates as previously described. This will offer the best opportunity to recover the specific antibody-secreting hybridoma. It is not uncommon for up to 20% of the initially identified hybridomas to "shut down" during the cloning process; thus, if possible, several antibody producing lines should be characterized and banked.

Once clonal hybridoma lines have been established and banked, the cultures should be shifted to medium devoid of HT supplement and slowly weaned off of any growth factor supplements employed. This may require multiple passages with twofold reductions in the growth factor. The rate of proliferation may decrease during this time; however, it is often accompanied by an increase in antibody production. After all of the supplements have been removed, FBS concentrations can also be reduced. Typically, hybridomas can be passaged in 20, 15, 12, and eventually 10% FBS. Some cell lines may even accommodate 5% FBS and eventually serum-free concentration. At each step, both growth characteristics and antibody production should be monitored. After clonal hybridomas have been established, the isotype of the immunoglobulin produced can be determined. Several kits and reagents are commercially available for analysis. Isotype identification is critical to select the proper method of antibody purification and may also impact the function of a given mAb.

The original positive culture and the top three cloned hybridomas derived from each line are expanded for cryopreservation. Individual culture wells are expanded slowly and maintained at densities between 1×10^5 and 1×10^6 cells/ml. Before cryopreservation hybridomas are kept in log phase growth. Cells are harvested, centrifuged at 4°C, and the pellet is resuspended at 5×10^6 cells/ml in cold growth medium containing 10% DMSO that has been filter-sterilized and chilled to 4°C. One ml vials (5×10^6 cells/vial) are stored at –70°C for several days and then transferred to liquid nitrogen for long-term storage. Cell lines should be preserved at each stage of identification and isolation: original fusion cultures, first round and second round subclones. In addition, multiple clones of each line should be frozen to ensure recovery of critical lines. In general, 10 vials of each cloned hybridoma will serve as an appropriate bank. Lines maintained in liquid nitrogen can survive for decades, although it is advisable to replenish the frozen stocks every few years.

5.3.6 HYBRIDOMA EXPANSION

After specific antibody-producing hybridomas have been isolated and cloned, and a frozen cell bank generated, the hybridomas may be expanded for antibody production. Hybridoma cultures in log phase growth may produce 3–20 μg/ml of antibody. Depending on the quantity of mAb needed, multiple methods of expansion may be used (see Chapter 6). Before expansion, hybridomas must be clonal so that non-producing cells do not overgrow the culture and reduce antibody production. For many applications, where less than 20 mg of the antibody are needed, simple expansion in flasks or roller bottles may be sufficient to generate the quantities of antibody needed. Production of 20–200 mg of antibody can be accomplished using either roller bottle cultures, specialized flasks capable of high-density cell growth (CeLLine flask, Integra), or small bioreactors. Large-scale output may require specialized equipment and expertise, or the accumulation of antibody from multiple rounds of smaller-scaled production methods.

Hybridomas may be adapted to different types of medium for expansion and antibody production. Many hybridomas can be adapted to lower concentrations of serum (2–5%) or specialized serum-free medium. Serum low in bovine immunoglobulin should be used to avoid any contaminating antibody in the final preparations. Defined serum-free medium and medium devoid of any animal proteins are available to facilitate antibody production and purification. Individual hybridomas may respond differently to alternate types of medium and should be assessed for growth rates and antibody production levels before large-scale expansion. An additional consideration for large-scale expansion is the quality of the medium used. All reagents should be monitored for endotoxin contamination; it is best to use components containing less than 0.2 EU/ml. Contaminating endotoxins can alter *in vitro* cellular responses and can render useless products designed for *in vivo* use. A variety of products do exist to remove endotoxin from antibody preparations; however, they often result in large protein losses as well. Antibody containing culture supernatants can be used directly in a number of assay systems (immunoprecipitation, Western blot analysis), but many applications will require the use of purified mAb. Protocols for antibody purification are described in Chapter 7. Purification procedures are dependent upon the species and isotype of the particular antibody. After mAb purification, the final antibody preparation should be monitored for both specificity and quality. Some parameters to be evaluated include sterility, concentration (>0.1 mg/ml), purity (assessed by sodium dodecyl sulfate–polyacrylamide gel electrophoresis (SDS–PAGE)), aggregation levels, endotoxin contamination (<0.2 EU/mg for *in vivo* use) and contamination by protein A leaching (<50 ng/ml).

5.4 CONCLUSIONS

Monoclonal antibodies serve as critical research tools, diagnostic reagents, and therapeutic treatments. Continued advancements in methods of antigen preparation, immunization strategies, high-throughput screening, and antibody production have facilitated the generation, identification, and production of novel antibody reagents. Today mAbs are used not only in research and diagnostic laboratories, but have also joined the arsenal of therapeutic reagents with broad usage in the areas of transplantation, oncology, infectious disease, and many other disciplines. Moreover, antibody modifications, including tagged, chimeric, humanized and engineered formats, have expanded their utility in numerous applications. Table 5.4 provides a list of web sites with information regarding the availability of previously generated antibodies and hybridomas, protocols, troubleshooting guides, suppliers and a variety of antibody-related links. Numerous commercial establishments can provide custom hybridoma technology to investigators; however, for those with basic tissue culture skills, the methods described here will provide the foundation for successful in-house hybridoma development. Careful design of immunization and screening strategies will provide for the identification and isolation of unique hybridomas that secrete antibodies with the novel, restricted specificity desired. Thus hybridoma development and antibody production continue to fulfill their promise as valuable tools for medical and industrial use.[1]

TABLE 5.4
Antibody-Related Web Sites

http://antibodyresource.com	Antibody-related links
http://linscottsdirectory.com	Antibody search engine
http://abcam.com	Antibody search engine
http://pathimm.wustl.edu	Washington University School of
Research Core Facilities → Hybridoma Center	Medicine Hybridoma Center
http://www.uiowa.edu/%7edshbwww	Developmental Studies Hybridoma Bank
http://www.drmr.com/abcon/index.html	Conjugation of Monoclonal Antibodies

REFERENCES

1. Kohler, G. and Milstein, C., Continuous cultures of fused cells secreting antibody of predefined specificity, *Nature,* 256, 495, 1975.
2. Velikovsky, C.A., Cassataro, J., Sanchez, M., Fossati, C.A., Fainboim, L., and Spitz, M., Single-shot plasmid DNA intrasplenic immunization for the production of monoclonal antibodies. Persistent expression of DNA, *J. Immunol. Methods,* 244 (1-2), 1–7, 2000.
3. Kilpatrick, K.E., Cutler, T., Whitehorn, E., Drape, R.J., Macklin, M.D., Witherspoon, S.M., Singer, S., and Hutchins, J. T., Gene gun delivered DNA-based immunizations mediate rapid production of murine monoclonal antibodies to the Flt-3 receptor, *Hybridoma,* 17, 569, 1998.
4. Kilpatrick, K.E., Kerner, S., Dixon, E.P., Hutchins, J.T., Parham, J.H., Condreay, J.P., and Pahel, G., *In vivo* expression of a GST-fusion protein mediates the rapid generation of affinity matured monoclonal antibodies using DNA-based immunizations, *Hybrid Hybridomics,* 21, 237, 2002.
5. Dunn, G.P., Bruce, A.T., Sheehan, K.C., Shankaran, V., Uppaluri, R., Bui, J.D., Diamond, M.S., Koebel, C M., Arthur, C., White, J.M., and Schreiber, R. D., A critical function for type I interferons in cancer immunoediting, *Nat. Immunol.,* 6, 722, 2005.
5a. Sheehan, K.C.F., Lai, K.S., Dunn, G.P., Bruce, A.T., Diamond, M.S., Heutel, J.D., Dungo-Arthur, C., Carrero, J.A., White, J.M., Hertzog, P.J. and Schreiber, R.D., Blocking monoclonal antibodies specific for mouse IFNα/β receptor subunit 1 (IFNAR1) from mice immunized by *in vivo* hydrodynamic transfection, *JICR,* in press.
6. Liu, F., Song, Y., and Liu, D., Hydrodynamics-based transfection in animals by systemic administration of plasmid DNA, *Gene Ther.,* 6, 1258, 1999.
7. Zhang, G., Budker, V., and Wolff, J. A., High levels of foreign gene expression in hepatocytes after tail vein injections of naked plasmid DNA, *Hum. Gene Ther.,* 10, 1735, 1999.
8. Yerganian, G., History and cytogenetics of hamsters, *Prog. Exp. Tumor Res.,* 16, 2, 1972.
9. Sheehan, K.C., Pinckard, J.K., Arthur, C.D., Dehner, L.P., Goeddel, D.V., and Schreiber, R.D., Monoclonal antibodies specific for murine p55 and p75 tumor necrosis factor receptors: identification of a novel *in vivo* role for p75, *J. Exp. Med.,* 181, 607, 1995.
9a. Sanchez-Madrid, F., Szklut, P. and Springer, T., Stable hamster-mouse hybridomas producing IgG and IgM hamster monoclonal antibodies of defined specificity, *J. Immunol.,* 130, 309, 1983.

10. Sheehan, K.C., Ruddle, N.H., and Schreiber, R. D., Generation and characterization of hamster monoclonal antibodies that neutralize murine tumor necrosis factors, *J. Immunol.,* 142, 3884, 1989.

11. Shalaby, M.R., Fendly, B., Sheehan, K.C., Schreiber, R.D., and Ammann, A.J., Prevention of the graft-versus-host reaction in newborn mice by antibodies to tumor necrosis factor-alpha, *Transplantation,* 47, 1057, 1989.

11a. Wall, D.A. and Sheehan, K.C.F., The role of tumor necrosis factor and interferon-gamma in graft-versus-host disease and related immunodeficiency, *Transplantation,* 57, 273, 1994.

12. Ferran, C., Sheehan, K., Dy, M., Schreiber, R., Merite, S., Landais, P., Noel, L.H., Grau, G., Bluestone, J., Bach, J.F., et al., Cytokine-related syndrome following injection of anti-CD3 monoclonal antibody: further evidence for transient *in vivo* T cell activation, *Eur. J. Immunol.,* 20, 509, 1990.

13. Fuhlbrigge, R.C., Sheehan, K.C., Schreiber, R.D., Chaplin, D.D., and Unanue, E., Monoclonal antibodies to murine IL-1 alpha. Production, characterization, and inhibition of membrane-associated IL-1 activity, *J. Immunol.,* 141, 2643, 1988.

14. Gronowski, A.M., Hilbert, D.M., Sheehan, K.C., Garotta, G., and Schreiber, R.D., Baculovirus stimulates antiviral effects in mammalian cells, *J. Virol.,* 73, 9944, 1999.

15. Hogquist, K.A., Nett, M.A., Sheehan, K.C., Pendleton, K.D., Schreiber, R.D., and Chaplin, D.D., Generation of monoclonal antibodies to murine IL-1 beta and demonstration of IL-1 *in vivo, J. Immunol.,* 146, 1534, 1991.

16. Rogers, H.W., Sheehan, K.C., Brunt, L.M., Dower, S.K., Unanue, E.R., and Schreiber, R.D., Interleukin 1 participates in the development of anti-*Listeria* responses in normal and SCID mice, *Proc. Natl. Acad. Sci. U. S. A.,* 89, 1011, 1992.

17. Spieker-Polet, H., Sethupathi, P., Yam, P. C., and Knight, K.L., Rabbit monoclonal antibodies: generating a fusion partner to produce rabbit-rabbit hybridomas, *Proc. Natl. Acad. Sci. U. S. A.,* 92, 9348, 1995.

18. Jat, P.S., Noble, M.D., Ataliotis, P., Tanaka, Y., Yannoutsos, N., Larsen, L., and Kioussis, D., Direct derivation of conditionally immortal cell lines from an H-2Kb-tsA58 transgenic mouse, *Proc. Natl. Acad. Sci. U. S. A.,* 88, 5096, 1991.

19. Green, L.L., Antibody engineering via genetic engineering of the mouse: XenoMouse strains are a vehicle for the facile generation of therapeutic human monoclonal antibodies, *J. Immunol. Methods,* 231, 11, 1999.

20. Mendez, M.J., Green, L.L., Corvalan, J.R., Jia, X.C., Maynard-Currie, C.E., Yang, X.D., Gallo, M.L., Louie, D.M., Lee, D.V., Erickson, K.L., Luna, J., Roy, C.M., Abderrahim, H., Kirschenbaum, F., Noguchi, M., Smith, D.H., Fukushima, A., Hales, J.F., Klapholz, S., Finer, M.H., Davis, C.G., Zsebo, K.M., and Jakobovits, A., Functional transplant of megabase human immunoglobulin loci recapitulates human antibody response in mice, *Nat. Genet.,* 15, 146, 1997.

21. Heeg, K. and Zimmermann, S., CpG DNA as a Th1 trigger, *Int. Arch. Allergy Immunol.,* 121, 87, 2000.

22. Murphy, K.M., Ouyang, W., Szabo, S.J., Jacobson, N.G., Guler, M.L., Gorham, J.D., Gubler, U., and Murphy, T.L., T helper differentiation proceeds through Stat1-dependent, Stat4-dependent and Stat4-independent phases, *Curr. Top. Microbiol. Immunol.,* 238, 13, 1999.

23. Harlow, E. and Lane, D., *Antibodies: A Laboratory Manual,* Cold Spring Harbor Laboratory, Cold Spring Harbor, NY, 1988.

24. Hunter, P. and Kettman, J.R., Mode of action of a supernatant activity from T-cell cultures that nonspecifically stimulates the humoral immune response, *Proc. Natl. Acad. Sci. U. S. A.,* 71, 512, 1974.

25. Kearney, J.F., Radbruch, A., Liesegang, B., and Rajewsky, K., A new mouse myeloma cell line that has lost immunoglobulin expression but permits the construction of antibody-secreting hybrid cell lines, *J. Immunol.,* 123, 1548, 1979.

26. Shulman, M., Wilde, C.D., and Kohler, G., A better cell line for making hybridomas secreting specific antibodies, *Nature,* 276, 269, 1978.

27. Galfre, G. and Milstein, C., Preparation of monoclonal antibodies: strategies and procedures, *Methods Enzymol.,* 73, 3, 1981.

28. Kilmartin, J.V., Wright, B., and Milstein, C., Rat monoclonal antitubulin antibodies derived by using a new nonsecreting rat cell line, *J. Cell. Biol.,* 93, 576, 1982.

6 Quantitative Production of Monoclonal Antibodies

David A. Fox and Elizabeth M. Smith

CONTENTS

6.1 INTRODUCTION: COMPARING METHODS

Monoclonal antibodies (mAbs) are among the most widely used tools in biomedical research, and are also of increasing value as diagnostic and therapeutic agents in clinical medicine. Production of mAbs for these diverse applications can be accomplished *in vivo*, usually as ascites, or *in vitro*, with a variety of systems. The choice of production method depends on several factors, including characteristics of the specific hybridoma, the quantity of antibody required (Table 6.1), the facilities available, local regulations concerning use of rodents for generation of ascites, cost considerations, and the specific skills or experience of the individual research team.

TABLE 6.1
Monoclonal Antibodies: Production Scale and Applications

Application	Production Scale per Batch
Research using *in vitro* systems	20–200 mg
Research using animal models	100 mg–1 g
Diagnostic kits and reagents	500 mg–10 g
Clinical therapeutics	>10 g (often >10 kg)

This chapter describes the advantages and disadvantages of both *in vivo* and *in vitro* approaches to mAb production and includes detailed protocols that are used in our facility. The emphasis will be on methods that can be used by an individual laboratory in a university setting and not on industrial-scale production of pharmaceutical-grade mAbs. The reader is also referred to a comprehensive report on this topic that was commissioned by the National Academy of Sciences and published in 1999.[1] In our view, the recommendations of this report remain current and valid.

6.2 OVERVIEW OF ANTIBODY PRODUCTION METHODS

Several original reports and reviews have considered the various factors that bear on the choice of method for producing a mAb.[1-8] The influence of these factors may change over time. Examples include animal use practices and regulations and the cost of both mice and the supplies needed for *in vitro* mAb production. However, all of these considerations remain not only valid, but important.

It must be emphasized that each hybridoma is a unique biologic system. Conclusions drawn from experiments that compare various mAb production methods in a small number of hybridomas may not be universally or even generally applicable. The process of sorting and shedding of chromosomes that occurs in fused cells is unlikely to result in retention in all hybridomas of the same repertoire of genes that affect cell growth and mAb production under various environmental conditions. Therefore, the investigator must examine the properties and preferences of each hybridoma line individually.

Furthermore, hybridomas can sometimes change their growth requirements and pattern of mAb production over time. It is essential, therefore, to cryopreserve ample stocks of hybridoma cells in a liquid nitrogen system, at several stages in the development and propagation of a cloned hybridoma line.

Table 6.2 summarizes the considerations involved in selecting a method of mAb production, and subsequent paragraphs will consider each of these factors in more detail. The favored method indicated for the various factors in Table 6.2 applies to production of smaller batches of mAbs for research uses. With production of larger batches for clinical use, most factors will favor *in vitro* production. A minority of

TABLE 6.2
In Vivo versus *in Vitro* **Monoclonal Antibodies: Important Factors to Consider**

Factor	Method Favored
Antibody yield	*In vivo*
Cost	*In vivo*
Facilities required	*In vivo*
Animal welfare	*In vitro*
Contamination with immunologically active molecules	*In vitro*
Microbial contamination	*In vivo*
Fidelity of mAb glycosylation and effector function	*In vivo*

hybridomas cannot be adapted to yield useful quantities of mAb by any *in vitro* method, and in such cases, *in vivo* production remains the only option.

6.2.1 YIELDS

The concentration of mAb in the supernatant of hybridoma cells in tissue culture wells or flasks is typically <0.01 mg/ml. In contrast, ascites usually contains 1–10 mg/ml of mAb, although some polyclonal mouse immunoglobulin is also present. This 2–3 log difference in mAb yield renders standard cell culture methods unsuitable for quantitative production of mAbs, and screening assays that select hybridomas of interest based on testing of mAbs in their supernatants must be sufficiently sensitive to low titers of mAb. High-density cell-culture methods (described in detail in section 6.4.2) have significantly closed the gap between the concentration of mAb in culture fluid versus ascites.[3,5,6] The time required to produce ascites is typically no more than 4 weeks, whereas batch production of mAb in high-density cell-culture can take up to 9 weeks.[6]

6.2.2 COST AND FACILITIES

Costs include capital costs, labor costs, and the cost of supplies, including mice and their maintenance.

In vivo mAb production requires an accredited animal care unit with appropriate environmental controls and veterinary services. *In vitro* production requires a cell-culture laboratory and a variety of disposable items, culture media, animal sera, and supplements. Although *in vitro* production of mAbs often involves adaptation of hybridomas to growth in low serum concentrations, the cost of animal sera has risen sufficiently to make this an important component of the total cost. It has been calculated that the labor costs of *in vitro* mAb production are at least twofold the labor costs of production of the same quantity of mAb *in vivo*, whereas the material costs may be similar.[6] A review of available information regarding costs of these procedures concluded that for small-scale mAb production, *in vitro* methods were

1.5- to 6-fold more costly, but for larger scale production *in vitro* methods were progressively more advantageous.[1]

6.2.3 ANIMAL WELFARE

Production of mAbs *in vivo* involves injection of malignant cells into the peritoneal cavity of an animal, usually a mouse, a procedure that will usually lead to the death of the animal unless euthanasia is administered when a preterminal level of distress is observed.

Attention has been given to the nature and degree of suffering experienced by mice that carry hybridoma-induced ascites.[1,3,9,10] Mice may have abdominal distension, decreased activity, decreased lean body mass, dehydration, difficulty walking, hunched posture, and respiratory distress. Anorexia, anemia, organ hypoperfusion, and, ultimately, shock can occur. Regarding pathologic findings, peritonitis, tumor infiltration, hemoperitoneum, and, abdominal adhesions have been documented.[3,9] These consequences of intraperitoneal injection of hybridomas have led to restrictions or prohibition of mAb production as ascites by many European countries and by some states in the United States.[3]

Jackson and colleagues reported a detailed clinical and pathological study of groups of 20 mice injected with five different hybridomas. Overall, survival was 98% to the first tap (draining of ascites fluid from the peritoneal cavity), 96% to the second tap, and 79% to the third tap. However, the rate of survival to the third tap varied from 35% to 100% depending on the hybridoma.[9] Therefore, the appropriate time of euthanasia must be individualized for each hybridoma. For four of five hybridomas, evidence of clinical dehydration was rare pretap but may have been present posttap.[9]

Using behavioral and hormonal measurements, Peterson analyzed the degree of stress experienced by mice up to 12 days after intraperitoneal injection of hybridomas.[10] He concluded that Pristane priming had little or no adverse effect and that the well-being of mice after hybridoma cell injection could be adequately monitored during the course of the development of ascites, at least up to Day 12. Based on these findings and a comprehensive review of other relevant literature, a report to the U.S. National Academy of Sciences advised against prohibition of *in vivo* mAb production, while favoring the use of *in vitro* methods whenever practical.[1] It should be noted that the *in vitro* alternatives to mAb production in ascites also involve animal sacrifice, whenever fetal calf serum is employed.

6.2.4 CONTAMINATION WITH IMMUNOLOGICALLY
ACTIVE MOLECULES

Either research or clinical applications of monoclonal antibodies can be confounded by the presence of contaminating proteins and other substances with biologic activities. Malignant ascites is an inflammatory fluid that contains polyclonal mouse immunoglobulin, other serum proteins, and cytokines, such as tumor necrosis factor, interleukin-1, and interleukin-6.[11] These cytokines are highly potent, even in low concentrations, and occasional hybridomas can even produce interleukin-6 themselves, either *in vitro* or *in vivo*.[11] Hybridoma ascites may also contain variable levels of an inhibitor of complement-mediated effector functions of mAbs.[12] These problems can generally be mitigated by purification of mAbs produced in ascites. Although contaminating proteins in diluted ascites will not typically interfere with assays in which antibodies are used to recognize their target antigen, such as flow cytometry, assays in which mAbs are included to modulate cellular functions are best performed with purified reagents.

mAbs produced *in vitro* suffer from a different problem: contamination by serum proteins, including immunoglobulins, from species other than the mouse. The level of immunoglobulin in fetal calf serum is relatively low, typically about 50 µg/ml, and this issue is usually a minor consideration if a low concentration of serum is used, However, if mAbs produced *in vitro* are employed for *in vivo* research studies in mice and are repeatedly injected, sensitization to foreign immunoglobulin could occur. Therefore, use of mAbs produced in mice is preferable in such situations.[1]

6.2.5 MICROBIAL CONTAMINATION

Cell-culture systems used for mAb production are potentially vulnerable to bacterial or fungal contamination. This may be a somewhat greater risk than for other short-term cell-culture procedures, in view of the complexity of some of the *in vitro* hybridoma culture systems. Although a theoretical risk exists of similar contamination of hybridoma ascites, hybridoma lines that become contaminated *in vitro* can occasionally be sterilized and rescued by *in vivo* passage as ascites because of the endogenous *in vivo* antimicrobial host defenses of the mouse.[1]

6.2.6 ANTIBODY GLYCOSYLATION

Glycosylation of mAbs is variable and can affect antibody recognition of target antigen as well as mAb effector function.[1,7,8,13–16] This parameter may affect the pharmacokinetics of mAbs injected into animals[7] and the efficacy of mAbs used as clinical therapeutics.[15] In general, mAbs produced *in vivo* are not glycosylated identically to those produced *in vitro*,[7,8,13–16] and notable differences may also exist when different *in vitro* production techniques are compared.[7,13,14,16] Available information suggests that mAb glycosylation and function is best preserved when hybridomas are cultured *in vitro* in the case of non-murine mAbs, but that ascites is the preferred production method for many murine mAbs. Caution is warranted in the application of these generalizations, because specific hybridomas may prove to be exceptions. Modest differences in oxygen tension[14] and pH[16] may alter the glycosylation patterns of mAbs produced *in vitro*, and during the course of generating a batch of mAb *in vitro*, the antibody glycosylation is likely heterogeneous between early and later stages of the same culture.[17] Also, subtle differences may exist between different batches of mAb made *in vitro*. These considerations emphasize that hybridomas are dynamic biological systems and that mAbs are, unfortunately, not entirely uniform or homogeneous glycoproteins.

6.3 PRODUCTION IN ASCITES

6.3.1 OVERVIEW OF METHODS

The production of a mAb *in vivo* should first be preceded by testing of hybridoma culture supernatant to clearly identify a mAb of interest, cryopreservation of primary cultures of that hybridoma, and typically one or two rounds of subcloning, rescreening, and cryopreservation of the hybridoma subclones.

The stages of *in vivo* production of a mAb in ascites are as follows.

* "Priming" of the mouse with Pristane (2,6,10,14-tetramethylpentadecane)
* Injection of hybridoma cells
* Collection of the ascites

Disagreement exists regarding the selection of mice, concerning gender, age, and strain. BALB/c mice are typically used to avoid alloreactivity against the hybridoma cells, because the myeloma line used in cell fusions is of BALB/c origin. Nevertheless, use of an F1 cross of BALB/c males × SW or MF1 females has been proposed because the mice are larger and can produce a greater volume of ascites.[18,19] It has also been argued that male mice should be used that are 6–11 weeks old and that hybridomas grow more rapidly in the presence of testosterone.[20] None of these procedures is widely followed in our experience.

Intervals of between 7 and 20 days are optimal between the injection of Pristane and injection of hybridoma cells.[20–23] The volume of Pristane injected is usually 0.5 ml.

Occasionally, solid intraperitoneal tumors form with insufficient ascites production when a standard Pristane priming schedule is followed. In the cases of two such hybridomas, a significant variation of the Pristane method was used. Adequate ascites was obtained when a 24-week interval elapsed between Pristane priming and hybridoma cell injection. Within 1 week of injection of 10 million hybridoma cells (an unusually high number of cells), ascites was present.[24]

The mechanism of the effect of Pristane is also disputed. According to one view, Pristane facilitates tumor growth by depressing the normal immunological function of the animal.[21] This concept was contested in a study that alluded to direct growth-stimulatory properties of Pristane on hybridoma cells.[25] Alternatively, an inflammatory response to Pristane in the peritoneum may induce a hospitable environment for hybridoma growth by engendering the production of cytokines that accelerate growth of these cells.

Alternative priming strategies involve the use of incomplete Freund's adjuvant.[26,27] Higher antibody yields and an interval between priming and hybridoma injection as short as 1 day have been reported.[26,27] This approach might be useful when immediate rescue of an infected or apoptotic cultured hybridoma line is required. One might, however, expect a higher level of contamination of the ascites by cytokines and other inflammatory mediators with this method of priming.

The optimal number of hybridoma cells for injection was defined in one study as 0.6–3.2 million, although injection of 5 million cells is also appropriate.[20] With a constant number of cells, the volume of ascites and the antibody concentration are variable.[23] If injection of a somewhat arbitrary selected number of cells proves suboptimal, either because of delayed and insufficient development of ascites or due to excessively rapid demise of the animal, subsequent batches of ascites should employ injection of a greater or lesser number of cells, respectively. In one study, intrasplenic injection of hybridoma cells yielded ascites after injection of a cell number 2–3 logs lower than is used for intraperitoneal injection.[28] This more difficult technique seems justified only in the very rare instance in which only a tiny number of hybridoma cells are available.

Ascites is harvested by insertion of a needle of sufficient caliber (e.g., 18 gauge) into the peritoneal cavity to allow drainage of ascites by gravity while avoiding persistent leakage. The issue of how many taps to perform requires balancing the need to sacrifice animals that are in notable distress against the fact that the more times ascites is harvested from each mouse, the fewer the total number of mice that will need to be injected and ultimately sacrificed.[1] Although it is uncertain how directly clinical observations in humans can be extrapolated to mice, it is clear that in humans with ascites, drainage of moderate to large volumes frequently leads to temporary symptomatic improvement.[1] With most hybridomas, two or three sequential taps are feasible and appropriate.

6.3.2 MODIFICATIONS FOR ALLOREACTIVITY OR XENOREACTIVITY

mAbs from species other than the mouse are often needed for a variety of research and clinical applications. In the research laboratory, mAbs of rat or hamster origin are often used to study antigens of mouse origin, which are usually not immunogenic in the mouse.

Human mAbs are sometimes required for their unique specificities or as therapeutic molecules that will not sensitize the human recipient to xenogeneic proteins. Although therapeutic mAbs are increasingly being produced in recombinant systems, non-mouse mAbs used for research are typically produced by heterohybridomas, in which lymphocytes of non-mouse origin are fused with a mouse myeloma line. Such heterohybridomas will typically be immunogenic in mice, because they carry xenogeneic determinants. Therefore, ascites production by standard methods will usually be unsuccessful, since the hybridoma cells will be immunologically rejected.

A variety of *in vivo* approaches can be used to circumvent these problems, including the use of immunodeficient mice (e.g., nude or SCID mice) or immunosuppressed mice (irradiated or pretreated with cyclophosphamide).[29–36] Ascites volumes and yields of mAbs are typically lower, whereas costs, especially animal purchase, are higher than ascites production of murine mAbs, making consideration of alternative *in vitro* mAb production methods more compelling.

It has generally been difficult to produce mAbs in rat ascites, but alternative approaches have been described. One method involves intravenous injection of hybridomas, which then proliferate in the liver and other organs. A suspension of such liver cells proliferated intraperitoneally and yielded ascites in Pristane-primed rats.[37] This approach was subsequently improved by the use of a mixture of Pristane and incomplete Freund's adjuvant for priming, which was immediately followed by injection of 10–15 million hybridoma cells.[38]

In our facility, a useful approach to generating mAbs against murine antigens has been the immunization of knockout mice, in which the antigen of interest is viewed as foreign. Such mice are almost always allogeneic to the BALB/c strain, and the allo-hybridomas will usually be rejected unless methods similar to those used to facilitate ascites production from hetero(xeno)hybridomas are employed.

6.3.3 IN VIVO PRODUCTION PROTOCOL: ASCITES IN BALB/C MICE

6.3.3.1 Priming of Mice and Injection of Hybridoma Cells

6.3.3.1.1 Materials

Mice: Two or four BALB/c female retired breeders
Pristane (Sigma T-7640)
Tuberculin syringes (1 cc)
Needles, sterile (21 gauge)
Hybridoma cells, 5×10^6 per mouse
Trypan blue solution, 0.4% (Sigma T8154)
Conical test tube, sterile (15 or 50 ml)
Dulbecco's phosphate-buffered saline (D-PBS), sterile

6.3.3.1.2 Equipment

Hemacytometer
Microscope with phase contrast optics
Centrifuge, bench top

6.3.3.1.3 Time required

Pristane priming of four mice requires about 10 min and should be done 10–14 days, but no more than 3 weeks before injection of hybridoma cells. Hybridoma cells should be grown ahead of time: preparation of the cells for injection requires about 30 min and injection of the cells takes about 10 min.

6.3.3.1.4 Procedure

1. Prime the mice with Pristane 10–14 days before planned injection of cells. Restrain the mouse by hand, tilting the head downward to move intestines out of the way and locate a point in the lower abdominal quadrant to the right of midline (Figure 6.1). Insert a 21-gauge needle, bevel up, angling (45°) away from the point of entry after the subcutaneous layers have been penetrated and inject 0.5 ml of Pristane into the peritoneal cavity. Some pressure may be required due to the viscosity of the agent. Care should be taken not to penetrate too deeply with the needle. After injection, hold the needle in place briefly without pressure on the syringe and then gently

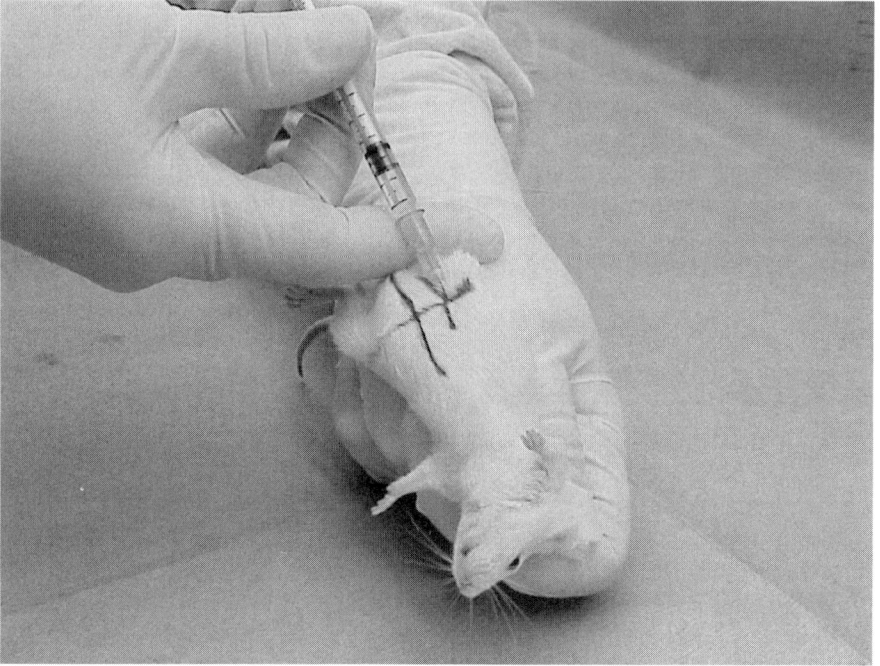

FIGURE 6.1 Priming a mouse.

withdraw the needle. This should avoid leakage of Pristane out of the puncture site.

2. The hybridoma cells should be split with fresh medium 1–2 days before injection to ensure that they will be in log growth phase. The cells should be at 90–100% viability on the day of injection as determined by trypan blue exclusion.

3. Mix the hybridoma cells well to create a suspension, dilute 1:10 in a 1:1 mixture of trypan blue in D-PBS and let stand 5–10 min. Transfer a drop of this dilution onto a hemacytometer, determine the viable cell count using a phase contrast microscope and calculate the number of viable cells per ml of cell suspension.

4. Transfer a cell suspension containing the desired number of viable cells to a 15- or 50-ml conical test tube and pellet at room temperature by centrifugation at $200 \times g$ for 5 min. Discard the supernatant and gently resuspend the cells in room temperature D-PBS at a concentration of 10×10^6 per ml. The cells should be injected into mice as quickly as possible to maintain viability.

5. Draw evenly suspended cells into a sterile syringe with a 21-gauge needle and inject 0.5 ml into the mouse intraperitoneally (i.p.) in the same location and manner as Pristane priming (see Step 1).

6. Monitor mice 24 h after injection of hybridoma cells and at least every other day thereafter for general health and development of ascites fluid.

6.3.3.2   Ascites Fluid Collection

6.3.3.2.1   Materials

Alcohol prep pads, sterile
Needles, sterile (18-gauge)
Conical test tubes, sterile (15 ml)
Conical test tube, sterile (50 ml)
Pasteur pipet, sterile

6.3.3.2.2   Time required

Ascites fluid usually builds up in the mouse within 7–10 days after injection of the hybridoma cells. Collection of the fluid and centrifugation takes approximately 30 min for each harvest (tap). A maximum of three taps should be harvested from each mouse over a 7-day period, with euthanasia performed at the time of the final collection.

6.3.3.2.3   Procedure

1. The first harvest of ascites fluid should be done when moderate distension of the abdomen is visible. The accumulation of ascites fluid can also be measured by weighing the mouse. Animal welfare experts recommend no more than a 20% increase in the baseline body weight before harvest. Yield will generally be reduced if harvesting is begun too early, whereas distress and mortality will be increased if too much fluid is allowed to collect without removal.
2. Have ready a sterile 15-ml test tube in a suitable rack to hold it securely. Restrain the mouse by hand, grasping as much skin as possible at the back and neck to stretch the abdomen taut. Swab the abdomen with a sterile alcohol prep pad.
3. Paracentesis should be performed by inserting an 18-gauge needle, bevel up, at a 45° angle into the lower abdominal cavity to the right of midline, avoiding organs in the higher abdomen and blood vessels near the midline. The needle hub should be directed toward the open mouth of the test tube as fluid may spurt out under high pressure. Alternatively, the ascites fluid may not flow easily and subtle, gentle repositioning of the needle (rotationally, by depth or by angle of insertion) may be needed to initiate or resume flow.
4. As much fluid as possible should be collected through the needle and then it should be carefully removed. Typically the yield from one harvest will be 4–6 ml per mouse but may vary from 2–10 ml. If the fluid flows around the needle or continues to drain significantly after needle removal, additional fluid may be collected by allowing it to drip from the needle insertion point into the test tube. Very gentle massaging of the abdomen away from the insertion point may assist the flow of ascites fluid. Special care should be taken to collect the fluid aseptically under these conditions and to avoid distress to the animal.

5. The ascites fluid collected from the paracentesis of a group of mice injected with the same hybridoma cell line may be pooled at each harvest. The fluid should be centrifuged at 1,500 g for 10 min to pellet cells. If the fluid clots, rim the clot by running a sterile pasteur pipet tip between the clot and the sides of the test tube before centrifugation. The ascites fluid supernatant should be transferred to a sterile 50-ml test tube and may be stored frozen at −20°C during the harvest period.
6. The mice should be returned to their cage and monitored daily for health changes that indicate distress. If the health does not decline significantly, additional taps may be performed. The fluid should be allowed to accumulate for approximately 48 h before repeating the tap procedure; a third tap may be done 24–48 h later at euthanasia. After centrifugation, the ascites fluid from all harvests may be pooled in a single test tube.
7. The fluid should be thawed, clarified by high-speed centrifugation, and lipids removed before purification. Ascites fluid may be used in this crude form or processed further by purification. For long-term storage, titer the antibody, aliquot the fluid into volumes that will avoid repeated freeze-thaw cycles during use and freeze at −70°C.

6.3.3.3 Results

Ascites volumes ranging from 5–10 ml per mouse may be collected. The concentration of monoclonal antibody in the fluid ranges from 1–10 mg/ml so that two BALB/c mice may produce 10–20 ml of ascites and 10–200 mg of antibody. The yield may vary between individual hybridomas and with each ascites batch of the same hybridoma cell line.

6.3.3.4 Problems and Troubleshooting

It is very important that the hybridoma cells injected into mice for ascites production be healthy and in a log growth phase. The culture supernatant should be assayed close to the date of injection to confirm desired antibody production. The number of cells injected per mouse can be as low as 1×10^6. Injecting more than 5×10^6 cells will usually reduce the yield of antibody because of increased distress and decreased survival time. When fewer cells are injected, it may take up to 21 days for sufficient ascites fluid to develop for harvest, but the mice may survive longer.

Some animals in a group of mice injected with the same hybridoma cell line may not produce any fluid. Ascites fluid can be transferred from one mouse of the group to a non-producing mouse by i.p. injection of 0.5 ml per mouse. Any ascites fluid that develops can be collected and pooled with the remainder of the batch. Some hybridomas or some mice may produce predominantly solid tumors. If even a small amount (0.2–1.0 ml) of ascites fluid develops in such a mouse, it can be transferred to a naive Pristane-primed mouse. If this fluid can be used to successfully initiate ascites production, cells may be recovered from the ascites harvest, cultured, assayed for antibody secretion, and stored in frozen aliquots for future ascites productions. If a hybridoma cell line consistently produces solid tumors but no ascites fluid or will not grow at all in the peritoneal cavity, the cell lines may grow and produce ascites in immunosuppressed BALB/c mice or athymic nude mice. Alternatively, such hybridomas may be more suitable for *in vitro* antibody production.

The size of the needles used for injections is important. Pristane will flow back out of the insertion hole, if the needle is too large. Injected cells may be damaged by passing through too small a needle. All needle insertions must be done carefully to avoid penetrating too deeply or hitting organs or major blood vessels. Aseptic techniques should be used to prevent introducing an infection to the mouse or contaminating the resulting ascites fluid. Ascites fluid should be centrifuged and stored at a temperature no higher than 4°C as soon as possible after collection to separate the fluid from cells and blood cell enzymes and to preserve the antibody.

It is also very important to carefully monitor the health of the mice for about 30 min after each tap, and daily thereafter, to assess the response to harvest and the suitability of each animal for continued harvesting after the first tap. Signs of distress include reduced activity and general lethargy, maintenance of a hunched posture, a ruffled hair coat, respiratory distress, and the animal's body feeling cold to the touch or appearing thin or severely dehydrated. Some of these signs of distress may be transient and mild immediately after the tap. In any group of mice, there may be some animals that become severely ill or show signs of distress earlier or later than others, so that the number of harvests possible may vary with each animal; the development of severe distress may also vary with different hybridoma cell lines. If a particular hybridoma cell line seems to cause more rapid deterioration of health and development of severe distress, the ascites production may be repeated using fewer cells at injection with the possibility of better results. In all cases when distress is persistent and severe, a final tap should be performed, immediately followed by euthanasia by an approved method.

At the time of writing, the availability of natural Pristane is limited, and substitution of other priming agents may be required. Unfortunately, there are few published data about synthetic Pristane or other suitable alternatives at the current time. Thus far we have encountered no disadvantages with the use of synthetic Pristane.

6.3.4 MONOCLONAL ANTIBODY PRODUCTION *IN VIVO*: ASCITES FROM XENOGENIC HYBRIDOMA CELL LINES

These variations of the protocol for ascites production in BALB/c mice may be followed to produce ascites from hetero-hybridomas resulting from fusions of spleen cells from a rat or a mouse strain other than BALB/c with standard mouse myeloma cells or from other xenogenic cell lines, such as rat-rat hybridomas.

6.3.4.1 Protocol 1: Ascites Production in Immunosuppressed BALB/c Mice

6.3.4.1.1 *Materials*

Mice: four BALB/c female retired breeder mice, Pristane-primed
Pristane (Sigma T-7640)

Tuberculin syringes (1 cc)
Needles, sterile (21 gauge)
Needles, sterile (30 gauge)
Hydrocortisone 21-hemisuccinate (Sigma H4881), 6 mg/ml solution in
  D-PBS, sterile
Hybridoma cells, 5×10^6 per mouse
Trypan blue solution, 0.4% (Sigma T8154)
Conical test tube, sterile (15 or 50 ml)
D-PBS, sterile

6.3.4.1.2 Equipment

Hemacytometer
Microscope with phase contrast optics
Centrifuge, bench top
Cesium gamma irradiator

6.3.4.1.3 Time Required

Pristane-priming of four mice takes 10 min. Preparing and administering hydrocortisone hemisuccinate takes 45 min, and gamma irradiation of four mice takes 10 min. Hybridoma cells should be grown ahead of time: preparation of the cells for injection requires about 30 min and injection of the cells takes about 10 min.

6.3.4.1.4 Procedure

1. Pristane-prime the mice 14 days before injection of cells (see 6.3.3.1).
2. Three days before injection of the cells inject 0.5 ml of the hydrocortisone hemisuccinate solution intramuscularly into the gluteal muscle of one leg of each mouse with a 30-gauge needle on a 1-cc tuberculin syringe.
3. Two days before the injection of hybridoma cells, gamma-irradiate the mice with a 550-rad sublethal dose.
4. See Sections 6.3.3.1 and 6.3.3.2 for procedures to prepare and inject hybridoma cells and for ascites fluid harvest procedures.

6.3.4.1.5 Results

The hybridoma cells usually grow more slowly in the allogeneic mice, even when they are immunosuppressed or immunodeficient than syngeneic hybridomas in intact BALB/c mice. Therefore, the ascites fluid builds up more slowly. Generally the mice are ready for the first ascites harvest (tap) 10–14 days after injection of the hybridoma cells, but if no ascites develops after 30 days, the procedure has failed.

Ascites volumes ranging from 2–10 ml per mouse may be collected. The concentration of monoclonal antibody in the fluid ranges from 1–10 mg/ml so that four immunosuppressed mice may produce 10–20 ml of ascites and 10–200 mg of antibody. The yield may vary both with individual hybridomas and with each ascites production of the same hybridoma cell line.

6.3.4.1.6   Problems and Troubleshooting

Situations similar to those described at the end of the "Ascites Production *in vivo* in BALB/c Mice" protocol may be observed.

The hydrocortisone hemisuccinate must be the sodium-salt form to be soluble in water, and the solution should be prepared within 48 h of use.

It is critical that a dose of gamma irradiation between 550–580 rad be administered. Lower doses will not be sufficient, and higher doses may be lethal within 2–5 days.

The mice are more prone to infection after immunosuppression, so special care should be taken to perform procedures to inject cells and harvest ascites by aseptic techniques. Feeding the mice sterilized food and acidified water will also reduce the chance of infection in the mice.

6.3.4.2 Protocol 2: Ascites Production in Athymic Nude Mice

6.3.4.2.1   Priming of Athymic Nude Mice and Injection
of Hybridoma Cells
6.3.4.2.1.1   Materials

> Mice, five 7- to 8-week-old female athymic nude (nu/nu) mice
> Pristane (Sigma T-7640)
> Tuberculin syringes (1 cc)
> Needles, sterile (25 gauge)
> Needles, sterile (21 gauge)
> Hybridoma cells, 5×10^6 per mouse
> Trypan blue solution, 0.4% (Sigma T8154)
> Conical test tube, sterile (15 or 50 ml)
> D-PBS, sterile

6.3.4.2.1.2   Equipment

> Hemacytometer
> Microscope with phase contrast optics
> Centrifuge, bench top

6.3.4.2.1.3   Time Required

The Pristane priming of five mice requires about 10 min. Two Pristane injections are done, with the first priming performed 12 days before injection of hybridoma cells. Hybridoma cells should be grown ahead of time: preparation of the cells for injection requires about 30 min and injection of the cells takes about 10 min.

6.3.4.2.1.4   Procedure

1. Twelve days before the planned injected of the hybridoma cells, prime the mice with 0.5 ml Pristane per mouse with a 25-gauge needle. The i.p. injection procedure is similar to that done with BALB/c mice. The Pristane is more likely to backflow from the needle opening in the smooth hairless

skin of the nude mouse: holding the needle in place briefly without pressure on the syringe, followed by slow, gentle withdrawal of the needle, is necessary to prevent this.

2. Two days before injection of the cells, repeat the Pristane priming procedure with 0.2 ml of Pristane per mouse.
3. The hybridoma cells should be pre-cultured, counted, and prepared for injection as with ascites production in BALB/c mice.
4. Evenly draw the suspended prepared cells into a sterile syringe with a 21-gauge needle and inject i.p. 0.5 ml into the mouse in the same location and manner as Pristane priming (see Step 1).
5. Monitor mice 24 h after injection of hybridoma cells and at least every other day thereafter for general health and development of ascites fluid.

6.3.4.2.2 Ascites Fluid Collection in Athymic Nude Mice
6.3.4.2.2.1 Materials

Alcohol prep pads, sterile
Needles, sterile (18 gauge)
Conical test tubes, sterile (15 ml)
Conical test tube, sterile (50 ml)

6.3.4.2.2.2 Time Required
The hybridoma cells usually grow more slowly in the athymic nu/nu mice than in BALB/c mice; therefore, the ascites fluid builds up more slowly. Generally, the animals are ready for the first ascites tap 10–14 days after injection of the hybridoma cells. Collection of the fluid and centrifugation takes approximately 30 min for each tap. A maximum of three taps should be harvested from each mouse over a 7-day period with euthanasia performed at the time of the final collection.

6.3.4.2.2.3 Procedure
See section 6.3.3.2 protocol for ascites fluid harvest procedures.

6.3.4.2.3 Results
Ascites volumes ranging from 2 to 5 ml per mouse may be collected. The concentration of monoclonal antibody in the fluid ranges from 1 to 10 mg/ml so that five athymic nu/nu mice may yield 10–20 ml of ascites and 10–100 mg of antibody. The yield may vary both with individual hybridomas and with each ascites production of the same hybridoma cell line.

6.3.4.2.4 Problems and Troubleshooting
Situations similar to those described at the end of the "Ascites Production *in Vivo* in BALB/c Mice" protocol may be observed.

Athymic nu/nu mice should be housed in a specific pathogen-free cage facility with laminar flow air circulation and handled with strict sterile technique to reduce the chance of infection developing in the animals.

Paracentesis is usually more difficult with athymic nu/nu mice as the fluid may drain more slowly and be more prone to flowing around the needle through the puncture site in the skin. Often, the fluid must be collected by removing the needle and allowing it to drip into a test tube but a best effort should be made to collect by drainage through the needle. If fluid has already wetted the skin of the abdomen, additional fluid may spread on the skin rather than dripping directly into the test tube. Quickly wiping the abdomen dry with a sterile alcohol prep pad may remedy this and increase the recovery of the ascites fluid at each tap.

Hybridoma cells may be more likely to the form solid tumors without yielding any ascites fluid when injected in athymic nu/nu mice. A solid tumor may be excised from the abdomen of the mouse in a laminar flow hood using careful aseptic technique. Cells should be released from the tumor into cold serum-free media, washed and centrifuged to pellet and then cultured under standard hybridoma conditions before reinjecting into naive Pristane primed nude mice. These *in vivo*–adapted cells may produce ascites without forming solid tumors.

6.4 PRODUCTION IN CELL CULTURE

6.4.1 *In Vitro* Production of Monoclonal Antibodies in Research Laboratories

In view of the limitations, disadvantages, and animal use constraints inherent in the production of mAbs in ascites, substantial effort has been devoted to improving techniques for producing mAbs in cell-culture systems. Standard culture in wells or flasks is not ideal, because the yield is typically <0.01 mg/ml. To obtain useful quantities of mAb from such cultures, it may be necessary to purify the mAb from several liters of culture fluid, a process that is tedious, costly, and inefficient.

Several systems have been described that improve the yield of mAb in cell culture and that vary in approach, scale, and complexity. None of these yield the same rate of mAb production per cell that occurs *in vivo* in ascites.[39] Some cell-culture systems are listed in Table 6.3. The term bioreactor has been rather broadly applied to many of these systems, including those in which oxygen and nutrients are distributed by mechanical pumping, and parameters such as pH, oxygen, nutrient, and waste concentrations are monitored and adjusted by automated sensing and control devices. Other simpler systems that lack pumping devices or automated controls, but that use specialized internal geometry of the cell culture devices to optimize cell density and mAb yield can also be thought of as bioreactors. These simpler systems are much more practical for smaller research laboratories, and a detailed protocol for use of one of them is included in this chapter.

Roller bottles and spinner flasks offer only modest advantages over standard tissue culture flasks in term of mAb yield. Use of gas permeable bags is an inexpensive approach and increased mAb yield by 70% in one study, compared to tissue culture flasks.[40] In a more comprehensive analysis of this method, involving 30 hybridoma lines, the mean yield of purified mAb from a 500 ml gas-permeable bag was 36.9 mg (range 1.8–102 mg).[41] Growth of hybridomas within dialysis tubing bags permits exchange of nutrients and metabolites, but not mAb, with a larger volume of culture medium in the culture vessel into which the dialysis bag is immersed.[42,43] This represents the simplest type of two-chamber

---

TABLE 6.3
**Systems for High-Density Hybridoma Growth and High-Yield
mAb Production in Cell Culture**

Roller bottles
Spinner flasks
Gas-permeable bags
Dialysis tubing within culture flasks or bottles
Two chamber flasks with a semipermeable membrane separating the chambers
 –Mini-PERM
 –CELLine
Hollow fiber bioreactors
 –Cellmax Quad
 –MiniMouse
 –Tecnomouse
 –Cellex Accusyst
 –Cell Pharm
 –Renal dialysis cartridges
 –Centrifugation-coupled bioreactors
Deep tank–stirred fermentors
Perfusion tank systems
Airlift bioreactors

---

culture system, that facilitates high-density cell growth and higher *in vitro* mAb production.

This approach has been used to develop self-contained two-chamber culture systems that are the simplest examples of the broad range of devices for mAb production that collectively are termed bioreactors.[44–47] The miniPERM system and the CELLine flask are two examples of different approaches to construction of such systems.[44–47]

The miniPERM bioreactor contains a 35-ml production module and a 550-ml supply module, separated by a dialysis membrane, and the entire assembly is rolled on an apparatus at 10 rpm within a tissue culture incubator.[45] With hybridomas adapted to low-serum conditions, yields of mAb >1 mg/ml can be achieved within a few weeks.[45] Typically, medium is exchanged three times weekly, and mAb harvested once or twice weekly.[44]

The CELLine flasks are even simpler, are fully disposable, contain two chambers separated by a dialysis membrane that allows passage of molecules <10 kDa, require no assembly and no equipment within the incubator, and yield similar mAb concentrations.[46,47] Procedures for medium exchange and mAb harvesting have been optimized[46,47] (see 5.2.2). In our experience it is easier to avoid microbial contamination with the use of the CELLine flask system. The final level of mAb contamination by bovine immunoglobulin from fetal calf serum is expected to be <1%.[46]

Hollow fiber bioreactors are more complex devices that involve high-density growth of cells within fibers designed to simulate the *in vivo* capillary system.[6,48–54] Such systems require mechanical devices to maintain constant perfusion of the fibers and are vulnerable to microbial contamination, cell death, and mechanical failures or fluctuations. Such systems are suitable for production of medium- to large-scale mAb batches in facilities with substantial expertise in tissue culture engineering. The use of supplemental natural and synthetic oxygen carrying molecules within bioreactor systems has been investigated as a method of enhancing mAb yield.[55]

Additional bioreactor designs incorporate packed-bed chambers, air-bubble technology, continuous centrifugation, filtration, controlled perfusion, and combinations of such approaches.[56-64] Sophisticated attention to nutrient and serum concentration, oxygen tension, pH, and conditions that minimize cell damage can optimize antibody yield.[39,65-83] These systems are most suitable for large-batch mAb production in the industrial setting, on a scale as large as several thousand liters. To produce 1 kg of a mAb, it has been estimated that 10 culture runs of a 1000 l bioreactor (or ascites production in thousands of mice) is required.[84] In contrast, use of the CELLine flasks is practical in smaller research laboratories, when more modest batches of mAb are required.

6.4.2 PROTOCOL: MONOCLONAL ANTIBODY PRODUCTION IN CELLINE CL-1000 FLASKS

6.4.2.1 Materials

CELLine CL-1000 bioreactor flask, Integra Biosciences/Argos Technologies
IMDM culture medium with 2 mM L-glutamine and 100 U/ml penicillin-streptomycin
Fetal bovine serum (FBS), heat-inactivated
Hybridoma cells, 50×10^6
Conical test tube, sterile (15 or 50 ml)
Serologic pipets (25 ml)
Spent-media waste reservoir, sterile (e.g., 1.5-l glass beaker or empty media bottles)
Trypan blue solution, 0.4% (Sigma T8154)
D-PBS, sterile

6.4.2.2 Equipment

Hemacytometer
Microscope with phase contrast optics
Centrifuge, bench top
CO_2 incubator

6.4.2.3 Time Required

Inoculation of the bioreactor flask requires 20 min. Each harvest and split back of the cell compartment and nutrient media replacement requires an additional 30 min. If handled carefully, a hybridoma cell line can be grown in a single bioreactor flask for 1–3 months with, at minimum, weekly nutrient medium replacement and cell compartment split back and antibody harvest.

6.4.2.4 Inoculation Procedure

1. Prewarm the nutrient media and cell compartment medium to 37°C before adding to the CELLine flask in all steps of the procedure. In this protocol, IMDM is specified: the hybridoma cells must be preadapted to grow in static culture in whatever medium (e.g., RPMI, DMEM) will be used in the bioreactor.

2. Remove the two-compartment CELLine bioreactor from the packaging in a laminar flow hood. The upper nutrient media compartment of the large T-style flask has a 1-l capacity and is accessible through a large screw-capped opening on the neck. A semipermeable cellulose acetate membrane separates the medium compartment from the lower cell compartment. This membrane allows free exchange of small molecules between the two compartments but prevents diffusion of larger molecules out of the cell compartment, ensuring that antibody secreted by the hybridoma cells is retained. The bottom of the cell compartment includes a silicone membrane that allows for the supply of oxygen and exchange of CO_2. The cell compartment is accessed separately by pipet through a capped silicone-sealed port on top of the flask.

3. Check that the small screw cap on the cell compartment access port is tightly closed. Equilibrate the semipermeable membrane by pipetting 50 ml of IMDM supplemented with 5% FBS into the nutrient medium compartment and let rest at least 5 min.

4. The hybridoma cells should be split with fresh medium 1–2 days before inoculation of the flask to ensure that they will be in logarithmic growth phase. The cells should be counted on the day of inoculation and should be at 90–100% viability as determined by trypan blue exclusion.

5. Transfer a volume of the cell-culture suspension containing 50×10^6 hybridoma cells into a sterile 50-ml conical test tube, and pellet at room temperature by centrifugation at $200g$ for 5 min. Discard the supernatant and gently resuspend the cells in 15 ml of IMDM supplemented with 15% FBS.

6. Loosen the large screw cap on the nutrient medium compartment to prevent an air lock and the backflow of liquid from the cell compartment port. It is essential to always perform this step before loosening the smaller cap to access the cell compartment.

7. Aspirate the cell suspension into a 25-ml pipet, remove the cap from the cell compartment, insert the pipet tip into the black cell compartment port, and gently inoculate the cells into the compartment. Keeping the pipet tip firmly pressed into the port to maintain a seal, immediately aspirate the cell suspension back out of the compartment into the pipet, and then gently draw up into the pipet until as many of the air bubbles in the compartment as possible are removed, allowing them to rise to the top of the pipet. Transfer the cell suspension back into the compartment, retaining the air bubbles in the pipet, and close the cap tightly.

8. In one hand, hold the bioreactor flask tilted on end in a semi-upright (45° angle) position with the neck opening facing the center of the

laminar flow hood. Carefully and aseptically pour 950 ml of IMDM supplemented with 5% FBS cell into the nutrient compartment. Close the nutrient compartment by completely tightening the cap and place the flask in an incubator under the optimal CO_2 and temperature conditions for the hybridoma cell line.

6.4.2.5 Cell-Compartment Harvest Procedure

The first harvest of antibody can generally be performed 7 days after inoculation and subsequent harvests can be performed at 5- to 7-day intervals. The viable cell population in the cell compartment should reach at least 350×10^6 before harvest.

1. Keeping the cell compartment tightly capped, remove the cap from the nutrient compartment and decant the spent medium into a sterile waste container by carefully pouring it out. Do this slowly to avoid drips and splashing.
2. Set the flask back down flat in the laminar flow hood, leaving the nutrient medium cap loose. Remove the cell compartment cap and gently aspirate the cell suspension into a 25-ml pipet. Mix the cell suspension well by slowly pipetting the liquid in and out of the compartment several times, note the total volume and return 4 ml or from $20–100 \times 10^6$ cells to the cell compartment. The volume returned to the cell compartment may need to be adjusted if the total volume of the suspension has increased from the day of inoculation due to osmotic flux, if the cells do not grow sufficiently or if the cell number increases at a rate significantly above average.
3. Transfer the harvest to a conical test tube, determine the cell count and viability, adjust the harvest volume by removing or adding to the cell compartment as necessary, and then tighten the cap on the cell compartment. Centrifuge the final harvest at 250 g for 10 min to pellet the cells. The harvest steps should be performed as rapidly as possible to maintain cell viability and minimize the time the nutrient compartment is empty and the semi-permeable membrane is dry.
4. While the harvest is being centrifuged, replace the nutrient medium by carefully pouring 1 l of IMDM supplemented with 1–2% FBS (serum can be reduced from the initial 5% depending on how well the cells grow during the first week) into the nutrient compartment. Leave the nutrient compartment cap loose and return the volume of the cell compartment to 15 ml by pipetting the appropriate volume of IMDM supplemented with 15% FBS into the cell compartment and removing the air bubbles as at inoculation. Tighten both the cell compartment and the nutrient caps well and return the CL-1000 bioreactor flask to the incubator.
5. Aspirate the harvest supernatant from the cell pellet and transfer to a 50-ml test tube. Store the antibody product frozen at –20 to –80°C.
6. Additional harvests should be performed at intervals of 5–7 days and can be pooled for storage. Generally, if the bioreactor is reseeded with $20–100 \times 10^6$ cells, the next harvest can be performed 7 days later.

6.4.2.6 Results

It is possible to achieve cell densities of up to 10^9 in the CELLine CL-1000 cell compartment, but it is best to try to maintain the density so that the viable cell number at each harvest is about 500×10^6 cells. Within 5–8 weeks of culture, a pooled harvest volume of 100 ml containing 0.5–2 mg/ml can be collected from a single CL-1000 bioreactor flask. The first week of culture generally produces less antibody than later weeks, but this depends on the cell density and the secretion rate of individual hybridoma cell lines. The antibody concentration in each harvest can be measured periodically to assess the progress of the monoclonal antibody production. At higher densities, the cell viability frequently decreases to between 50% and 70%, but the best antibody production often occurs at this reduced viability.

Well-established cultures can be maintained with serum-free IMDM (or other medium) in the nutrient compartment or a serum-free medium designed specifically for hybridoma cell culture (for example, HyQSFM4mAb-Utility [Hyclone #SH30382]) can be substituted in the nutrient compartment at a 1:1 ratio with the original nutrient medium (either with 1% FBS or serum-free). The serum in the cell compartment can sometimes be reduced to 5% if osmotic flux does not dilute the cell compartment medium too much and if the cells maintain high-density proliferation at greater than 50% viability and continue to secrete antibody. The CELLine bioreactor can also be set up and maintained through the entire production in a hybridoma specific serum-free medium if the hybridoma cell line of interest has been successfully grown and demonstrated to secrete antibody in the same medium in static culture conditions.

6.4.2.7 Problems and Troubleshooting

The membrane in the CL-1000 flask is thin and somewhat delicate. It can withstand normal handling as described in this protocol, but especially when the nutrient compartment is empty, it is very important not to shake the flask, bang it to "loosen" cells, or pipet from the cell compartment with too much force. If the membrane tears, it will be obvious that cells have entered the nutrient compartment when clouding of the medium is observed, especially if they are allowed to grow there. To reduce the chance of disrupting the membrane due to drying, add 50 ml of fresh culture medium to the nutrient compartment immediately after the spent medium is decanted, but before harvesting.

The Integra CELLine bioreactor flask is prone to contamination because of frequent handling with multiple rounds of removal and addition of cells and medium. All procedures should be performed using strict sterile technique in a laminar flow hood, the surface of which has been thoroughly wiped down with 70% alcohol. A vacuum system (Integra VACUSAFE) for aspirating spent nutrient medium is available from the manufacturer. If the medium is poured out, the waste receptacle should be sterile, and great care should be taken to avoid drops of medium on the neck of the flask. If these do occur, they can be removed from the interior of the neck with a sterile Pasteur pipette or from the exterior by carefully wiping with a sterile alcohol prep pad. Loosening the screw cap on the nutrient compartment before opening the cap on the cell compartment port is critical; this prevents backflow of the cell suspension and possible contamination of the cell compartment.

As antibody is secreted into the cell compartment, the protein gradient between the nutrient and cell compartment medium may produce osmotic flux, driving water into the cell compartment and increasing the volume. The compartment seems to be able to withstand at least 25 ml of suspension without bursting the membrane. If the volume has increased significantly over the initial 15 ml at inoculation, the harvest volume should be adjusted to retain sufficient cells in the bioreactor for continued proliferation.

If the hybridoma cells do not grow well after the first 7 days of culture, several steps can be taken. First, be sure that the cells are in log growth at the time of inoculation and check for mycoplasma contamination, which will accumulate in the cell compartment and which may cause effects not seen in static culture. Inoculate with a minimum of 40×10^6 viable cells. Starting out with larger cell numbers $(100–200 \times 10^6)$ in the CL-1000 flask cell compartment may result in better or more rapid adaptation to growth in the bioreactor. Use 5% FBS-containing medium in the nutrient chamber at inoculation and until high-density growth is established. Maintain a static culture of the hybridoma line so that additional cells can be added after the first week if the initial inoculation does not grow well. The serum in the cell compartment can also be increased to 20% if the hybridoma cells do not reach high density. If the density of the hybridoma cells approaches or reaches 10^9 or if the viability decreases below 50%, it may be necessary to harvest the cell compartment every 3–5 days or to reseed the flask with fewer cells after the harvest.

The properties of individual hybridoma cell lines can vary in the CELLine bioreactor flask. Therefore, the growth conditions must often be adjusted for optimal proliferation and antibody production. Some hybridoma cell lines do not seem to grow or produce antibody well in these flasks no matter what variations of the procedures are attempted.

6.4.3 RECOMBINANT AND TRANSGENIC SYSTEMS FOR PRODUCING ANTIBODIES

Recombinant DNA approaches to production of antibodies or antibody-like molecules are discussed in Chapter 8 of this volume and involve expression of the gene constructs that encode the antibody reagent in mammalian or microbial cell cultures. Large-scale production can be undertaken in bacteria, mammalian cell lines, transgenic plants, or in milk of transgenic animals, such as goats.[85–91] As with conventional mAbs, issues of contamination with other biologically active molecules and consistency of antibody glycosylation and function require attention.

6.5 CONCLUSION

In the three decades since the methods for making mAbs were first described, a wide range of techniques have been developed for generation of mAbs in quantities useful for research and for clinical applications. The choice of methodology for mAb production must take into account the desired use of the reagent, the quantity required, and the individual properties of each hybridoma. It is expected that techniques for mAb production will continue to be optimized, but for the foreseeable future, no single approach will be universally suitable for all hybridomas, and availability of both *in vivo* and *in vitro* techniques will be required.

ACKNOWLEDGMENT

The authors would like to thank Stephanie Taylor for her assistance with the literature search for this chapter.

REFERENCES

1. Institute of Laboratory Animal Resources (U.S.). Committee on Methods of Producing Monoclonal Antibodies, *Monoclonal Antibody Production*, National Academy Press, Washington, D.C., 1999.
2. Marx, U. and Merz, W., *In vivo* and *in vitro* production of monoclonal antibodies, In *Methods in Molecular Biology, vol. 45: Monoclonal antibody protocols*, Davis, W.C., ed., Humana Press, Totowa, N.J., 1995.
3. Hendriksen, C.F. and de Leeuw, W., Production of monoclonal antibodies by the ascites method in laboratory animals, *Res. Immunol.*, 149, 535, 1998.
4. Falkenberg, F.W., Monoclonal antibody production: problems and solutions, *Res. Immunol.*, 149, 542, 1998.
5. Nagel, A. et al., Membrane-based cell culture systems—an alternative to *in vivo* production of monoclonal antibodies, *Dev. Biol. Stand.*, 101, 57, 1999.
6. Jackson, L.R. et al., Evaluation of hollow fiber bioreactors as an alternative to murine ascites production for small scale monoclonal antibody production, *J. Immunol. Methods*, 189, 217, 1996.
7. Maiorella, B.L. et al., Effect of culture conditions on IgM antibody structure, pharmacokinetics and activity, *Biotechnology. (N. Y.)*, 11, 387, 1993.
8. Weitzhandler, M. et al., Analysis of carbohydrates on IgG preparations, *J. Pharm. Sci.*, 83, 1670, 1994.
9. Jackson, L.R. et al., Monoclonal antibody production in murine ascites. I. Clinical and pathologic features, *Lab. Anim. Sci.*, 49, 70, 1999.
10. Peterson, N.C., Behavioral, clinical, and physiologic analysis of mice used for ascites monoclonal antibody production, *Comp. Med.*, 50, 516, 2000.
11. Gearing, A.J. et al., Presence of the inflammatory cytokines IL-1, TNF, and IL-6 in preparations of monoclonal antibodies, *Hybridoma*, 8, 361, 1989.
12. Appelmelk, B.J. et al., Murine ascitic fluids contain varying amounts of an inhibitor that interferes with complement-mediated effector functions of monoclonal antibodies, *Immunol. Lett.*, 33, 135, 1992.
13. Patel, T.P. et al., Different culture methods lead to differences in glycosylation of a murine IgG monoclonal antibody, *Biochem. J.*, 285, 839, 1992.
14. Kunkel, J.P. et al., Comparisons of the glycosylation of a monoclonal antibody produced under nominally identical cell culture conditions in two different bioreactors, *Biotechnol. Prog.*, 16, 462, 2000.

15. Kemminer, S.E. et al., Production and molecular characterization of clinical phase I anti-melanoma mouse IgG3 monoclonal antibody R24, *Biotechnol. Prog.*, 17, 809, 2001.
16. Muthing, J. et al., Effects of buffering conditions and culture pH on production rates and glycosylation of clinical phase I anti-melanoma mouse IgG3 monoclonal antibody R24, *Biotechnol. Bioeng.*, 83, 321, 2003.
17. Schenerman, M.A. et al., Comparability testing of a humanized monoclonal antibody (Synagis) to support cell line stability, process validation, and scale-up for manufacturing, *Biologicals*, 27, 203, 1999.
18. Brodeur, B.R. and Tsang, P.S., High yield monoclonal antibody production in ascites, *J. Immunol. Methods*, 86, 239, 1986.
19. Stewart, F., Callander, A., and Garwes, D.J., Comparison of ascites production for monoclonal antibodies in BALB/c and BALB/c-derived cross-bred mice, *J. Immunol. Methods*, 119, 269, 1989.
20. Brodeur, B.R., Tsang, P., and Larose, Y., Parameters affecting ascites tumour formation in mice and monoclonal antibody production, *J. Immunol. Methods*, 71, 265, 1984.
21. Hoogenraad, N.J. and Wraight, C.J., The effect of Pristane on ascites tumor formation and monoclonal antibody production, *Methods Enzymol.*, 121, 375, 1986.
22. Hoogenraad, N., Helman, T., and Hoogenraad, J., The effect of pre-injection of mice with Pristane on ascites tumour formation and monoclonal antibody production, *J. Immunol. Methods*, 61, 317, 1983.
23. Jackson, L.R. et al., Monoclonal antibody production in murine ascites. II. Production characteristics, *Lab. Anim. Sci.*, 49, 81, 1999.
24. Hasegawa, N. et al., Production of monoclonal antibody in mouse ascitic fluid with two solid tumor-forming hybridoma cell lines, *Hybridoma*, 10, 647, 1991.
25. Ruiz-Bravo, A., Perez, M., and Jimenez-Valera, M., The enhancement of growth of a syngeneic plasmacytoma in BALB/c mice by Pristane priming is not due to immunosuppressive effects on antibody-forming cell or mitogen-responsive splenocytes, *Immunol. Lett.*, 44, 41, 1995.
26. Jones, S.L., Cox, J.C., and Pearson, J.E., Increased monoclonal antibody ascites production in mice primed with Freund's incomplete adjuvant, *J. Immunol. Methods*, 129, 227, 1990.
27. Mueller, U.W., Hawes, C.S., and Jones, W.R., Monoclonal antibody production by hybridoma growth in Freund's adjuvant primed mice, *J. Immunol. Methods*, 87, 193, 1986.
28. Witte, P.L. and Ber, R., Improved efficiency of hybridoma ascites production by intrasplenic inoculation in mice, *J. Natl. Cancer Inst.*, 70, 575, 1983.
29. Truitt, K.E. et al., Production of human monoclonal antibody in mouse ascites, *Hybridoma*, 3, 195, 1984.
30. Abrams, P.G. et al., Production of large quantities of human immunoglobulin in the ascites of athymic mice: implications for the development of anti-human idiotype monoclonal antibodies, *J. Immunol.*, 132, 1611, 1984.
31. Ware, C.F., Donato, N.J., and Dorshkind, K., Human, rat or mouse hybridomas secrete high levels of monoclonal antibodies following transplantation into mice with severe combined immunodeficiency disease (SCID), *J. Immunol. Methods*, 85, 353, 1985.

32. Weissman, D. et al., Methods for the production of xenogeneic monoclonal antibodies in murine ascites, *J. Immunol.,* 135, 1001, 1985.

33. Witt, S. et al., [Optimal production of murine monoclonal antibodies in ascites of syngeneic mice by a single whole body irradiation], *Allerg. Immunol. (Leipz).,* 33, 259, 1987.

34. Matsumoto, M., Mochizuki, K., and Kobayashi, Y., Productive ascites growth of heterohybridomas between Epstein-Barr virus-transformed human B cells and murine P3X63Ag8.653 myeloma cells, *Microbiol. Immunol.,* 33, 883, 1989.

35. Pistillo, M.P., Sguerso, V., and Ferrara, G.B., High yields of anti-HLA human mono-clonal antibodies can be provided by SCID mice, *Hum. Immunol.,* 35, 256, 1992.

36. Kan-Mitchell, J. et al., Altered antigenicity of human monoclonal antibodies derived from human-mouse heterohybridomas, *Hybridoma,* 6, 161, 1987.

37. Hirsch, F. et al., Rat monoclonal antibodies. III. A simple method for facilitation of hybridoma cell growth *in vivo, J. Immunol. Methods,* 78, 103, 1985.

38. Kints, J.P., Manouvriez, P., and Bazin, H., Rat monoclonal antibodies. VII. Enhancement of ascites production and yield of monoclonal antibodies in rats following pretreatment with Pristane and Freund's adjuvant, *J. Immunol. Methods,* 119, 241, 1989.

39. Kundu, P.K. et al., Getting higher yields of monoclonal antibody in culture, *Indian J. Physiol. Pharmacol.,* 42, 155, 1998.

40. Peterson, N.C., Considerations for *in vitro* monoclonal antibody production, *Res. Immunol.,* 149, 553, 1998.

41. Lipski, L.A. et al., Evaluation of small to moderate scale *in vitro* monoclonal antibody production via the use of the i-mAb gas-permeable bag system, *Res. Immunol.,* 149, 547, 1998.

42. Kasehagen, C. et al., Metabolism of hybridoma cells and antibody secretion at high cell densities in dialysis tubing, *Enzyme Microb. Technol.,* 13, 873, 1991.

43. Mathiot, B. et al., Increase of hybridoma productivity using an original dialysis culture system, *Cytotechnology,* 11, 41, 1993.

44. Marx, U., Membrane-based cell culture technologies: a scientifically and economically satisfactory alternative to malignant ascites production for monoclonal antibodies, *Res. Immunol.,* 149, 557, 1998.

45. Falkenberg, F.W., Production of monoclonal antibodies in the miniPERM bioreactor: comparison with other hybridoma culture methods, *Res. Immunol.,* 149, 560, 1998.

46. Trebak, M. et al., Efficient laboratory-scale production of monoclonal antibodies using membrane-based high-density cell culture technology, *J. Immunol. Methods,* 230, 59, 1999.

47. Bruce, M.P. et al., Dialysis-based bioreactor systems for the production of monoclonal antibodies—alternatives to ascites production in mice, *J. Immunol. Methods,* 264, 59, 2002.

48. Gorter, A. et al., Production of bi-specific monoclonal antibodies in a hollow-fibre bioreactor, *J. Immunol. Methods,* 161, 145, 1993.

49. Goodall, M., A simple hollow-fiber bioreactor for the "in-house" production of monoclonal antibodies, *Methods Mol. Biol.,* 80, 39, 1998.

50. Kreutz, F.T. et al., Production of highly pure monoclonal antibodies without purification using a hollow fiber bioreactor, *Hybridoma,* 16, 485, 1997.

51. Manzke, O. et al., Single-step purification of bispecific monoclonal antibodies for immunotherapeutic use by hydrophobic interaction chromatography, *J. Immunol. Methods,* 208, 65, 1997.

52. Lipman, N.S. and Jackson, L.R., Hollow fibre bioreactors: an alternative to murine ascites for small scale (<1 gram) monoclonal antibody production, *Res. Immunol.,* 149, 571, 1998.

53. Dowd, J.E. et al., Predictive control of hollow-fiber bioreactors for the production of monoclonal antibodies, *Biotechnol. Bioeng.,* 63, 484, 1999.

54. Evans, T.L. and Miller, R.A., Evaluation of hollow-fiber bioreactor systems for large-scale production of murine monoclonal antibodies, *Targeted Diagn. Ther.,* 3, 25, 1990.

55. Shi, Y., Sardonini, C.A., and Goffe, R.A., The use of oxygen carriers for increasing the production of monoclonal antibodies from hollow fibre bioreactors, *Res. Immunol.,* 149, 576, 1998.

56. Wang, G. et al., Modified CelliGen-packed bed bioreactors for hybridoma cell cultures, *Cytotechnology,* 9, 41, 1992.

57. Gudermann, F., Lutkemeyer, D., and Lehmann, J., Design of a bubble-swarm bioreactor for animal cell culture, *Cytotechnology,* 15, 301, 1994.

58. Tan, W.S., Dai, G.C., and Chen, Y.L., Quantitative investigations of cell-bubble interactions using a foam fractionation technique, *Cytotechnology,* 15, 321, 1994.

59. Banik, G.G. and Heath, C.A., Partial and total cell retention in a filtration-based homogeneous perfusion reactor, *Biotechnol. Prog.,* 11, 584, 1995.

60. Persson, B. and Emborg, C., A comparison of simple growth vessels and a specially designed bioreactor for the cultivation of hybridoma cells, *Cytotechnology,* 8, 179, 1992.

61. Deo, Y.M., Mahadevan, M.D., and Fuchs, R., Practical considerations in operation and scale-up of spin-filter based bioreactors for monoclonal antibody production, *Biotechnol. Prog.,* 12, 57, 1996.

62. Johnson, M. et al., Use of the Centritech lab centrifuge for perfusion culture of hybridoma cells in protein-free medium, *Biotechnol. Prog.,* 12, 855, 1996.

63. Yang, J.D. et al., Achievement of high cell density and high antibody productivity by a controlled-fed perfusion bioreactor process, *Biotechnol. Bioeng.,* 69, 74, 2000.

64. Petrossian, A. and Cortessis, G.P., Large-scale production of monoclonal antibodies in defined serum-free media in airlift bioreactors, *Biotechniques,* 8, 414, 1990.

65. Tiebout, R.F., Tissue culture in hollow-fibre systems: implications for downstream processing and stability analysis, *Dev. Biol. Stand.,* 71, 65, 1990.

66. Hagedorn, J. and Kargi, F., Coiled tube membrane bioreactor for cultivation of hybridoma cells producing monoclonal antibodies, *Enzyme Microb. Technol.,* 12, 824, 1990.

67. Ozturk, S.S. and Palsson, B.O., Effects of dissolved oxygen on hybridoma cell growth, metabolism, and antibody production kinetics in continuous culture, *Biotechnol. Prog.,* 6, 437, 1990.

68. Batt, B.C., Davis, R.H., and Kompala, D.S., Inclined sedimentation for selective retention of viable hybridomas in a continuous suspension bioreactor, *Biotechnol. Prog.,* 6, 458, 1990.

69. Handa-Corrigan, A. et al., Controlling and predicting monoclonal antibody production in hollow-fiber bioreactors, *Enzyme Microb. Technol.,* 14, 58, 1992.

70. Franek, F. and Dolnikova, J., Hybridoma growth and monoclonal antibody production in iron-rich protein-free medium: effect of nutrient concentration, *Cytotechnology,* 7, 33, 1991.

71. Abu-Reesh, I. and Kargi, F., Biological responses of hybridoma cells to hydrodynamic shear in an agitated bioreactor, *Enzyme Microb. Technol.,* 13, 913, 1991.

72. Ozturk, S.S. and Palsson, B.O., Growth, metabolic, and antibody production kinetics of hybridoma cell culture: 1. Analysis of data from controlled batch reactors, *Biotechnol. Prog.,* 7, 471, 1991.

73. Ozturk, S.S. and Palsson, B.O., Growth, metabolic, and antibody production kinetics of hybridoma cell culture: 2. Effects of serum concentration, dissolved oxygen concentration, and medium pH in a batch reactor, *Biotechnol. Prog.,* 7, 481, 1991.

74. Schurch, U., Cryz, S.J., Jr., and Lang, A.B., Scale-up and optimization of culture conditions of a human heterohybridoma producing serotype-specific antibodies to Pseudomonas aeruginosa, *Appl. Microbiol. Biotechnol.,* 37, 446, 1992.

75. Mercille, S. and Massie, B., Induction of apoptosis in oxygen-deprived cultures of hybridoma cells, *Cytotechnology,* 15, 117, 1994.

76. Xie, L. and Wang, D.I., Applications of improved stoichiometric model in medium design and fed-batch cultivation of animal cells in bioreactor, *Cytotechnology,* 15, 17, 1994.

77. Mancuso, A. et al., Effect of extracellular glutamine concentration on primary and secondary metabolism of a murine hybridoma: an *in vivo* ^{13}C nuclear magnetic resonance study, *Biotechnol. Bioeng.,* 57, 172, 1998.

78. Cherlet, M. and Marc, A., Intracellular pH monitoring as a tool for the study of hybridoma cell behavior in batch and continuous bioreactor cultures, *Biotechnol. Prog.,* 14, 626, 1998.

79. Iyer, M.S., Wiesner, T.F., and Rhinehart, R.R., Dynamic reoptimization of a fed-batch fermentor, *Biotechnol. Bioeng.,* 63, 10, 1999.

80. Tatiraju, S., Soroush, M., and Mutharasan, R., Multi-rate nonlinear state and parameter estimation in a bioreactor, *Biotechnol. Bioeng.,* 63, 22, 1999.

81. Seifert, D.B. and Phillips, J.A., The production of monoclonal antibody in growth-arrested hybridomas cultivated in suspension and immobilized modes, *Biotechnol. Prog.,* 15, 655, 1999.

82. Wen, Z.Y., Teng, X.W., and Chen, F., A novel perfusion system for animal cell cultures by two step sequential sedimentation, *J. Biotechnol.,* 79, 1, 2000.

83. Dhir, S. et al., Dynamic optimization of hybridoma growth in a fed-batch bioreactor, *Biotechnol. Bioeng.,* 67, 197, 2000.

84. Kretzmer, G., Industrial processes with animal cells, *Appl. Microbiol. Biotechnol.,* 59, 135, 2002.

85. Inoue, Y. et al., Production of recombinant human monoclonal antibody using ras-amplified BHK-21 cells in a protein-free medium, *Biosci. Biotechnol. Biochem.,* 60, 811, 1996.

86. Larrick, J.W. et al., Production of antibodies in transgenic plants, *Res. Immunol.,* 149, 603, 1998.

87. Chadd, H.E. and Chamow, S.M., Therapeutic antibody expression technology, *Curr. Opin. Biotechnol.,* 12, 188, 2001.

88. Yazaki, P.J. et al., Mammalian expression and hollow fiber bioreactor production of recombinant anti-CEA diabody and minibody for clinical applications, *J. Immunol. Methods,* 253, 195, 2001.

89. Cruz, H.J. et al., Process development of a recombinant antibody/interleukin-2 fusion protein expressed in protein-free medium by BHK cells, *J. Biotechnol.,* 96, 169, 2002.

90. Ehsani, P. et al., Expression of anti human IL-4 and IL-6 scFvs in transgenic tobacco plants, *Plant Mol. Biol.,* 52, 17, 2003.
91. Kipriyanov, S.M. and Le Gall, F., Generation and production of engineered antibodies, *Mol. Biotechnol.,* 26, 39, 2004.

7 Purification and Characterization of Antibodies

Joseph P. Chandler

CONTENTS

7.1 INTRODUCTION

Antibodies have been a biologic workhorse for well over a century. Since the first description of immunoglobulins in the 1870s by Emil von Behring and Shibasaburo Kitasato (reviewed in reference 1), extraordinary advances have been made in the use of antibodies in a wide range of research and commercial applications. Antibodies were first used as crude antiserum for viral diseases. Today, highly engineered therapeutic antibodies combat a variety of life-threatening illnesses.

This chapter will define several methods for purifying antibodies, in particular monoclonal antibodies (mAbs), which can later be used for a number of applications. Methodologies to characterize the antibodies before and after purification will also be described. Throughout the chapter, data generated from this laboratory will illustrate the results with the various techniques that are used to purify and characterize antibodies.

The approaches described in this chapter are designed for small to moderate scale purifications that can be performed manually or with the use of semiautomated equipment. No one method will work for all antibodies, and any one antibody can be purified by a number of different procedures.

Most laboratories should consider a platform of choices, each serving a desired outcome. In addition, there may be instances when no purification is required, such as when performing an immunoelectrophoresis or radial diffusion immunoassay. Crude extracts of antibody can work quite well in these instances as long as appropriate controls are included. However, for antibodies in preclinical trials, advanced multistep purification procedures must be used to generate highly purified antibody that is low in endotoxin. The greater the repertoire of purification schemes, the more efficient and economical the purification laboratory will become.

7.2 ANTIBODY PURIFICATION

The number of approaches for purifying antibodies is vast, too extensive for a single chapter. The author's intent is to describe in detail a few procedures that will work successfully for most applications. Notwithstanding, an overview of the general purification schemes, mentioning their strengths and limitations is presented. Table 7.1 illustrates the most common purification methods and their characteristics.[2]

7.2.1 PARTIAL PURIFICATION BY PRECIPITATION

Precipitation of protein using ammonium sulfate is one of the oldest methods for partial purification of antibodies. A saturated solution of ammonium sulfate is added to serum or ascites to precipitate the antibody. The precipitate is separated from the nonprecipitated proteins by centrifugation. When the pellet is resuspended in an aqueous medium, the resulting solution is enriched for antibodies.

In contract, after precipitation using caprylic acid, most of the proteins that precipitate are not immunoglobulins. This leaves the antibodies in solution at the end of the procedure.

Preparation of standard reagents (e.g., 1 × phosphate-buffered saline [PBS]) are described elsewhere (see Chapters 8, 9, and 11).

7.2.1.1 Ammonium Sulfate Precipitation

1. Prepare a saturated solution (4.1 M) of ammonium sulfate (AS). (Adding an equal volume of saturated AS to serum or ascites will yield a 50% saturated solution that will precipitate antibodies.)
2. Centrifuge the ascites or serum at 2,500 rpm for 15–30 min to clarify the starting material. Transfer the serum/ascites to a clean beaker with a stir bar and place on a magnetic stirrer.
3. Add a volume of saturated AS solution equal to the volume of the starting material. Slowly add the ammonium sulfate while it is blended into the ascites/serum. Allow precipitates to redissolve before adding more AS. When you have added nearly all of the AS, irreversible precipitation will occur. Continue to add the AS until all of it has been added to the preparation. Allow the solution to continue to mix for 6–24 h at 4°C.
4. When precipitation is complete, centrifuge the solution at 3,000 g for 30 min. Decant the supernatant. (Save a sample of the discard for later testing.) Slowly add 150 mM PBS to the pellet and gently stir with a pipette. Add sufficient PBS equal to 25–50% of the original volume of the starting material.
5. Dialyze the dissolved antibody preparation against PBS in a 20:1 (volume/volume) ratio of PBS to antibody preparation. Because of the high salt concentration of the AS, at least three changes of buffer for a minimum of 2 h each are recommended.
6. Protein concentration determination, zonal electrophoresis, densitometry scan, and specificity testing of the final preparation can then be performed

TABLE 7.1
Overview of Antibody Purification Methods

Procedure	Ease of Use	Cost	Versatility	Species Specificity	Best Use	Limitations	Scalable
Ammonium sulfate/caprylic acid precipitation	Very easy	Inexpensive	Works for all serum and ascites. Less useful for cell culture supernatants	None	Increase the concentration of an antibody	Does not completely purify	Yes, but limited
Anion exchange chromatography	Somewhat difficult—user must know pI of antibody	Moderately inexpensive	Good for most antibodies and isotypes	None	Good for removal of non-antibody proteins, protein A, DNA, endotoxin, retroviruses. IgG is soluble during loading and elution	Some capacity issues with some IgGs. DNA irreversibly binds to matrix	Very scalable
Cation exchange chromatography	Somewhat difficult—user must know pI of antibody	Moderately inexpensive	Good for most antibodies and isotypes	None	Good removal of non-antibody proteins, protein A. Has a high capacity for IgG	Moderately good removal of DNA and endotoxin. IgG can precipitate under conditions best suited for binding. Buffers can be corrosive.	Very scalable
Size exclusion	Packing and maintaining column can be difficult	Moderately inexpensive	Good for all antibodies	None	Can isolate fragments, aggregates	Can only purify small highly concentrated quantities	Not highly scalable

Method	Ease of use	Cost	Species/isotype	Denaturing	Role in purification	Comments	Scalability
Hydroxyapatite	Somewhat easy, matrix requires special care during packing and has a shorter use life	Moderately inexpensive	Good for most antibodies and isotypes	None	Polishing step—good for removal of non-antibody proteins, protein A, DNA, and endotoxins	Media is unstable below pH 6.25. Media binds to metal contaminants	Very scalable
Hydrophobic interaction columns	Somewhat difficult—user must know pI of antibody	Moderately inexpensive	Good for most antibodies and isotypes	None	Polishing step—good for removal of non-antibody proteins, protein A, DNA and endotoxin	High salt give better yields but is corrosive. Low salt easier to use but lowers capacity	Very scalable
Protein A	Some difficulty depending on species and isotype	Expensive	Best for mouse and rabbit antibodies. IgG1 isotypes require high salt	Yes	Best first step in antibody purification	May have fragments and aggregates formed during purification	Scalable, but at a high cost
Protein G	Easier than Protein A but *not* for all species	Expensive	Best for human, goat, sheep. All isotypes will bind	Yes	Best first step in antibody purification	May have fragments and aggregates formed during purification	Scalable, but at a high cost

Ig, immunoglobulin; pI, isoelectric point.

(see Section 7.3). These test results will indicate the level of purity and specific activity of the antibody.

7. The antibody preparation should be divided into aliquots and stored at –70°C.

7.2.1.2 Caprylic Acid Precipitation

1. Measure and record the volume of ascites or serum and transfer it to a clean beaker or flask. Add stir bar and place the flask on a magnetic stirrer.
2. Add two volumes of 60 mM acetate buffer, pH 4.0. Adjust the pH to 4.8.
3. Slowly add the caprylic acid (octanoic acid undiluted) dropwise to the ascites or serum while slowly stirring the solution. Add 0.4 ml caprylic acid for each 10 ml diluted ascites or serum. Let the solution stir for 30 min at room temperature.
4. Centrifuge the preparation for 10 min at 5,000 g. Decant and save the supernatant.
5. The antibody-enriched solution should be dialyzed against 20 volumes PBS as described previously before storage.

Yields by these methods are not more than 50%, and purity is not usually more than 70%. Because neither procedure yields a highly purified preparation of antibody, an additional purification step may be needed. One strategy is to perform a double precipitation on ascites to generate a moderately pure antibody preparation. In our laboratory, we preform the ammonium sulfate precipitation first, followed by caprylic acid precipitation. After the AS precipitation, the precipitated antibody can be reconstituted in PBS and dialyzed against 60 mM acetate buffer. After dialysis, the pH should be adjusted to 4.8 before performing the caprylic acid purification. The antibody preparation resulting from either AS or caprylic acid or both yields an antibody that is useful in its own right. Antibodies purified by precipitation can be successfully used in a number of procedures such as enzyme immunoassays, immunoprecipitation, Western blot, and dot blot.

Purity of antibody after AS and caprylic acid precipitation can be checked by zonal electrophoresis gels (Figure 7.1). For this analysis, we purified mAb from ascites by AS (lane 1), caprylic acid, and the combination of the two precipitation steps (lane 3). The concentration of MAb increased from 16.3% in the starting material to 23.0% after the AS precipitation and to 32.75% after the double precip-itation. AS precipitation alone enhanced the concentration of this mAb, but the combined procedures produced the most favorable enhancement of antibody purity. However, even with both treatments, the antibody is only about 75–80% pure.

7.2.2 PROTEIN A AND PROTEIN G

In our hands, protein A and protein G provide the best means of purifying mAbs from sera, ascites, or cell-culture supernatants. The end product tends to be at least 95% pure and can be used for full range of assays and procedures. Further processing of protein A– or protein G–purified mAbs will yield antibodies that can be used for

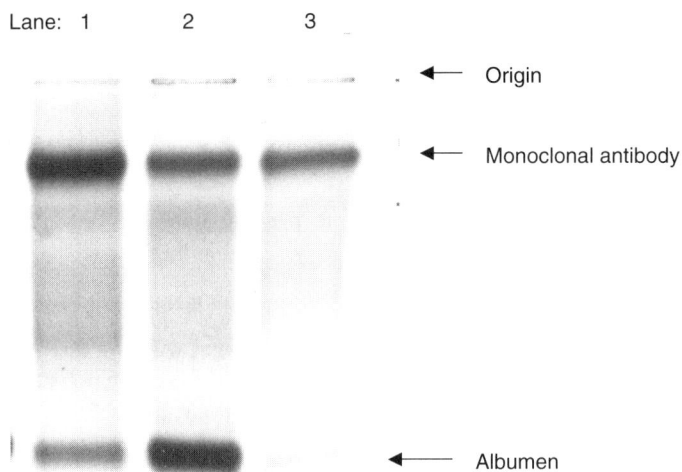

Lane 1 – Ammonium sulfate precipitate

Lane 2 – Original sample

Lane 3 – Amm. sulfate + caprilic acid Ppts.

FIGURE 7.1 Zonal electrophoresis gels showing the purity of a serum sample (S) after precipitation by ammonium sulfate alone (A) or with caprylic acid (A+C).

preclinical or clinical trials. Also, there is a variety of purification equipment that allows the process to be automated.

7.2.2.1 Protein A

Protein A is a constituent of bacterial cell walls and has a high binding affinity for the F(c) region. Protein A is commercially available from a number of suppliers, either in pure form or coupled to a resin. Some suppliers provide prepacked columns; however, preparing one's own column is not difficult. Depending on the binding and eluting conditions, protein A can be used to purify mAbs based on isotype. For example, mouse ascites has host-contributed immunoglobulin (primarily IgG1 and IgM) as well as the hybridoma-secreted mAbs. Protein A does not bind IgM and, in low salt conditions, binds poorly to IgG1. Thus, if the isotype of the mAb is IgG2a, IgG2b, or IgG3, a relatively clean preparation of these mAbs can be obtained. Protein A can be used successfully for IgG1 if a high-salt (~3 M) binding buffer is used. Under these conditions, there will be some host contributed Ig when ascites is the source material. If there is concern for any exogenous Ig, then the mAb should be prepared in cell culture (see Chapter 5).

7.2.2.1.1 Reagents

10X PBS: 1.5 M PBS, pH 7.6
NaCl (solid)

Binding buffer: 3.0 M NaCl in100 mM citrate/phosphate buffer; pH 9.0
Eluting buffer: 100 mM citrate/phosphate buffer, pH 3.0
Neutralizing buffer: 1.0 M Tris
Transport buffer: 150 mM PBS; pH 7.6

7.2.2.1.1.1 Procedure

1. The size of the column you need to pack or purchase depends on the amount of antibody you plan to purify. The binding capacity of protein A ranges from 5 to 20 mg Ig/ml resin. If you will be purifying 10–100 mg, therefore, a 5- to 10-ml gel column will be sufficient. When you need to purify gram quantities, then 50–100 ml of gel will be best.

2. Wash and dry a glass column. (If you purchased a column prepacked with protein A-agarose, then disregard this step.) Pour the protein A coupled to agarose slurry into the column. Let the beads settle. Start a flow of binding buffer over the column at about 5 ml per min. Do not allow the column to run dry. Let 3–5 volumes of binding buffer pass through the column to wash the storage solution from the beads. After the column has been washed, it is ready for use or can be stored at 4°C.

3. Clarify the source material before applying it to any column. Centrifugation and filtration through glass-fiber filters (e.g., Whatman GF/C) will remove lipids, microclots, and other unwanted constituents from ascites and serum.

4. In preparing cell-culture supernatants, the source of the culture medium is important. Media from conventional culture, including T-flasks, spinner flasks, roller bottles and perfusion culture systems, should be concentrated 10-fold. Media from advanced systems, such as hollow-fiber and stirred tank bioreactors, do not need to be concentrated unless the volumes are exceedingly large. Then concentration of the medium is for handling reasons. In either case, after the medium has been harvested and concentrated, it must be clarified by centrifugation and filtration.

5. After the ascites, serum or supernatant has been clarified, add 1 ml of 10× PBS to every 9 ml of starting material. Determine the final volume of the starting material and add NaCl according to this equation: Volume starting material (ml) × 0.1753 = g NaCl to be added. Add the salt and stir until dissolved.

6. Before adding the starting material, the column should be removed from storage, allowed to come to room temperature, and flushed with binding buffer. It is advisable that the effluent pass through an ultraviolet monitor so that proteins passing through and out of the column can be watched and recorded. When the flow of binding buffer reaches base line and is steady, the column is ready for the starting material.

7. The amount of starting material applied to the column depends on the size of the column and the estimated amount of antibody per ml in the starting material. Most protein A has a binding capacity of at least 4–5 mg Ab/ml.

Thus, if you have a 10-ml column, apply no more than 40–50 mg of antibody. Typically, serum has 5–10 mg/ml antibody, ascites contains 3–12 mg/ml, and cell-culture supernatant, if concentrated, contains 0.2–2 mg/ml of antibody.

8. When the trailing edge of binding buffer approaches the upper surface of the column, slowly add the volume of starting material. Allow the starting material to pass completely into the column. After all the starting material has entered the column, wash the walls of the column with a few ml of binding buffer. Let that enter into the column. Then add binding buffer to continue the flow of starting material through the column. Continue to add binding buffer while monitoring the passage of protein out of the column using absorbance at 280 nm (A_{280}).

9. Have a clean glass vessel ready. To the vessel, add enough 1 M Tris so that at the end of the elution, the Tris will be diluted 10-fold with the eluting antibody solution. For example, if you expect to harvest about 10 ml of antibody, add 1 ml of Tris solution to the vessel.

10. When the chart recorder returns to baseline, add the eluting buffer. When the chart recorder indicates that protein is beginning to pass out of the column, place your vessel with the Tris under the column to collect the purified antibody. Harvesting the eluting antibody is complete when the chart recorder returns close to baseline and flattens out. In all likelihood, the chart recorder will not return to the original baseline because the eluting buffer and binding buffer differ in their salt concentration and have slightly different absorbances at 280 nm. A new baseline will be reached when the antibody is completely eluted.

11. After all the antibody is collected, add binding buffer. Wash the column with at least two volumes of binding buffer or until the chart recorder returns to the original baseline.

12. The column is then ready for another pass of starting material. (Passing some of the void volume, the first broad A_{280} peak, back over the column should be done to assure that you did not overload the column.) Purification of antibody from starting material can continue until the antibody has been harvested. The eluted antibody can be stored at 4°C, and each harvest can be added to it until the entire purification is complete. It is best to check the pH of the harvested antibody to ensure that it is pH 7.2–7.6.

13. When the entire amount of antibody has been collected from the starting material, it should be dialyzed against PBS. It can also be further concentrated as it may become diluted during dialysis. Most antibodies can be safely concentrated to 1–10 mg/ml. Be aware that mAbs may precipitate during concentration. If you are concerned about loosing antibody, only concentrate it to 2–5 mg/ml.

14. One consideration when eluting mAbs at such low pH (pH 3.0) is for coeluting IgG1 antibodies. IgG1s can be eluted at pH 6.0. This may be of value when planning your purification strategy. If the mAb is unstable at very low pH, then using elution buffer at pH 4.5–6.0 may be important

for antibody stability. Also, consider a gradient elution approach keeping the citrate/phosphate buffer concentration the same but decreasing the pH. In this way, the antibody will elute at the best pH.

15. This protocol will work for all isotypes of mouse antibodies. However, milder conditions can be used for IgG2a, IgG2b, or IgG3 isotypes. Instead of a high-salt binding buffer (e.g., 3 M NaCl in citrate/phosphate buffer), use a low-salt binding buffer (0.5 M NaCl in buffer) Likewise, use the following formula to "spike" the starting material with NaCl: volume of starting material (ml) $\times$ 0.0292 = g of NaCl to be added. Adjust the pH to 9.0. This modification will serve to reduce the amount of host-contributed IgG1 resident in ascites from copurifying with the mAbs of the other isotypes. Because these isotypes bind quite well to protein A, they must be eluted at pH 3.0–4.5. In all cases, be sure to neutralize the eluting antibody in Tris to prevent Ab decomposition or degradation.

Figure 7.2 illustrates the purity of mAbs generated in both ascites and cell culture after protein A purification.

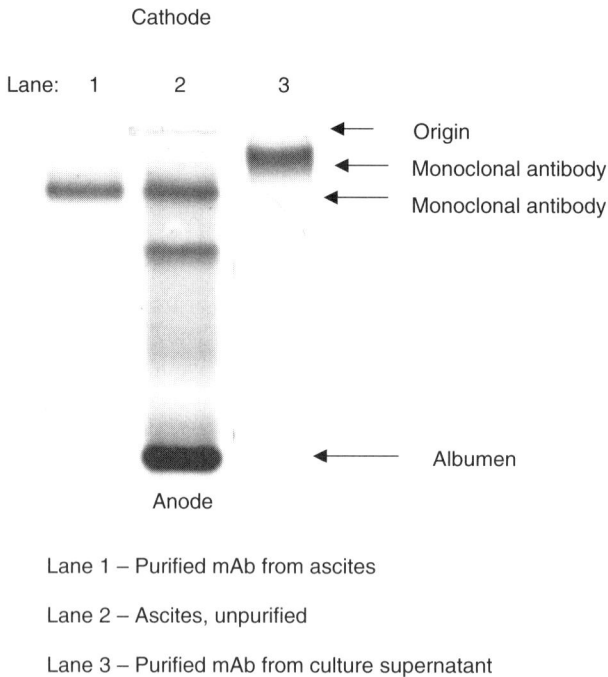

Lane 1 – Purified mAb from ascites

Lane 2 – Ascites, unpurified

Lane 3 – Purified mAb from culture supernatant

FIGURE 7.2 Zonal electrophoresis gels showing the improved purity of antibody in an ascites solution (lane 2) after purification by protein A (lane 1).

7.2.2.2 Protein G

Protein G is another bacterial cell wall protein with a high affinity for the Fc region of immunoglobulin. In our experience, protein G works best on human and animal sera other than rabbit. Although it can be used for purifying murine mAbs, using protein A is preferable for mouse and rabbit antibodies.

7.2.2.2.1 Reagents

> 10X sodium acetate buffer: 1.0 M sodium acetate; pH 5.0
> Binding buffer: 100 mM sodium acetate buffer, pH 5.0
> Elution buffer: 100 mM glycine; pH 2.5
> Neutralizing buffer: 1.0 M Tris
> Transport buffer: 150 mM PBS; pH 7.6

7.2.2.2.1.1 Procedure

The use of protein G is similar to that of protein A. Thus protein G is coupled to a resin and can be purchased in prepacked columns or as a slurry. The processing of the starting material falls under the same consideration as was described above for protein A. Therefore, steps 1–4 described above apply here as well and will not be reiterated. The protocol begins with the handling of the starting material.

1. To the clarified starting material, add 1 ml 10× sodium acetate buffer for every 9 ml starting solution. Adjust the pH to 5.0.
2. Apply to the protein G column and allow the starting material to enter into the matrix. Continue the flow with binding buffer. When all the nonbinding material has passed through the column and the monitor has returned to baseline, apply the elution buffer.
3. Collect the eluting antibody into a vessel containing 1M Tris so that the Tris is diluted 10-fold with the harvested antibody.
4. Repeat the purification as described for protein A until all the starting material has been processed. Dialyze the collected antibody against PBS.

Figure 7.3 shows the results of a protein G purification of polyclonal serum. As can be seen, purified polyclonal antibody (pAb) is a blend of different antibodies with a range of electrophoretic mobilities. pAbs are revealed as a smear as opposed to a discrete band as seen with mAbs. The two lanes represent two lots of the same purified pAb.

7.2.2.2.1.2 Rejuvenation and Reactivation of Columns

Protein A and G columns are reusable. However, there are a few considerations regarding the reuse of either of these purification matrices. It would be best to use one column for one antibody. But that may not be feasible for every laboratory. If a column is going to be reused for a different antibody, then it should be rejuvenated. Check with the manufacturer, but protein A can be rejuvenated with 2 M urea, 1 M LiCl, or 100 mM glycine. If appropriate, columns can be depyrogenated with 100 mM NaOH. Usually at least two column volumes of these solutions should be

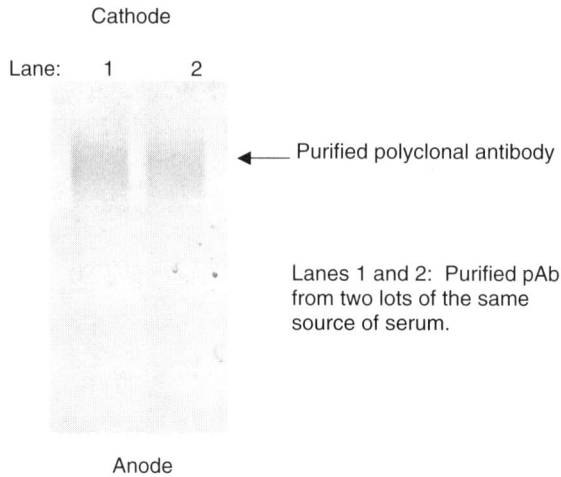

FIGURE 7.3 Polyclonal antibodies are a mixture of different antibodies with a range of electrophoretic mobilities. Consequently, they are seen as a smear after electrophoresis rather than a discrete band as seen with monoclonal antibodies. The two lanes represent two lots of the same purified polyclonal antibody.

passed over the column. Once rejuvenated, pass PBS over the column to remove these solutions. The column should be stored at 4°C with or without NaN_3.

For protein G, the column is rejuvenated with 100 mM glycine, pH 2.5. As with protein A, the column should be stored at 4°C, with or without NaN_3.

7.2.2.3 Other Purification Methods for IgG Antibodies

There are other methods for purifying antibodies. Most notably are size exclusion chromatography (SEC) and ion exchange chromatography (IEC). For IEC, there are numerous ion exchange resins. These work better for mAbs than they do for pAbs because knowing the pI (isoelectric point) of the antibody is advisable for maximum yields of the antibody you wish to purify. Clearly, mAbs have a narrow pI range as opposed to the broad pI range found for pAbs. The following chart illustrates the range of molarity of NaCl at the peak of elution of various mAbs using either a cation or anion exchange column (Figure 7.4).

Because pAbs have a wide range of pIs, not all of the antibodies in the serum will remain bound to the matrix before elution at a particular buffer pH and ionic strength. It would be best then to use a gradient elution system. Eluted IgG from serum can later be selected and pooled. Other serum protein may elute under the same conditions as for some of the pAbs. The primary advantage of IEC is the cost. Both anion and cation exchange matrices cost less than proteins A and G. But, in comparison to proteins A and G, this method is far more variable and must be defined for each antibody. Purity ranges from 60% to 95%.

The greatest advantage for IEC is that is will "polish" purified mAbs. IEC will serve to separate shed protein A or G, DNA, retroviruses and endotoxin from the

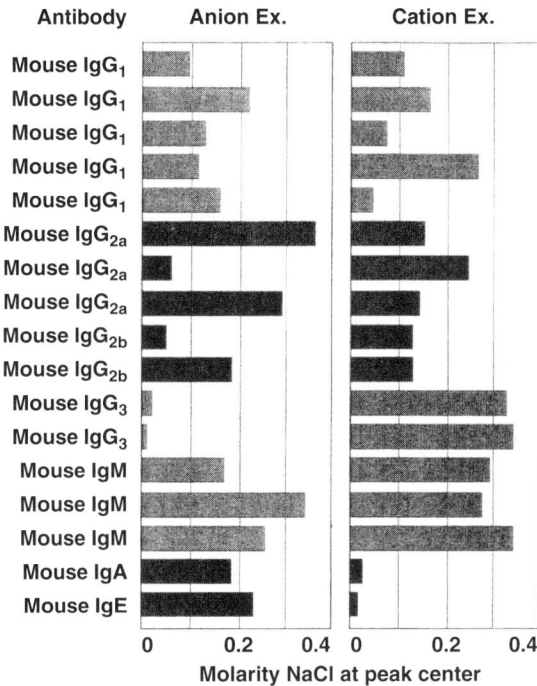

FIGURE 7.4 Paired anion and cation exchange elution characteristics of 17 monoclonal antibodies. Anion exchange results were obtained with 50 mM Tris, pH 8.6. Cation exchange results were obtained with 50 mM MES, pH 5.6.

mAb. The resulting mAb solution can be considered >99% pure. This is of great value when using mAbs for animal studies and clinical trials.

SEC, one of the oldest techniques for purifying proteins, is also a gentle method because physiologic buffers can be used at the most optimal pH for the antibody. There are many resins from a number of manufacturers. For antibodies, Sephadex G150, G200, and G300 (Pharmacia/Pfizer) are commonly used matrices. (The numerical value represents the porosity of the beads, with the higher value indicating larger pores. Large molecules are excluded from the pores and therefore pass through the column at a faster rate than smaller molecules. The smaller molecules are retarded by passing through the pores into the column matrix and therefore elute later than the larger molecules.)

After protein A/G purification, there may be some leached protein A/G, cell-contributed DNA, small antibody fragments, and large protein aggregates. Use of SEC after protein A/G will eliminate these minor constituents resulting in an antibody preparation that is nearly >99% pure. SEC can also be used as an inexpensive means of analyzing the purity of an antibody preparation. A small column, to which a small amount of concentrated antibody has been applied, can provide

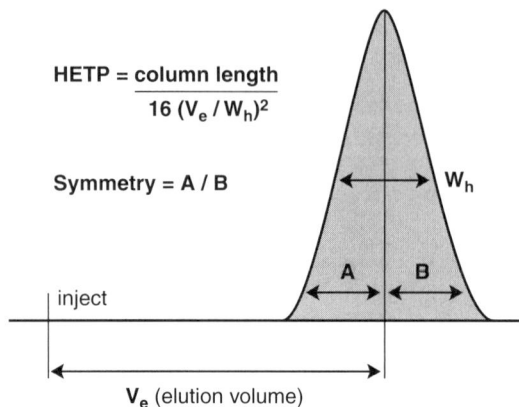

$$\text{HETP} = \frac{\text{column length}}{16\,(V_e / W_h)^2}$$

$$\text{Symmetry} = A / B$$

W_h

A B

inject

V_e (elution volume)

FIGURE 7.5 A relationship between height and width of a gel filtration column and the maximum sample size that can be applied to that column. Height equivalent of a theoretical plate (HETP). W_h refers to the peak width at half height. A and B are measured at 10% of the peak height.

a chromatograph illustrating the level of purity and the size of the contaminants. The results are comparable to HPLC (high-performance liquid chromatography).

Depending on which SEC column you prepare, there is a minimal absolute relationship between height and width of the column and the maximum sample size that can be applied to the column. Refer to the chart above on the calculations of column dimensions (Figure 7.5).

Unless the column is two or more inches in diameter and several feet tall, large samples cannot be applied to a column. In our hands, the volume of antibody we apply to a column is 5% of the total volume of the column. In most cases, only small milligram quantities can be purified at a time. Thus purification of a large amount of antibody can be a tedious process. Nevertheless, SEC is a can be a valuable method for increasing the purity of an antibody preparation as well as an inexpensive means of analyzing the purity of a final antibody preparation.

7.2.3 Purification of IgM and F(ab) Antibody Fragments

IgM antibodies are notoriously difficult to purify because of its pentameric structure. To purify IgM, a multistep approach is possible combining salt precipitation with IEC or SEC. This is a satisfactory way of preparing purified IgM, but it is tedious for large amounts of antibody and low yields can be expected. In our laboratory, we use two other affinity columns to purify IgM. These are protein L affinity column and goat anti-mouse IgM affinity column.

7.2.3.1 Anti-Mouse IgM Affinity Chromatography

The preferred technique for purifying IgM in our laboratories is to use a goat anti-mouse IgM affinity column. We purchase the antibody already coupled to agarose and prepare our own column.

7.2.3.1.1 Reagents

Binding buffer: 10 mM phosphate buffer with 0.5 M NaCl, pH 7.2
2X binding buffer: 20 mM phosphate buffer with 1 M NaCl, pH 7.2
Elution buffer: 100 mM glycine with 150 mM NaCl, pH 2.4
Neutralizing buffer: 1.0 M Tris
Transport buffer: 150 mM PBS, pH 7.6

Preparation of this column should follow the procedures as described for protein A/G. Pour the goat anti-mouse IgM-conjugated agarose beads into clean glass column. Flush with binding buffer after the beads have settled. Wash the slurry with several void volumes of binding buffer. When the A_{280} monitor indicates a level base line for the absorbance of the eluate, the column is ready to use. Although murine IgM from any source can be applied to the column, we recommend media supernatants over ascites. Low levels of albumen contamination of the purified IgM may be seen when ascites is the source material. If there is no choice, a further purification of the IgM with IEC or SEC may be necessary.

7.2.3.1.2 Procedure

1. Dilute the clarified starting material 1:1 with the 2× binding buffer. Only prepare as much sample as will be purified that day. Leave the remainder of the starting material undiluted until it will be purified.
2. For the first passage of starting material over the column, we recommend that you apply an excess amount of your preparation. Save the void volume because there will still be antibody in this preparation. The idea is to determine the capacity of the column for your particular IgM.
3. After the sample has been applied, collect and save the starting material and the first wash off the column. Wash the column with binding buffer until the monitor returns to baseline.
4. Elute the bound antibody with elution buffer. Determine the amount of antibody eluted using an ultraviolet spectrophometer set at OD = 280 nm (see "Determination of Antibody Concentration"). Using this information, you can then proceed with the rest of the purification, knowing the antibody-binding capacity of the column you have prepared.
5. Following the procedures outlined earlier, apply all or a portion of the void volume back over the column depending the estimated amount of antibody in the starting material.
6. Wash the column free of unbound material with binding buffer. When the monitor returns to baseline, elute the bound IgM directly into the Tris neutralizing buffer.
7. When all of the antibody has been eluted and the monitor has stabilized, apply binding buffer to the column. When the monitor reaches the original baseline, the column is ready for the next application of starting material.
8. Store the eluted and neutralized antibody at 4°C to be combined with the remainder of the purified IgM.

9. Continue to apply the starting material and elute the bound antibody until all of the antibody has been purified.
10. Pool the eluates and dialyze against PBS. The antibody can be further concentrated but IgM has a tendency to precipitate. We recommend keeping the final antibody preparation to 2–5 mg/ml.

The purity of IgM preparation purified over a goat anti-mouse IgM affinity column can be determined using gel electrophoresis (polyacrylamide gel electrophoresis [PAGE] or isoelectric focusing [IEF]). As is typical for most IgM mAbs, a diffuse banding pattern is seen on a zonal electrophoresis gel. IgM is a large molecule, ~750 kDa, and has a large number of glycosylization sites. With so many sites, IgM has a much broader range of charges within a single mAb population. Purification of IgM from supernatant leaves the antibody free of contaminating murine albumen from ascites (Figure 7.6).

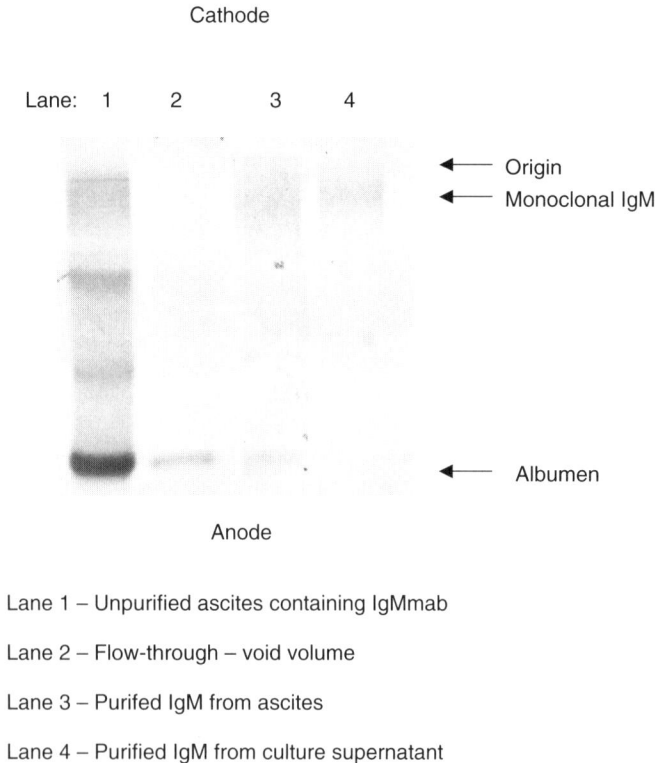

Lane 1 – Unpurified ascites containing IgMmab

Lane 2 – Flow-through – void volume

Lane 3 – Purifed IgM from ascites

Lane 4 – Purified IgM from culture supernatant

FIGURE 7.6 The purity of immunoglobulin (Ig)M preparation purified over a goat anti-mouse IgM affinity column can be determined using gel electrophoresis (PAGE or IEF). As is typical for most IgM mAbs, a diffuse banding pattern is seen on a zonal electrophoresis gel. IgM is a large molecule, ~750 kDa, and has a large number of glycosylization sites. With so many sites, IgM has a much broader range of charges within a single mAb population. Purification of IgM from supernatant leaves the antibody free of contaminating murine albumen from ascites.

7.2.3.2 Protein L affinity chromatography

Protein L has a high affinity for lambda light-chains of Igs and will purify intact and fragmented antibodies. Thus IgM and F(ab) and F(ab)$'_2$ fragments can be purified using this matrix. Despite these properties, we have had inconsistent results and find that the matrix has a short half-life. However, protein L purification is an alternative "broad spectrum" antibody purification matrix. Protein L binds to kappa light chains of immunoglobulins and is useful for purification of antibody fragments and intact IgG, IgM, and IgA.

7.2.3.2.1 Reagents

Binding buffer: 10 mM PBS, pH 7.4
2X binding buffer: 20 mM PBS, pH 7.4
Elution buffer: 100 mM citrate buffer, pH 3.0
Neutralizing buffer: 1.0 M Tris
Transport buffer: 150 mM PBS, pH 7.6

7.2.3.2.2 Procedure

1. Dilute the clarified starting material 1:1 with the 2× binding buffer. Only prepare as much sample that will be purified that day. Leave the remainder of the starting material undiluted until you will be purifying it.
2. For the first passage of starting material over the column, we recommend that you apply an excess amount of your preparation. Save the void volume because there will still be antibody in this preparation. The idea is to determine the capacity of the column for your particular intact or fragmented antibody.
3. After the sample has been applied, collect and save the starting material and the first wash off the column. Wash the column with binding buffer until the monitor (A_{280}) returns to baseline.
4. Elute the bound antibody with elution buffer. Determine the amount of antibody eluted using an ultraviolet spectrophometer (see "Determination of Antibody Concentration"). Using this information, you can then proceed with the rest of the purification, knowing the antibody binding capacity of the column you have prepared.
5. After the procedures outlined earlier, apply all or a portion of the void volume back over the column, depending on the estimated amount of antibody in the starting material.
6. Wash the column free of unbound material with binding buffer. When the monitor returns to baseline, elute the bound antibody directly into the Tris neutralizing buffer.
7. When all of the antibody has been eluted and the monitor has stabilized, apply binding buffer to the column. When the monitor reaches the original baseline, the column is ready for the next application of starting material.
8. Store the eluted and neutralized antibody at 4°C to be combined with the remainder of the purified antibody.

9. Continue to apply the starting material and elute the bound antibody until all of the antibody has been purified.
10. Pool the eluates and dialyze against PBS. The antibody can be further concentrated to a final antibody preparation to 2–10 mg/ml.

(In our hands, the matrix has a short half-life and can only be used a maximum of 10 times before the binding capacity of the column is cut in half.)

Protein A, protein G, and IEC are the most scalable of the above procedures. Large columns can be prepared: the greater the volume of purification matrix that you purchase, the less expensive it becomes per unit volume. There is essentially no limit to the size of a column one with any of these three matrices. Many commercially available columns will accommodate liters to kiloliters of purification products.

7.2.3.2.3 Economic Considerations

SEC can only be scaled to the height of the ceiling in the laboratory or purification facility. Certainly, very large columns can accommodate larger volumes of starting material, although packing problems may reduce the efficiency of the column. Cost of the goat anti-mouse IgM and protein L matrix does not decrease significantly with increased volumes. It might be more cost effective to purchase high-grade goat anti-mouse IgM and couple it to activated (e.g., CNBr) agarose beads if a large column is required to purify gram quantities of IgM mAb.

In addition to these methods, others are also available, including polyethylene glycol adsorption, hydroxyapatite chromatography, and hydrophobic interaction chromatography.

The final use for the antibody will dictate the procedure or combination of procedures that should be employed for purification. For rapid, inexpensive purification, precipitation is adequate. For diagnostics, protein A/G provides a highly reproducible method. For preclinical studies, protein A/G followed by IEC, SEC, or hydrophobic interaction chromatography will provide a highly purified product. For any preclinical or clinical application, antibodies must be purified using aseptic and low endotoxin conditions and therefore conform to local (e.g., Food and Drug Administration) regulatory protocols. Therefore, for each laboratory and for each antibody, a purification strategy should be developed.

7.2.3.3 High-Performance Chromatography

As inferred, use of non–gravity-based chromatographic technology, such as HPLC, can also be used to purify small quantities of antibodies to extremely high purity using only a single column of small proportions.[3] Although not practiced in this laboratory, these techniques are of particular use when isolating and characterizing small quantities of starting material (e.g., isolate from biopsy samples), when affinity binding must be measured, used for therapy, when a high throughput system is required (e.g., antibody screening), or to remove endotoxin contaminants.[4] Here we describe briefly an number of examples used by others.

Josic et al. demonstrated HPLC purification of crude rabbit anti-transferrin antiserum using Affi-Gel Blue columns.[3] Similar results were obtained using mouse ascites fluid. In a first round of purification, albumin was eluted at around 0.4 M

NaCl using a reverse sodium chloride gradient, the IgG eluting at much lower concentrations of NaCl. According to Josic et al., this increased considerably the column capacity for subsequent HPLC separations.

Bowles et al. developed a large scale method of purifying intact IgG monoclonal Abs using either high performance hydroxylapatite chromatography (HPHT) or DEAE 5PW (Waters Protein Pak) HPLC.[5] For HPHT of clarified ascites fluid, Bowles et al. loaded samples of ascites fluid that had been dialyzed, diluted, and filtered through 0.22-μm Millipore membranes onto a BioRad MAPS (monoclonal antibody purification system) column; the IgG1 was eluted at about 250 mM phosphate using a gradient from 20 mM to 300 mM phosphate buffer (pH 6.8). The IgG peaks (~100 mg) represented about 25–30% of the total protein eluted from the column. However, some antibodies appeared to reduce the column lifetime significantly, probably because of antibody aggregation during the dilution phase.[6] Fab fragments obtained by papain digestion of the pooled IgG HPHT fractions were purified on the MAPS (monoclonal antibody purification system) HPHT column using a linear phosphate gradient and eluted from about 120 to 150 mM phosphate. Bowles et al also purified Ab using DEAE 5PW HPLC and noted that fewer protein contaminants were observed in the DEAE HPLC-purified antibody when analyzed using SDS-PAGE.[5]

Manzke et al. described a large-scale production and single-step purification of bispecific Abs using hydrophobic interaction HPLC that could yield between 8 and 12 g of IgG per month. Bispecific Abs are of use particularly for therapy, such as treatment of human cancers and, possibly, against viral infections.[7] Manzke et al. used Abs prepared from hybrid hybridomas (tetradomas) that typically contain a heterogeneous combination of heavy and light chains from both parental hybridomal cells.[7] The tetradomas were expanded using hollow fiber bioreactors and the media were harvested, filtered (0.2 μm), adjusted to 1.0 M AS in 100 mM phosphate buffer (pH 7.9), refiltered, and applied to an 8 ml phenyl-Superose HR10/10 column (Pharmacia). The Abs were eluted using a descending AS gradient (1.0 M to 0 M AS in phosphate buffer). The column sufficiently resolved the bispecific bi-isotypic Abs (IgG1/IgG2a) from the mono-isotypic parental Abs, the bispecific Abs eluting at between about 40 and 25 mM AS.

7.3 CHARACTERIZATION OF ANTIBODIES

Antibodies can be characterized by a wide variety of physiochemical and functional analyses. The procedures to follow are the most standard ones and will provide information on an antibody's uniqueness, integrity, purity, and functionality. These include: zonal electrophoresis on cellulose acetate, A280 concentration determination, sodium dodecyl sulfate polyacrylamide gel electrophoresis (SDS-PAGE), Western blot, isoelectric focusing, and enzyme-linked immunoassay (enzyme immunoassays).

7.3.1 ZONAL (CELLULOSE ACETATE) ELECTROPHORESIS

Zonal electrophoresis reveals a pattern of separation among the proteins based on their charge properties. We use commercially available kits from Helena Laboratories (Beaumont, TX), which include the wicks, cellulose acetate gel, blotter, templates, and buffer mix. Also available from Helena are the chambers, power supply units, fixative and Coomassie stains that need to be purchased and are not parts of each kit. The procedure is well detailed with the product insert from Helena.

7.3.1.1 Procedure

1. Fill the two middle troughs of the chamber with water and place in the freezer.
2. Prepare the buffer, fixative stain, and destain as described by the manufacturer.
3. Remove the gel from its protective package and blot one end with the blotter paper. Lay the template onto the tamped surface of the gel.
4. Pipet 2–5 µl of sample into the wells of the template. Up to seven samples can be applied to each gel. One well should be reserved for a standard or control, such as normal serum. Allow the samples to be absorbed into the gel for 5 min. Reblot the gel and remove the template.
5. Remove the chamber from the freezer and fill the outer troughs with the buffer. Lay the gel onto the middle of the chamber with the sample end adjacent to the cathode (−). Place the wicks onto the ends of the gel and insert the wicks into the buffer trough. They should immediately get wet.
6. Apply the top of the chamber into position and connect the chamber to the power supply with the lead wires.
7. Turn on the power and adjust to 250 V. Leave the power on for no more than 20 min. Turn off the power supply and disconnect the wires.
8. Remove the gel from the chamber. Wash the chamber and refill the middle troughs with water. Place the chamber back into the freezer for future use.
9. Stain the gel with Coomassie blue (0.1% in 10% glacial acetic acid/40% MeOH volume/volume) for 15 min. Destain with acetic acid/methanol (10%/40%) solution.
10. The gel can then be dried with a hair dryer or vacuum gel dryer to leave a permanent record of the gel.

The pattern revealed by zonal electrophoresis has a number of applications. Individual serum proteins will migrate to a specific location depending on the species. That is, albumen, which has the highest concentration in serum, will have the largest band and will migrate furthest from the point of origin because of its high negative charge. Antibodies from polyclonal serum (pAbs) will be revealed as a diffuse band close to the origin. mAbs will have a clear discrete band also close to the origin and different mAbs will migrate to different locations (see Figure 7.7 below). This can be a useful tool to differentiate between mAbs or to demonstrate lot-to-lot consistency. As seen in Figure 7.7, this gel has three different mAb samples from ascites. In the first three lanes, the mAbs are different; each of the uppermost bands illustrates the differences in the electrophoretic mobility based on the charge of the mAb. In the last two lanes, the bands are similar. These are the same mAb but from different lots of ascites.

The gel can also be scanned and the area under each band can be determined. Each band can be defined as a percentage of the total with different software programs. If the protein concentration of the ascites is determined by modified Lowry

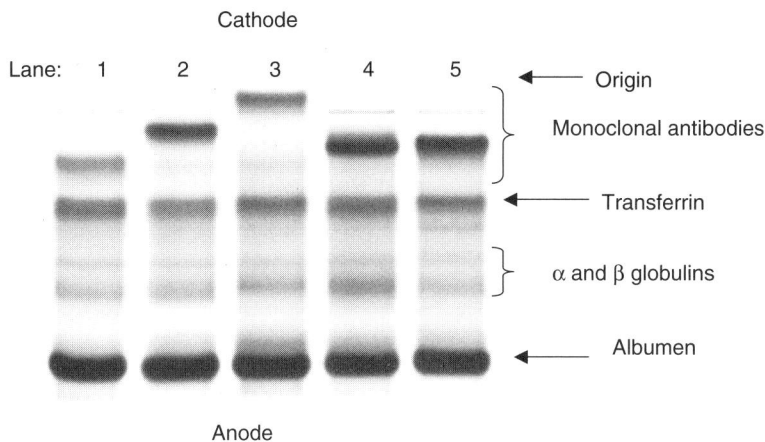

Lane 1 – mAb 1
Lane 2 – mAb 2
Lane 3 – mAb 3
Lane 4 – mAb 4, Lot #1
Lane 5 – mAb 4, Lot #2

FIGURE 7.7 Zonal electrophoresis can both differentiate monoclonal antibodies (mAbs) as well as demonstrate lot-to-lot consistency. In the first three lanes, the different mAbs samples from ascites are shown; each of the uppermost bands illustrates the differences in the electrophoretic mobility based on the charge of the mAb. In the last two lanes, the bands are similar. These are the same mAb but from different lots of ascites.

assay,[8] for example, an approximate concentration of the mAb can be calculated (percent of total from gel scan × total mg/ml protein = mg/ml mAb).

7.3.2 ANTIBODY CONCENTRATION BY ABSORBANCE AT 280 NM

One of the fastest methods for determining purified antibody concentration is by absorption at 280 nm (A_{280}). For this, an ultraviolet-visible spectrophotometer is needed to detect light transmitted in the ultraviolet range. Proteins that contain the aromatic amino acid residues tryptophan, phenylalanine, and tyrosine absorb light energy at a wavelength of about 280 nm. Each protein has its own extinction coefficient based on the known number of these three amino acids. Here, in the case of immunoglobulins, the compositional differences can largely be ignored. The greater the absorbance, the higher the concentration of protein. To obtain satisfactory results, the preparation must be >95% pure.

7.3.2.1 Procedure

1. Turn on spectrophotomer and adjust setting to 280 nm.
2. Dilute sample so that the protein concentration is estimated to be 0.5–1.5 mg/ml. If the sample is highly concentrated, prepare two different dilutions (e.g., 1:5 and 1:10) to confirm results.

TABLE 7.2
Extinction Coefficients of Various Immunoglobulins

Protein Moiety	Absorbance at 1.0 mg/ml
IgG	1.36
IgM	1.18
IgA	1.32
IgE	1.53
IgG F(ab)	1.50
IgG F(ab)$'_2$	1.48

Ig, immunoglobulin.

3. Fill a cuvette with diluent. When the spectrophotometer has had ample time to warm up, insert the cuvette with the diluent and adjust the reading to "0."
4. Empty the cuvette and tamp dry. Transfer enough of the more diluted sample to the cuvette, mix briefly and discard this sample. Transfer another volume of the same sample to the cuvette. Insert into the spectrophotometer and record the value.
5. Discard the sample. If there is a second, more-concentrated sample, transfer a volume of that to the cuvette. Again, mix briefly and discard. Transfer a volume of the same sample and insert into the spectrophotomer. Record the value.
6. Determine the concentration of the diluted sample. The absorbance reading of the sample is divided by the proper value displayed in Table 7.2. The chart indicates what the absorbance reading should be (extinction coefficient) if the antibody concentration is at 1.0 mg/ml.
7. Thus if an IgG sample absorbance is 0.75, then the concentration is 0.75 ÷ 1.36 = 0.55 mg/ml. If this happens to be a 1:5 dilution of the original sample, then the concentration of the original sample is 0.55 × 5 = 2.75 mg/ml.

7.3.3 SDS–PAGE

SDS is an anionic detergent which denatures proteins by "wrapping around" the polypeptide backbone. SDS binds to proteins fairly specifically in a mass ratio of 1.4:1.[10] In so doing, SDS confers a negative charge to the polypeptide in proportion to its length (i.e., the denatured polypeptides become "rods" of negative charge cloud with equal charge or charge densities per unit length). It is usually necessary to reduce disulfide bridges in proteins before they adopt the random-coil configuration necessary for separation by size. This is done with 2-mercaptoethanol or dithiothre-itol. In denaturing SDS-PAGE separations, therefore, migration is determined not by intrinsic electrical charge of the polypeptide, but by molecular weight. Sample are usually run as both native and denatured.

A number of companies manufacture electrophoresis chambers, gels, and systems. Our general recommendation is the simpler and the more leak-proof the setup, the better. The following is an example, not an endorsement.

7.3.3.1 Equipment

Biorad Criterion cell
Biorad gel holder
Fisher Biotech electrophoresis system

7.3.3.2 Reagents

Run buffer: 50 mM Tris base, 50 mM Tricine, 0.1% SDS
5× nonreducing sample buffer: 62.5 mM Tris-HCl, 10% glycerol, 2% SDS, 0.025% Bromophenol blue
5× reducing sample buffer: 62.5 mM Tris-HCl, 10% glycerol, 2% SDS, 0.7 M 2-mercaptoethanol; 0.025% Bromophenol blue
Antibody diluting buffer: 150 mM PBS

7.3.3.3 Procedure

1. Prepare a native (nonreduced) and reduced sample of the material to be tested. Label two microvessels as reduced and nonreduced. Transfer 2 µg of antibody to each of the tubes.
2. To each of the tubes, add enough PBS to bring the volume to 20 µl.
3. To the non-reduced sample, add 5 µl of nonreducing sample buffer; to the reduced sample, add 5 µl of reducing sample buffer. Mix each sample well on a vortex mixer.
4. To break the disulfide bonds between the H and L chains of the antibody, the reduced sample needs to be boiled for 2 min before applying it to the gel. The nonreduced sample should not be boiled.
5. Before applying the samples to the gel, follow all the manufacturers recommending procedures. This may include washing the gel in deionized water (DI H$_2$0) and cutting the bottom of the gel.
6. Place the gel and its holder inside the chamber of the electrophoresis device. Secure the gel into the chamber with a tight seal. Fill the chamber with SDS running buffer.
7. Apply samples to the wells of the gel. The gel should be thought of as having two halves. The first half is for nonreduced samples; the second half for reduced. The first well in each half should be for the molecular mass markers. Apply each pair of samples, non-reduced/reduced, into the

wells of each half in the same sequence. Record in your notebook or standard operating procedure the sequence for applying the samples.

8. After the samples have been applied to the well, secure the top of the chamber and attach the chamber to the appropriate electrodes. Turn on the power supply and adjust the current to 60 mA. Allow the gel to run for approximately 30 min. The leading edge of the blue dye should approach the bottom of the gel but not run off. Turn off the power at the end of the run and disconnect the electrodes.
9. Remove the lid to the chamber and remove the gel. Discard the running buffer.
10. Transfer the gel to a shallow plastic container. Wash the gel three times in DI H_2O. After the third wash, add the preferred stain for revealing proteins, such as Coomassie Blue or Gel Blue. Allow the gel to stain for 15 min.
11. Destain with acetic acid/methanol or DI H_2O depending on the stain used. Destaining may require up to three changes of acetic acid/methanol or water.
12. After destaining, apply the gel to a hard glass surface for observation. Gels can be sealed or dried for long-term storage. It is best to make some permanent record of the gel by using an imaging system. Gels can also be scanned for further analysis.

SDS-PAGE (Figure 7.8) can reveal the integrity and stability of an antibody. It is also essential when for analyzing F(ab) and F(ab)$'_2$ fragments. Intact antibody will migrate at approximately coincident with the 150 kDa marker. If there are antibody fragments, you will see minor bands below the strong antibody band. If there are aggregates, you will see these revealed as bands between the major antibody band and the base of the well. On the reduced side, there should be two distinct bands from samples of purified antibody. One band, at the 50 kDa marker, represents each half of the heavy chain, and the other at about 25 kDa is the light chain.

7.3.4 WESTERN BLOT

Western blot analysis can detect a protein moiety in a mixture of any number of proteins while giving you information about the size of the protein. This method is, however, dependent on the use of a high-quality antibody directed against a desired protein. It is a methodology that combines the separation of proteins by molecular weight using SDS-PAGE and direct immunoassay procedures. With Western blot, one can determine not only the presence or absence of a particular protein or peptide in a mixture but also confirm its molecular mass.

7.3.4.1 Reagents

Transfer buffer: 25 mM Tris, 192 mM glycine, 20% methanol, pH 7.8–8.2
Wash buffer: 150 mM PBS, 0.05% Tween
Blocking buffer: 150 mM PBS, 1.0% nonfat dry milk
Primary antibody specific to the target ligand

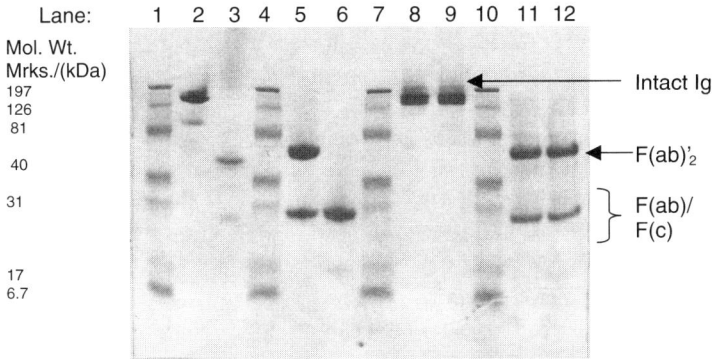

Lanes 1, 4, 7, and 10 – Molecular Weight Markers

Lane 2 – Non-reduced intact pAb

Lane 3 – Non-reduced pAb F(ab)'$_2$ fragment

Lane 5 – Reduced intact pAb (complement to Lane 2)

Lane 6 – Reduced pAb F(ab)'$_2$ fragment (complement to Lane 3)

Lanes 8, 9 – Non-reduced intact mAb

Lanes 11, 12 – Reduced intact mAb (complement to lanes 8, 9)

FIGURE 7.8 SDS-PAGE of various antibody samples, including nonreduced intact pAb (2), nonreduced pAb F(ab) fragment (3), reduced intact pAb (5), reduced pAb F(ab) fragment (6), nonreduced intact mAb (8, 9), reduced intact mAb (11, 12), and molecular mass markers (1, 4, 7, 10).

Secondary antibody specific for the species of the primary antibody and conjugated to horseradish peroxidase (HRP)

TMB membrane peroxidase substrate

7.3.4.2 Equipment and Other Supplies

Western blot apparatus
Cooling unit
Plate rotator
Gel bond film/gel blot paper/filter paper
Magnetic stir plate
Transfer sponges
Transblot transfer medium

7.3.4.3 Procedure

1. Perform an SDS-PAGE as described previously. Apply to the gel, samples containing the target molecule both reduced and nonreduced. After performing the electrophoresis, remove one of the glass plates covering the gel and notch a corner of the gel.

2. Rinse the gel in DI H_2O in a plastic container. Pour off the water and add transfer buffer. Let the gel incubate for about 30 min.

3. Cut one piece of nitrocellulose paper and two pieces of filter paper to the same size as the gel.

4. To another plastic container, add transfer buffer. Place the gel, the three pieces of paper, and two transfer sponges to the plastic container containing the transfer buffer. Let the items stand for 2–3 min.

5. Prepare the WB apparatus according to manufacturer's instruction. Insert the electrode into the buffer chamber. Place a 1-inch stir bar into the bottom of the chamber.

6. Open the gel holder cassette. Layer the sponges, gels, and paper onto the black surface of the cassette holder in the following sequence: sponge, filter paper, gel, nitrocellulose paper, filter paper, and sponge. Make sure that the gel side faces the nitrocellulose and that there are no air bubbles between the two. Close the cassette.

7. Pour in about 400 ml transfer buffer into the blot apparatus. Insert the gel holder cassette with the black side of the cassette facing the black side of the apparatus. Insert the cooling unit.

8. Set the tank on top of a magnetic stirrer and turn the stirrer on to slow speed. Apply the lid to the chamber. Connect the wires to the appropriate electrodes.

9. Turn the power on and run the system at 100 volts. The current should be at about 250 mA at the beginning of the run and end at about 350 mA. Run the transfer for about 1 h.

10. After the run, remove the cassette and gently transfer the gel to a plastic container containing wash buffer. Wash the gel three times for about 5 min each.

11. Incubate the gel in 10 ml of blocking buffer for a minimum of 30 min. Blocking can done overnight, if necessary. After blocking, wash the gel in wash buffer three times for 5 min each.

12. Prepare primary antibody by diluting the antibody in blocking buffer as described below:

13. Purified antibody is best used at 0.5 to 1.0 µg/ml.

14. Serum is best used if diluted to 1:200 to 1:5,000.

15. Ascites fluid is best used of diluted to 1:500 to 1:10,000.

16. Cell-culture supernatant containing mAb is best used neat or diluted 1:10 to 1:100.

17. Prepare enough of the diluted antibody to cover the gel in a plastic container, 15–20 ml. Pour the diluted antibody over the gel which already is in a plastic container. Place the container on a rocker platform and incubate the gel for 30 min. Set the rocker to a slow speed.

18. After the incubation step, wash the gel in wash buffer three times for 5 min each.

19. Prepare the HRP-conjugated secondary antibody. Depending on the source, follow the recommended dilutions from the supplier. The antibody should be diluted in wash buffer from 1:500 to 1:10,000. If the conjugate

is too concentrated, there can be background problems and if it is too dilute, you may miss minor positive sites.

20. Transfer the diluted conjugate to the plastic container containing the gel. Again, place the container on a rocker platform and let the gel incubate for about 30 min. After the incubation step, wash the gel three times with wash buffer for 5 min each time.

21. After decanting the last wash, add the TMB membrane substrate. Incubate for 2–10 min. Bluish bands will begin to appear where specific antibody-antigen complexes have formed.

22. Stop the reaction by decanting off the substrate solution and adding DI H_2O. Scan the gel and save by whatever means available. For example, the data can be saved in digital format; the membrane can be vacuum-dried, then sealed in a bag and stored at 4°C.

7.3.5 ISOELECTRIC FOCUSING

Isoelectric focusing is a highly useful procedure that determines the pI of an antibody. Knowing the pI of an antibody will determine the correct purification condition when using IEC as well as defining the optimal storage conditions. The results are also a "fingerprint" for a particular antibody as there is an array of bands that form depending on the mode of production. Alterations in the "fingerprint" of an antibody will be evident when changes in the storage buffer occur, if there has been any deamination, or if production changes from *in vivo* to *in vitro* techniques.

7.3.5.1 Reagents

Note: Because we use the BioRad Criterion Cell system, we use their reagents. Again, this is an example and not an endorsement.

Antibody: 1–2 mg/ml in sample buffer
Sample buffer: 50% glycerol in DI H_2O
10X anode buffer
10X cathode buffer
IEF standards: pI 4.45–9.6
IEF gel stain
IEF destaining solution

7.3.5.2 Equipment

Criterion cell system
Criterion precast gel cassette

7.3.5.3 Procedure

1. Add 10 μl of antibody to a capped sample tube. Add 5 μl of glycerol solution.
2. Prepare 10 μl of pI standards.

3. Prepare 60 ml of cathode buffer from 10× stock solution. Prepare 400 ml of anode buffer from 10× stock solution.
4. Remove the gel cassette and remove the comb from the gel. Rinse the gel in DI H_2O.
5. Insert the gel into the Criterion apparatus and fill the upper chamber with 60 ml of the cathode buffer.
6. Load the samples and standard into the wells left from having removed the comb.
7. Fill the lower chamber with 400 ml of anode buffer.
8. Apply the lid and connect the leads to the power source.
9. Turn on the power and set the voltage to 100 V for 1 h, 200 V for 1 h, and 500 V for 30 min.
10. After the 2.5-h run, turn off the power and disconnect the leads. Remove the lid and remove the cassette from the apparatus. Following the manufacturer's instructions, open the cassette to expose the gel.
11. Stain the gel in Coomassie Blue gel stain for 45 min. Destain with 100 ml of destain solution. This may take several washes.
12. Preserve the gel. The gel can be scanned by eye or electronically.

7.3.6 ENZYME IMMUNOASSAY

Enzyme immunoassays, performed as an ELISA, are the mainstay of immunoassays. Understanding how ELISAs work leads to an understanding of all antibody-antigen based assays. Depending on how the assay is performed, results from ELISAs define specificity, affinity, concentration, and sensitivity for any given antibody. The following is the procedure for performing either a direct or an indirect ELISA. The procedure for performing a direct versus an indirect is essentially the same. The difference is whether or not the primary antibody which binds to the target ligand is conjugated to the enzyme label. In the direct assay, the primary antibody is conjugated; in the indirect assay, the primary antibody is not conjugated. For the indirect approach, a secondary antibody specific for the species of the primary is used that is conjugated to an enzyme.

7.3.6.1 Reagents and Supplies

Diluent: 150 mM PBS, pH 7.6
Coating buffer: 200 mM carbonate/bicarbonate buffer, pH 9.5
Wash buffer: 150 mM PBS with 0.05% Tween
Blocking buffer: 150 mM PBS with 1% nonfat dry milk
Primary antibody, conjugated to an enzyme such as HRP, alkaline phosphatase, or biotin, or primary antibody, unconjugated
Secondary antibody conjugated to an enzyme such as HRP, alkaline phosphatase, or biotin
TMB
Stop reagent: 0.6 N HCL

7.3.6.2 Procedure

1. Before running the assay, a strategy for the plate design should be established. There should always be blank wells for all the negative controls. For example, we often coat the antigen in just the odd-numbered columns if we are going to perform a titration on serum or media supernatant. If we are screening for positive hybridomas, we will coat the entire plate except for one or two wells.

2. It is best to coat the plates the night before you wish to run the assay. In this way, there is ample time for the target ligand to bind to the wells. Plates can be coated and left at 4°C for up to 2 weeks. Dilute the target ligand in coating buffer to 2–5 µg/ml. Coat the wells with 0.1 ml diluted ligand per well.

3. When the assay is to be run, remove the plates from the refrigerator. Discard the contents of the well and wash the entire plate three times with wash buffer. Tamp the plates between each wash.

4. Add 300 µl of blocking buffer to all the wells. Let the plates stand at room temperature for at least 30 min. Then, wash the plates one time with wash buffer.

5. While the plates are being blocked, make dilutions of the primary antibody in PBS. To screen antiserum, we often prepare fivefold serial dilutions starting with a 1:50 or 1:100. If medium is to be tested, twofold serial dilutions may be adequate. For hybridoma supernatants, do not make any dilutions.

6. If the primary antibody has been conjugated, prepare a range of dilutions from 1:100 to 1:10,000 dilution. Apply several of these dilutions to the antigen-coated wells.

7. Apply 100 µl of the diluted antibody to the wells. For serially diluted antibodies, it is best to apply each dilution in duplicate sets of wells (four wells in total). Let the plates incubate at 37°C for 30 min.

8. After the incubation, wash the plates three times with wash buffer. Tamp the plates between each wash.

9. If conjugated primary antibody was applied to the plate, skip to Step 10. For indirect assays, apply 100 µl of diluted, conjugated secondary antibody to all the wells. Let the plates incubate for 30 min at 30°C.

10. Wash the plates five times with wash buffer. Tamp the plates between each wash.

11. Add 100 µl of TMB substrate to all the wells. Watch for color development. Let the color develop until the negative controls show slight color. This may not necessarily happen. If not, stop the reaction after 5–10 min.

12. Stop the reaction with 50 µl of stop solution.

13. Read the plates on an automatic plate reader at the appropriate absorbance to get a numerical assignment of color intensity. For TMB, this will be 450 nM. The larger the value the more the antibody is present.

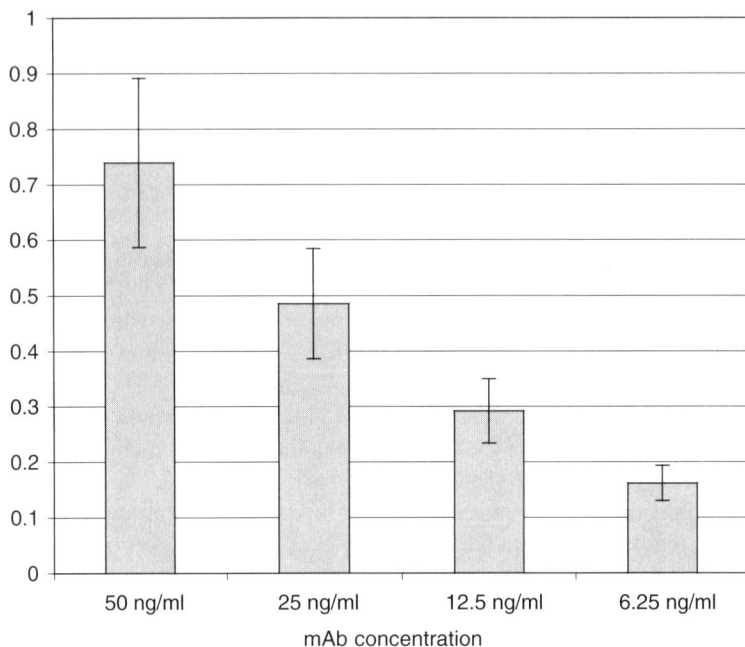

FIGURE 7.9 Titration results.

Figure 7.9 is a histogram illustrating the numerical results from a titration. A purified mAb was diluted and applied to the plate coated with its specific ligand. As can be seen, even when the mAb was diluted to 6.25 ng/ml, specific activity could still be detected.

7.4 CONCLUDING COMMENTS

It is the hope of this contributor that this chapter has provided a platform of procedures for purifying and characterizing antibodies. Although the focus was on mAbs, all of the procedures will work for either monoclonal or pAbs. Clearly, no one purification procedure will accommodate all the different antibodies being used in one laboratory. Thus having a repertoire of methodologies will best serve the laboratory. Likewise, depending on the type of information needed to characterize an antibody, one needs to have an array of techniques available. Hopefully, this chapter will provide the reader a sufficient collection of techniques for purification and characterization of most antibodies.

ACKNOWLEDGMENTS

Special thanks go to Nathalie Forster, Cheryl Steenstra, Trudi Church, Skip Graf, and Adam Curtis for their assistance in helping me prepare this chapter.

REFERENCES

1. Janeway, C.A., et al., *Immunobiology: The Immune System in Health and Disease*, 4th ed., Current Biology Publications, New York, 1999.
2. Gagnon, P., The Quest for a Generic IgG Purification Procedure, Presentation given at the Waterside Conference, Bal Harbour, FL, May 3, 2005.
3. Josic et al., *J. Chromatogr.*, 353, 13, 1986.
4. Zola, *Monoclonal Antibodies: A Manual of Techniques*, CRC Press, Boca Raton, 1987.
5. Bowles et al., *Int. J. Immunopharmacol.*, 10, 537, 1988.
6. Zola and Neoh, *Biotechniques* 7, 802, 1989.
7. Manzke et al., *J. Immunol. Methods,* 208, 65, 1997.
8. Lowry et al., *J. Biol. Chem.*, 193, 265, 1951.

8 Making Antibodies in Bacteria

Frederic A. Fellouse and Sachdev S. Sidhu

CONTENTS

8.1 INTRODUCTION

The bacterium *Escherichia coli* is a well-characterized and extremely robust host for recombinant protein expression. Among other advantages, expression in *E. coli* permits efficient mutagenesis and DNA manipulation, rapid establishment and

(a)

Heavy chain

Light chain

Fab

Fc

(b)

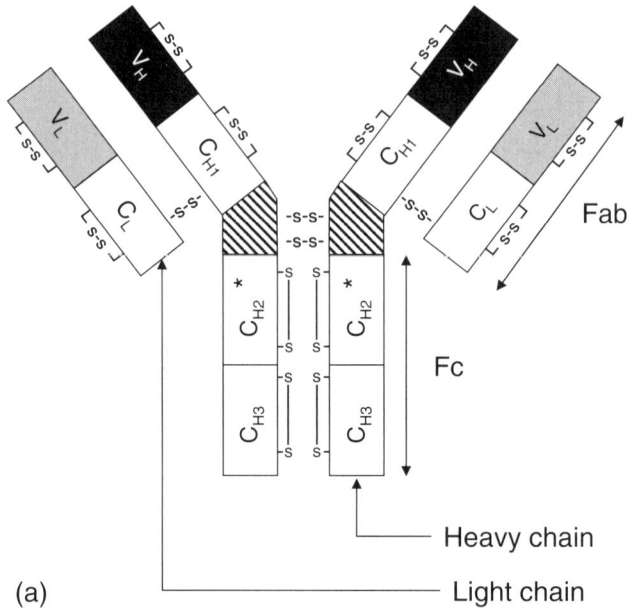

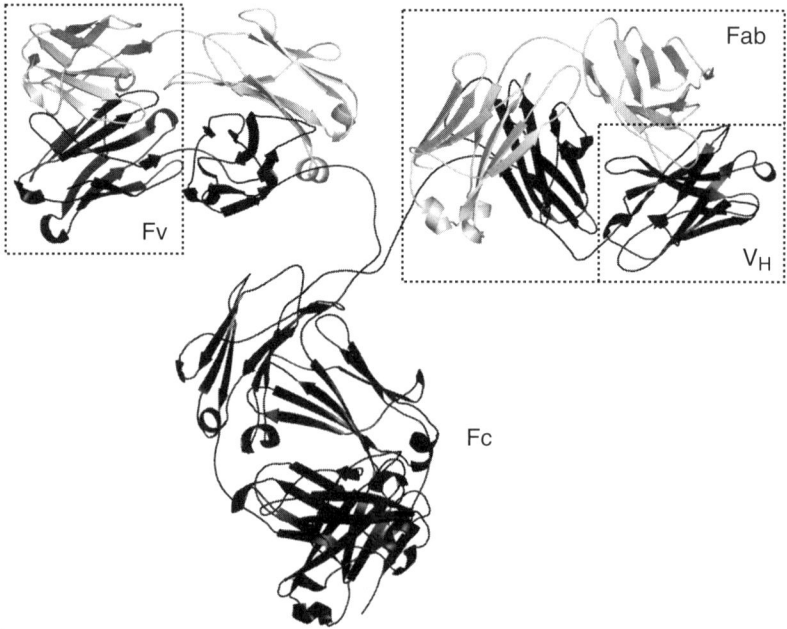

FIGURE 8.1 (*Caption on facing page*)

growth of expression cultures, and in ideal cases, high yields of purified proteins. Thus, the ability to express and purify antibody fragments from *E. coli* has been of great utility for both basic research into antibody structure and function and for the development of antibody reagents and therapeutics. However, because of the complex nature of the immunoglobulin molecule, the development of efficient expression protocols in bacteria faced significant challenges, which have nonetheless been gradually surmounted through the efforts of many researchers.

Immunoglobulin G (IgG) is the most common class of recombinant antibody, and the full-length molecule is a heterotetramer of two heavy chains and two light chains (Figure 8.1). The complexity of the molecule is increased further by the presence of several disulfide bonds (both intermolecular and intramolecular) and by glycosylation on the heavy chain. Proper assembly of the heterotetramer requires transport through the secretion pathway and correct formation of the disulfide bonds. Under most conditions, this process is highly inefficient in a bacterial host,[1,2] and bacteria also lack glycosylation mechanisms. Fortunately, the full-length immunoglobulin is not required for antigen recognition, as the antigen-binding site is contained within the nonglycosylated antigen-binding fragment (Fab), whereas the glycosylated crystallizable fragment (Fc) is responsible for mediating biologic effector functions.

A major breakthrough in antibody research was the development of methods for expressing Fabs in *E. coli*, because this provided a means of purifying the antigen-binding activity of antibodies from a bacterial host.[3] The binding entity was simplified further by the development of the single-chain variable fragment (scFv) format,[4,5] which consists of only the heavy- and light-chain variable domains joined together by a short linker. The surprising discovery of functional cameloid antibodies that lack light chains lead to the expression of heavy-chain variable (V_H) domains,[6–8] which represent the simplest natural antigen-binding entities. All three of these antibody fragments can be produced from *E. coli,* and this has greatly facilitated the study of antibodies, particularly for mutagenic and structural analysis. More recently, the high-level expression of full-length, unglycosylated IgG has also been achieved in *E. coli*.[9] As a result of these advances, it has been possible to rapidly modify and improve antibody binding properties, and this has led to the development of better

FIGURE 8.1 The immunoglobulin G (IgG) molecule. **(A)** Schematic view of an IgG heterotetramer composed of two light chains and two heavy chains. The constant domains are shown in white and the hinge region is cross-hatched. The light and heavy chain variable domains are colored grey and black, respectively. The antigen binding fragment (Fab) and the crystallizable fragment (Fc) are labeled. Intrachain and interchain disulfide bonds are shown (–S–S–) and a glycosolation site in C_{H2} is denoted by an asterisks (*). **(B)** The crystal structure of an IgG. The main chain traces of the IgG (PDB ID code 1IGT) are depicted as cartoons, and the light and heavy chains are colored gray or black, respectively. The dashed boxes delineate the three antibody fragments that are used for phage-displayed libraries: the V_H domain, the Fab, and the Fv, which is displayed in the form of a scFv. The Fc is also labeled. The structure was generated by the computer program PyMOL (DeLano Scientific, San Carlos, CA).

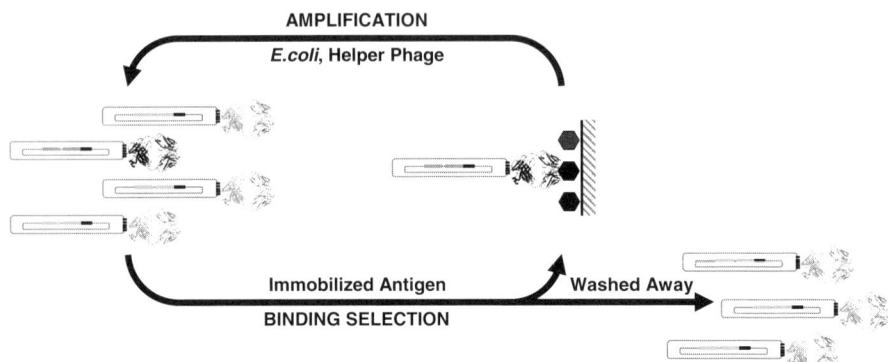

FIGURE 8.2 The phage display cycle. Fabs are displayed on phage particles that also contain the encoding DNA. Specific binding clones can be selectively enriched from library pools by binding to immobilized antigen and washing away nonbinding particles. Subsequently, the bound phage can be eluted and amplified for additional rounds of selection by passage through an *Escherichia coli* host coinfected by helper phage.

reagents and, most importantly, has greatly facilitated the development of antibody therapeutics.

A second major breakthrough in antibody research came with the development of phage display technology,[10] which made it possible to produce large libraries of antibody fragments *in vitro*.[11-25] Phage display relies on the fact that, if gene fragments encoding antibody fragments (or other proteins) are fused to filamentous phage coat protein genes, these "fusion genes" can be incorporated in phage particles that also display the antibody fragments on their surfaces (Figure 8.2).[10,26-28] In this way, a physical linkage is established between phenotype and genotype. Phage-displayed antibodies with a particular binding specificity can be selected from library pools by binding to an immobilized antigen, and the selected phage can be amplified for additional rounds of selection by passage through a bacterial host. Most important, the sequence of any displayed antibody can be readily deduced by sequencing the encoding DNA encapsulated within the phage particle.

In essence, phage display technology has allowed for the development of *in vitro* antibody libraries that mimic the key features of the natural immune system by providing for clonal selection of antigen-specific antibodies from large, diverse pools. Efficient systems have been developed for the display of antigen-binding sites in either the Fab,[12,29] scFv,[30,31] or V_H[19,21,32] domain format. The earliest phage-displayed antibody libraries relied on repertoires obtained from immunized animals, and thus provided a method for the enrichment of binding activities that were already evolved by the natural repertoire.[24,33] Subsequently, the technology was advanced significantly by the construction of naive libraries that still relied on the natural repertoire, but were large and diverse enough to be used as sources of a naive antibody response against antigens to which the natural repertoire had not been exposed.[25,34,35] Most recently, the need for a natural repertoire has been circumvented completely by the development of so-called "synthetic" antibody libraries in which synthetic DNA is used to introduce diversity within antigen-binding sites built in the context of defined frameworks.[12,18,22,31,36,37]

Synthetic antibody libraries hold great promise for high-precision antibody engineering, because defined scaffolds are used and diversity is tailored in a site-specific manner. In this chapter, we describe principles and methods for the construction and use of phage-displayed libraries that obviate the need for natural immune repertoires and enable the development of synthetic antibodies against essentially any antigen of interest.

8.2 REQUIRED MATERIALS

1. 0.2-cm gap electroporation cuvette (BTX, San Diego, CA)
2. 1.0 M H_3PO_4
3. 1.0 M Tris base, pH 8.0
4. 1.0 mM Hepes, pH 7.4 (4.0 ml of 1.0 M Hepes, pH 7.4 in 4.0 l of ultrapure irrigation USP water, filter sterilize)
5. 3,3′,5,5′-tetramethylbenzidine/H_2O_2 peroxidase (TMB) substrate (Kirkegaard & Perry Laboratories, Gaithersburg, MD)
6. 10% (v/v) ultrapure glycerol (100 ml ultrapure glycerol in 900 ml ultrapure irrigation USP water, filter sterilize)
7. 10 mM adenosine 5′-triphosphate (Amersham-Pharmacia, Piscataway, NJ)
8. 10x polymerase chain reaction buffer (600 mM Tris-Hcl, pH 8.3, 250 mM KCl, 15 mM $MgCl_2$, 1% Triton X100, 100 mM β-mercaptoethanol)
9. 10x TM buffer (0.1 M $MgCl_2$, 0.5 M Tris, pH 7.5)
10. 100 mM dNTP mix (solution containing 25 mM each of dATP, dCTP, dGTP, dTTP) (Amersham-Pharmacia)
11. 96-well Maxisorp immunoplates (NUNC, Roskilde, Denmark)
12. 96-well microtubes (VWR, Chicago, IL)
13. 100 mM HCl
14. 100 mM dithiothreitol (DTT) (Sigma, St. Louis, MO)
15. 2YT medium (10 g bacto-yeast extract, 16 g bacto-tryptone, 5 g NaCl. Add water to 1.0; adjust pH to 7.0 with NaOH; autoclave.)
16. 2YT/carb/cmp medium (2YT, 50 µg/ml carbenicillin, 5 µg/ml chloramphenicol)
17. 2YT/carb/kan medium (2YT, 50 µg/ml carbenicillin, 25 µg/ml kanamycin)
18. 2YT/carb/kan/uridine medium (2YT, 50 µg/ml carbenicillin, 25 µg/ml kanamycin, 0.25 µg/ml uridine)
19. 2YT/carb/KO7 medium (2YT, 50 µg/ml carbenicillin, 10^{10} phage/ml of M13KO7)
20. 2YT/tet medium (2YT, 5 µg/ml tetracycline)
21. 2YT/carb/tet medium (2YT, 50 µg/ml carbenicillin, 10 µg/ml tetracycline)
22. 2YT/carb/tet/KO7 (2YT, 50 µg/ml carbenicillin, 10 µg/ml tetracycline, 10^{10} phage/ml of M13KO7)
23. 2YT/kan medium (2YT, 25 µg/ml kanamycin)

24. 2YT/kan/tet medium (2YT, 25 µg/ml kanamycin, 5 µg/ml tetracycline)
25. 2YT top agar (16 g tryptone, 10 g yeast extract, 5 g NaCl, 7.5 g granulated agar. Add water to 1.0 l and adjust pH to 7.0 with NaOH, heat to dissolve, autoclave)
26. AmpliTaq DNA polymerase (Applied Biosystems, Hayward, CA)
27. Carbenicillin (5 mg/ml in water, filter sterilize)
28. Chloramphenicol (50 mg/ml in ethanol)
29. Complete C.R.A.P. medium (3.57 g $(NH_4)_2SO_4$, 0.71 g NaCitrate-$2H_2O$, 1.07 g KCl, 5.36 g yeast extract, 5.36 g Hycase SF-Sheffield. Adjust pH to 7.3 with KOH, add water to 1.0 l, autoclave, cool to 55°C. Add 110 ml 1.0 M MOPS, pH 7.3, 11 ml 50% glucose, 7 ml 1.0 M $MgSO_4$.)
30. ECM-600 electroporator (BTX)
31. *E. coli* 34B8 (Stratagene, La Jolla, CA)
32. *E. coli* CJ236 (New England Biolabs, Beverly, MA)
33. *E. coli* MC1061 (Bio-Rad, Hercules, CA)
34. *E. coli* XL1-blue (Stratagene)
35. EZ-Link NHS-SS-biotin (Pierce, Rockford, IL)
36. Filtration Units (Nalgene Nunc, Rochester, NY)
37. Homogenizer, model T-50 Ultra-Turrax (Janke & Kunkel, Staufen, Germany)
38. Horseradish peroxidase/anti-M13 antibody conjugate (Amersham-Pharmacia)
39. Kanamycin (5 mg/ml in water, filter sterilize)
40. LB/carb plates (LB agar, 50 µg/ml carbenicillin)
41. LB/tet plates (LB agar, 5 µg/ml tetracycline)
42. M13KO7 helper phage (New England Biolabs)
43. Magnetic stir bars (2 inch) soaked in ethanol
44. Microfluidizer processor, model M-110Y (Microfluidics, Newton, MA)
45. Neutravidin (Pierce)
46. Phosphate-buffered saline (PBS) (137 mM NaCl, 3 mM KCl, 8 mM Na_2HPO_4, 1.5 mM KH_2PO_4. Adjust pH to 7.2 with HCl, autoclave)
47. PBS, 0.2% bovine serum albumin (BSA)
48. PBS/BSA/Tween buffer (PBS, 0.05% Tween 20, 0.2% BSA)
49. PT buffer (PBS, 0.05% Tween 20)
50. PCR clean up mix (exonuclease I and shrimp alkaline phosphatase) (United States Biochemical, Cleveland, OH)
51. PEG/NaCl (20% PEG-8000 (w/v), 2.5 M NaCl. Mix and autoclave.)
52. Protein A-Sepharose resin (Amershan-Pharmacia)
53. QIAprep Spin M13 Kit (Qiagen, Valencia, CA)
54. QIAquick Gel Extraction Kit (Qiagen)
55. Resuspension buffer (PBS, 25 mM EDTA, protease inhibitor cocktail) (Roche, Indianapolis, IN)
56. SOC medium (5 g bacto-yeast extract, 20 g bacto-tryptone, 0.5 g NaCl, 0.2 g KCl. Add water to 1.0 l and adjust pH to 7.0 with NaOH, autoclave; add 5.0 ml of autoclaved 2.0 M $MgCl_2$ and 20 ml of filter sterilized 1.0 M glucose.)

57. Superbroth medium (12 g tryptone, 24 g yeast extract, 5 ml glycerol; add water to 900 ml, autoclave, add 100 ml of autoclaved 0.17 M KH_2PO_4, 0.72 M K_2HPO_4)
58. Superbroth/tet/kan medium (Superbroth medium, 5 µg/ml tetracycline, 25 µg/ml kanamycin)
59. T4 polynucleotide kinase (New England Biolabs)
60. T4 DNA ligase (Invitrogen, Carlsbad, CA)
61. T7 DNA polymerase (New England Biolabs)
62. TAE buffer (40 mM Tris-acetate, 1.0 mM EDTA; adjust pH to 8.0; autoclave)
63. TAE/agarose gel (TAE buffer, 1.0% (w/v) agarose, 1:5000 (v/v) 10% ethidium bromide)
64. Tetracycline (5 mg/ml in water, filter sterilize)
65. Ultrapure irrigation USP water (Braun Medical, Irvine, CA)
66. Uridine (25 mg/ml in water, filter sterilize)
67. Ultrapure glycerol (Invitrogen)

8.3 METHODS

The following sections detail all of the methods necessary for the construction and use of phage-displayed libraries for the isolation of antigen-specific antibodies. First, a phagemid is designed for the display of the parental antibody framework, and subsequently, the library is constructed by introducing appropriate genetic diversity into the complementarity determining regions. The genetic library is converted to a phage-displayed protein library by passage through an *E. coli* host, and the library phage pool can be used for selection experiments to obtain antigen-specific clones. These clones can be assayed for specific binding directly as phage particles, and the protein sequences can be deduced by sequencing the encapsulated DNA. Finally, antibodies of interest can be purified as free proteins for subsequent analysis or use.

8.3.1 PHAGEMID DESIGN

Phage-displayed antibody libraries are usually constructed using a specialized vector known as a phagemid (Figure 8.3), and if proper considerations are taken into account, the same vector can be readily modified for the production of free protein for purification and analysis. A phagemid vector contains a double-stranded DNA (dsDNA) origin of replication (dsDNA ori) to enable replication as a plasmid in *E. coli* and, also, a single-stranded DNA (ssDNA) filamentous phage origin of replication (f1 ori) to allow packaging into phage particles. We describe a versatile phagemid that has been used for the display of Fabs,[18] scFvs,[37] and V_H domains,[21] and also, for the display of other proteins and peptides.[26,27] The vector also contains a β-lactamase (*bla*) gene conferring resistance to ampicillin and carbenicillin (Amp[r]), but alternatively, other resistance markers can be used. The final component is a DNA cassette under the control of the alkaline phosphatase promoter (*PphoA*), which is responsible for expression of the antibody fragment to be displayed. We have used the alkaline phosphatase promoter for Fab display,[18] because this promoter is well-suited for the expression and purification of Fab protein. However, other promoters

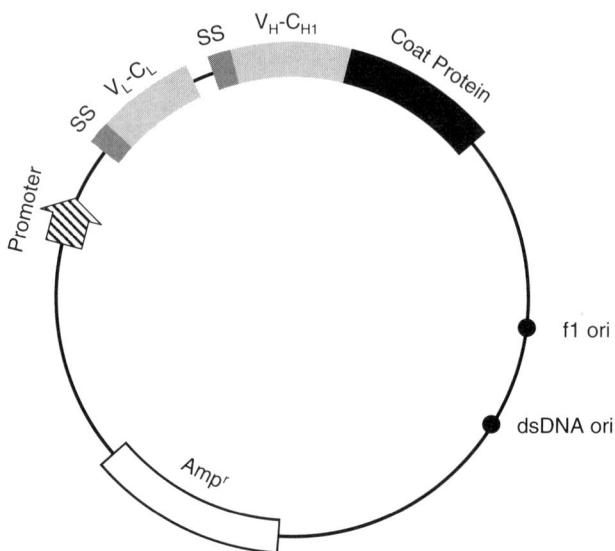

FIGURE 8.3 A phagemid vector designed for Fab display. The phagemid vector contains origins of single-stranded (f1 ori) and double-stranded (dsDNA ori) DNA replication and a selective marker, such as the β-lactamase gene (Amp^r), which confers resistance to ampicillin antibiotics. For Fab display, the phagemid also contains a cassette consisting of a promoter that drives transcription of a bi-cistronic message that encodes for the light chain (V_L-C_L) and the variable and first constant domains of the heavy chain (V_H-C_{H1}). Each chain is fused to an N-terminal secretion signal (SS), and in addition, the C terminus of the heavy chain is fused to a phage coat protein. In the periplasm, the light and heavy chains associate to assemble functional Fab, which is incorporated into phage particles that are secreted from the *Escherichia coli* host.

can also be used. For example, we have used the isopropyl 1-thio-β-D-galactopyr-anoside-inducible P$_{tac}$ promoter for the display of scFv and V$_H$ domain libraries.[21,37] In the case of monomeric scFv or V$_H$ domain display, the DNA cassette encodes for a fusion protein consisting of a secretion signal, followed by the protein to be displayed and the C-terminal domain of the M13 bacteriophage gene-3 minor coat protein (Protein-3, P3). For Fab display, the situation is complicated somewhat by the heterodimeric nature of the protein, which necessitates the use of a bicistronic expression system. The heavy chain is fused to the phage coat protein, whereas the light chain is expressed independently. Each chain is also fused to a secretion signal so that both chains are directed to the periplasm where they assemble to form the heterodimeric Fab, which is subsequently displayed on phage.

In an *E. coli* host, the phagemid replicates as a double-stranded plasmid. After coinfection with a helper phage (such as M13KO7), ssDNA replication is initiated, and the phagemid ssDNA is preferentially packaged into phage particles. The ssDNA can be easily purified from phage and used for efficient sequencing, mutagenesis, or library construction (see the following section). Although the helper phage provides all of the proteins necessary for phage assembly, copies of phagemid-encoded coat protein are also incorporated into the assembling virions. In this way, polypeptides fused to phagemid-encoded coat protein are displayed in linkage to their encoding DNA, and phage particles can be used for library selections and screens.

8.3.2 Library Construction

Very large phage-displayed antibody repertoires ($>10^{10}$ members) can be constructed quite rapidly by using optimized procedures[18,37,38] that are based on the classical oligonucleotide-directed mutagenesis method of Kunkel et al. (Figure 8.4).[39] First, mutagenic oligonucleotides are used to introduce stop codons at the sites to be randomized, and the resulting "stop template" phagemid can be used as the template for library construction because the presence of stop codons eliminates wild-type (wt) protein display. Uracil-containing ssDNA (dU-ssDNA) stop template (purified from an *E. coli dut⁻/ung⁻* host) is then annealed with mutagenic oligonucleotides designed to replace the stop codons with the appropriate degenerate codons. For example, an NNK degenerate codon (N = A/G/C/T, K = G/T) encodes for all 20 amino acids. Alternatively, other degenerate codons can be used to allow only

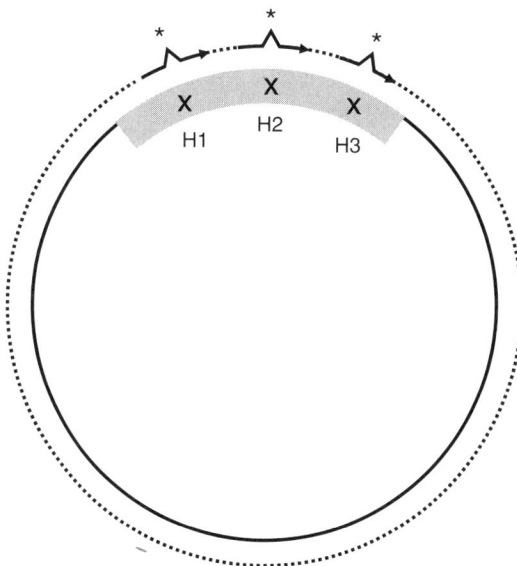

FIGURE 8.4 Library construction by oligonucleotide-directed mutagenesis. Synthetic oligonucleotides (arrows) are annealed to the dU-ssDNA template (solid circle). In this example, three different oligonucleotides are annealed to mutate the three complementarity determining regions (H1, H2, and H3) within the heavy chain variable domain (gray box). Each oligonucleotide is designed to encode mutations (*) in the mismatched diversity region that is flanked by perfectly complementary sequences. Heteroduplex covalently closed circular, double-stranded DNA is enzymatically synthesized by T7 DNA polymerase and T4 DNA ligase (dashed circle), and is introduced into an *Escherichia coli* host where the mismatched region is repaired to either the wild-type sequence or the mutant sequence. The template contains stop codons in the regions to be mutated (X), and thus Fab proteins are only displayed by clones in which all the stop codons have been repaired by the mutagenic oligonucleotides.

particular subsets of the amino acids, to favor amino acids that are common in natural antibodies[18,37] or are particularly well suited for antigen recognition.[40,41] The mutagenic oligonucleotides are used to prime the synthesis of a complementary DNA strand that is ligated to form a covalently closed circular, double-stranded DNA (CCC-dsDNA) heteroduplex. To complete the library construction, the CCC-dsDNA heteroduplex is introduced into an *E. coli dut⁺/ung⁺* host by electroporation and the mismatch is repaired to either the wt or mutant sequence. In an *ung⁺* strain, the uracil-containing template strand is preferentially inactivated, and the synthetic, mutant strand is replicated, thus resulting in efficient mutagenesis (>50%). The use of a template with stop codons at all of the sites to be randomized ensures that only fully mutagenized clones are displayed on phage, because only these clones contain open reading frames that produce full-length proteins fused to the phage coat protein. The library members can be packaged into phage particles by coinfection of the *E. coli* host with a helper phage.

8.3.2.1 Purification of dU-ssDNA Template

Mutagenesis efficiency depends on template purity, and thus the use of highly pure dU-ssDNA is critical for successful library construction. We use the Qiagen QIAprep Spin M13 Kit for dU-ssDNA purification, and the following is a modified version of the Qiagen protocol. It yields at least 20 µg of dU-ssDNA for a medium copy number phagemid (e.g., pBR322 backbone), and this is sufficient for the construction of one library.

1. From a fresh LB/carb plate, pick a single colony of *E. coli* CJ236 (or another *dut⁻/ung⁻* strain) harboring the appropriate phagemid into 1 ml of 2YT medium supplemented with M13KO7 helper phage (10¹⁰ pfu/ml) and appropriate antibiotics to maintain the host F' episome and the phagemid. For example, 2YT/carb/cmp medium contains carbenicillin to select for phagemids that carry the β-lactamase gene and chloramphenicol to select for the F' episome of *E. coli* CJ236.
2. Shake at 200 rpm and 37°C for 2 h and add kanamycin (25 µg/ml) to select for clones that have been coinfected with M13KO7, which carries a kanamycin resistance gene.
3. Shake at 200 rpm and 37°C for 6 h and transfer the culture to 30 ml of 2YT/carb/kan/uridine medium.
4. Shake 20 h at 200 rpm and 37°C.
5. Centrifuge for 10 min at 15 krpm and 4°C in a Sorvall SS-34 rotor (27,000*g*). Transfer the supernatant to a new tube containing 1/5 volume of PEG/NaCl and incubate for 5 min at room temperature.
6. Centrifuge 10 min at 10 krpm and 4°C in an SS-34 rotor (12,000*g*). Decant the supernatant; centrifuge briefly at 4 krpm (2000*g*) and aspirate the remaining supernatant.
7. Resuspend the phage pellet in 0.5 ml of PBS and transfer to a 1.5-ml microcentrifuge tube.
8. Centrifuge for 5 min at 13 krpm in a microcentrifuge, and transfer the supernatant to a 1.5 ml microcentrifuge tube.

9. Add 7.0 µl of buffer MP (Qiagen) and mix. Incubate at room temperature for at least 2 min.
10. Apply the sample to a QIAprep spin column (Qiagen) in a 2-ml microcentrifuge tube. Centrifuge for 30 sec at 8 krpm in a microcentrifuge. Discard the flow-through. The phage particles remain bound to the column matrix.
11. Add 0.7 ml of buffer MLB (Qiagen) to the column. Centrifuge for 30 sec at 8 krpm and discard the flow-through.
12. Add 0.7 ml of buffer MLB. Incubate at room temperature for at least 1 min.
13. Centrifuge at 8 krpm for 30 sec. Discard the flow-through. The DNA is separated from the protein coat and remains adsorbed to the matrix.
14. Add 0.7 ml of buffer PE (Qiagen). Centrifuge at 8 krpm for 30 sec and discard the flow-through.
15. Repeat step 14. Residual proteins and salt are removed.
16. Centrifuge at 8 krpm for 30 sec in a fresh 1.5-ml microcentrifuge tube to remove residual PE buffer.
17. Transfer the column to a fresh 1.5-ml microcentrifuge tube.
18. Add 100 µl of buffer EB (Qiagen; 10 mM Tris-Cl, pH 8.5) to the center of the column membrane. Incubate at room temperature for 10 min.
19. Centrifuge for 30 s at 8 krpm. Save the eluant, which contains the purified dU-ssDNA.
20. Analyze the DNA by electrophoresing 1.0 µl on a TAE/agarose gel. The DNA should appear as a predominant single band, but faint bands with lower electrophoretic mobility are often visible. These are likely caused by secondary structure in the dU-ssDNA.
21. Determine the DNA concentration by measuring absorbance at 260 nm (A_{260} = 1.0 for 33 ng/µl of ssDNA). Typical DNA concentrations range from 200 to 500 ng/µl.

8.3.2.2 *In Vitro* Synthesis of Heteroduplex CCC-dsDNA

A three-step procedure is used to incorporate the mutagenic oligonucleotides into heteroduplex CCC-dsDNA, using dU-ssDNA as a template. The protocol described here is an optimized, large-scale version of a published method.[39] The oligonucleotide is 5′-phosphorylated and annealed to a dU-ssDNA template. The oligonucleotide is enzymatically extended and ligated to form heteroduplex CCC-dsDNA (Figure 8.4), which is then purified and desalted. The protocol below produces ~20 µg of highly pure, low conductance CCC-dsDNA. This is sufficient for the construction of a library containing more than 10^{10} unique members.

8.3.2.2.1 *Oligonucleotide Phosphorylation with T4 Polynucleotide Kinase*

1. In a 1.5-ml microcentrifuge tube, combine 0.6 µg of the mutagenic oligonucleotide, 2.0 µl 10× TM buffer, 2.0 µl 10 mM ATP, and 1.0 µl 100 mM DTT. Add water to a total volume of 20 µl.
2. Add 20 units of T4 polynucleotide kinase. Incubate for 1.0 h at 37 °C and use immediately for annealing.

8.3.2.2.2 Annealing of the Oligonucleotides to the Template

1. To 20 µg of dU-ssDNA template, add 25 µl 10x TM buffer, 20 µl of each phosphorylated oligonucleotide, and water to a final volume of 250 µl. In cases where more than one region of the DNA are to be mutated (e.g., mutagenesis of multiple complementarity determining regions), two or more mutagenic oligonucleotides can be added simultaneously, as long as no sequences within the oligonucleotides overlap with each other. These DNA quantities provide an oligonucleotide:template molar ratio of 3:1, assuming that the oligonucleotide:template length ratio is 1:100.
2. Incubate at 90°C for 3 min, 50°C for 3 min, and 20°C for 5 min.

8.3.2.2.3 Enzymatic Synthesis of CCC-dsDNA

1. To the annealed oligonucleotide/template mixture, add 10 µl 10 mM ATP, 10 µl 100 mM dNTP mix, 15 µl 100 mM DTT, 30 Weiss units T4 DNA ligase, and 30 units T7 DNA polymerase.
2. Incubate overnight at 20°C.
3. Affinity purify and desalt the DNA using the Qiagen QIAquick DNA purification kit. Add 1.0 ml of buffer QG (Qiagen) and mix.
4. Apply the sample to two QIAquick spin columns placed in 2-ml micro-centrifuge tubes. Centrifuge at 13 krpm for 1 min in a microcentrifuge. Discard the flow-through.
5. Add 750 µl buffer PE (Qiagen) to each column, and centrifuge at 13 krpm for 1 min.
6. Transfer the column to a fresh 1.5-ml microcentrifuge tube, and centrifuge at 13 krpm for 1 min.
7. Transfer the column to a fresh 1.5-ml microcentrifuge tube, and add 35 µl of ultrapure irrigation USP water to the center of the membrane. Incubate at room temperature for 2 min.
8. Centrifuge at 13 krpm for 1 min to elute the DNA. Combine the eluants from the two columns. The DNA can be used immediately for *E. coli* electroporation, or it can be frozen for later use.
9. Electrophorese 1.0 µl of the eluted reaction product alongside the ssDNA template. Use a TAE/agarose gel with ethidium bromide for DNA visualization (Figure 8.5).

A successful reaction results in the complete conversion of ssDNA to dsDNA, which has a lower electrophoretic mobility. Usually, at least two product bands are visible and there should be no remaining ssDNA (Figure 8.5). The product band with higher electrophoretic mobility represents the desired product: correctly extended and ligated CCC-dsDNA, which transforms *E. coli* efficiently and provides a high mutation frequency (~80%). The product band with lower electrophoretic mobility is a strand-displaced product resulting from an intrinsic, unwanted activity of T7 DNA polymerase.[42] Although the strand-displaced product provides a low mutation frequency (~20%), it also transforms *E. coli* at least 30-fold less

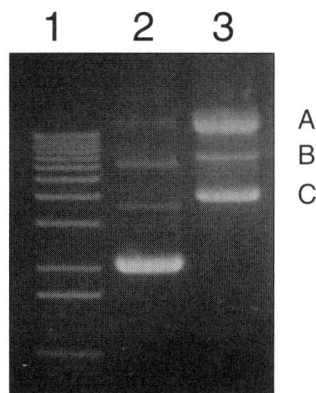

FIGURE 8.5 *In vitro* synthesis of heteroduplex covalently closed circular, double-stranded DNA (CCC-dsDNA). The reaction products were electrophoresed on a 1.0% TAE/agarose gel containing ethidium bromide for DNA visualization. Lane 1: DNA markers; Lane 2: the uracil-containing single-stranded DNA template; Lane 3: reaction product from the heteroduplex CCC-dsDNA synthesis reaction. The lower band (C) is correctly extended and ligated CCC-dsDNA, the middle band (B) is knicked, dsDNA and the upper band (A) is strand-displaced dsDNA.

efficiently than CCC-dsDNA. If a significant proportion of the single-stranded template is converted to CCC-dsDNA, a highly diverse library with high mutation frequency will result. Sometimes a third band is visible, with an electrophoretic mobility between the two product bands described previously. This intermediate band is correctly extended but unligated dsDNA (knicked dsDNA), which results from either insufficient T4 DNA ligase activity or from incomplete oligonucleotide phosphorylation.

8.3.2.3 Conversion of CCC-dsDNA into a Phage-Displayed Library

To complete the library construction, the heteroduplex CCC-dsDNA is introduced into an *E. coli* host that contains an F′ episome to enable M13 bacteriophage infection and propagation. Phage-displayed library diversities are limited by methods for introducing DNA into *E. coli*, with the most efficient method being high-voltage electroporation.

We have constructed an *E. coli* strain (SS320) that is ideal for both high-efficiency electroporation and phage production.[43] Using a standard bacterial mating protocol,[44] we transferred the F′ episome from *E. coli* XL1-blue to *E. coli* MC1061. The progeny strain was selected for double resistance to streptomycin and tetracycline, because *E. coli* MC1061 carries a chromosomal marker for streptomycin resistance and the F′ episome from *E. coli* XL1-blue confers tetracycline resistance. *E. coli* SS320 retains the high electroporation efficiency of *E. coli* MC1061, while the presence of an F′ episome enables infection by M13 phage.

The following optimized protocols allow for the large-scale preparation of electrocompetent *E. coli* SS320 and, subsequently, for electroporation to produce high-diversity libraries.

8.3.2.4 Preparation of Electrocompetent *E. coli* SS320

The following protocol yields approximately 12 ml of highly concentrated, electro-competent *E. coli* SS320 (~3×10^{11} cfu/ml) infected by M13KO7 helper phage. The cells can be stored indefinitely at $-70°C$ in 10% glycerol. The use of *E. coli* infected by helper phage ensures that, once transformed with a phagemid, each cell will be able to produce phage particles.

1. Inoculate 25 ml 2YT/tet medium with a single colony of *E. coli* SS320 from a fresh LB/tet plate. Incubate at 37°C with shaking at 200 rpm to mid-log phase ($OD_{550} = 0.8$).
2. Make 10-fold serial dilutions of M13K07 by diluting 20 μl into 180 μl of PBS (use a new pipette tip for each dilution).
3. Mix 500 μl of *E. coli* SS320 at exponential phase with 200 μl of each M13K07 dilution and 4 ml of 2YT top agar.
4. Pour the mixtures onto prewarmed LB/tet plates and grow overnight at 37°C.
5. Pick a well-separated, single plaque and place in 1 ml of 2YT/kan/tet medium. Incubate 8 h at 37°C.
6. Transfer the culture to 250 ml of 2YT/kan medium in a 2-l baffled flask. Grow overnight at 37°C with shaking at 200 rpm.
7. Inoculate six 2-l baffled flasks containing 900 ml of superbroth/tet/kan medium with 5 ml of the overnight culture. Incubate at 37°C with shaking at 200 rpm to mid-log phase ($OD_{550} = 0.8$).
8. Chill three of the flasks on ice for 5 min with occasional swirling. The following steps should be done in a cold room, on ice, with prechilled solutions and equipment.
9. Centrifuge at 5.5 krpm ($5,000g$) and 4°C for 10 min in a Sorvall GS-3 rotor.
10. Decant the supernatant and add culture from the remaining flasks (these should be chilled while the first set is centrifuging) to the same tubes.
11. Repeat the centrifugation and decant the supernatant.
12. Fill the tubes with 1.0 mM Hepes, pH 7.4, and add sterile magnetic stir bars to facilitate pellet resuspension. Swirl to dislodge the pellet from the tube wall and stir at a moderate rate to resuspend the pellet completely.
13. Centrifuge at 5.5 krpm ($5,000g$) and 4°C for 10 min in a Sorvall GS-3 rotor. Decant the supernatant, being careful to retain the stir bar. To avoid disturbing the pellet, maintain the position of the centrifuge tube when removing from the rotor.
14. Repeat steps 12 and 13.
15. Resuspend each pellet in 150 ml of 10% ultrapure glycerol. Use stirbars and do not combine the pellets.
16. Centrifuge at 5.5 krpm ($5,000g$) and 4°C for 15 min in a Sorvall GS-3 rotor. Decant the supernatant and remove the stir bar. Remove remaining traces of supernatant with a pipette.

17. Add 3.0 ml of 10% ultrapure glycerol to one tube and resuspend the pellet by pipetting. Transfer the suspension to the next tube and repeat until all of the pellets are resuspended.
18. Transfer 350 µl aliquot into 1.5 ml microcentrifuge tubes.
19. Flash freeze with liquid nitrogen and store at –70°C.

8.3.2.5 *E. coli* Electroporation and Phage Propagation

1. Chill the purified CCC-dsDNA (20 µg in a minimum volume) and a 0.2 cm gap electroporation cuvette on ice.
2. Thaw a 350 µl aliquot of electrocompetent *E. coli* SS320 on ice. Add the cells to the DNA and mix by pipetting several times (avoid introducing bubbles).
3. Transfer the mixture to the cuvette and electroporate. For electroporation, follow the manufacturer's instructions, preferably using a BTX ECM-600 electroporation system with the following settings: 2.5 kV field strength, 129 ohms resistance, and 50 µF capacitance. Alternatively, a Bio-Rad Gene Pulser can be used with the following settings: 2.5 kV field strength, 200 ohms resistance, and 25 µF capacitance.
4. Immediately rescue the electroporated cells by adding 1 ml SOC medium and transferring to 10 ml SOC medium in a 250-ml baffled flask. Rinse the cuvette twice with 1 ml SOC media. Add SOC medium to a final volume of 25 ml.
5. Incubate for 30 min at 37°C with shaking at 200 rpm.
6. To determine the library diversity, plate serial dilutions on LB/carb plates to select for the phagemid.
7. Transfer the culture to a 2-l baffled flask containing 500 ml 2YT medium, supplemented with antibiotics for phagemid and M13KO7 helper phage selection (e.g., 2YT/carb/kan medium).
8. Incubate overnight at 37°C with shaking at 200 rpm.
9. Centrifuge the culture for 10 min at 10 krpm and 4°C in a Sorvall GSA rotor ($16,000g$).
10. Transfer the supernatant to a fresh tube and add 1/5 volume of PEG/NaCl solution to precipitate the phage. Incubate 5 min at room temperature.
11. Centrifuge for 10 min at 10 krpm and 4°C in a GSA rotor. Decant the supernatant. Spin again briefly and remove the remaining supernatant with a pipette.
12. Resuspend the phage pellet in 20 ml of PBT buffer.
13. Pellet insoluble matter by centrifuging for 5 min at 15 krpm and 4°C in an SS-34 rotor ($27,000g$). Transfer the supernatant to a clean tube.
14. Estimate the phage concentration spectrophotometrically ($OD_{268} = 1.0$ for a solution of 5×10^{12} phage/ml).
15. The library can be used immediately for selection experiments. Alternatively, the library can be frozen and stored at –80°C, following the addition of glycerol to a final concentration of 10%. In general, it is best to use

libraries immediately, because levels of displayed proteins can be reduced over time from denaturation or proteolysis.

8.3.3 SELECTION OF ANTIGEN-SPECIFIC ANTIBODIES

Phage-displayed antibody libraries can be used to select antigen-specific antibodies by a variety of strategies, and we describe the two most common methods. In the first method, the phage-displayed library is incubated with the antigen immobilized on an immunoplate. In the second method, it is incubated with the biotinylated antigen in solution and, subsequently, bound phage are captured in immunoplates coated with Neutravidin (Pierce/Invitrogen, Carlsbad, CA).

In the first method, the avidity effect produced by the immobilization of the antigen on a solid surface allows for the selection of ligands with weak affinity. In the second method, the stringency of the selection can be adjusted to favor high-affinity clones by adjusting the concentration of biotinylated antigen incubated with the library. The first method is useful for selection from a naive library, and the second is more appropriate for the affinity improvement of an existing antibody.

8.3.3.1 Selection against Immobilized Antigen

1. Coat Maxisorp immunoplate wells with 100 μl of antigen solution (5 μg/ml in coating buffer) for 2 h at room temperature or overnight at 4°C. The number of wells required depends on the diversity of the library. Ideally, the phage concentration should not exceed 10^{13} phage/ml, and the total number of phage should exceed the library diversity by 1000-fold. Thus, for a diversity of 10^{10}, 10^{13} phage should be used, and using a concentration of 10^{13} phage/ml, 10 wells will be required.
2. Remove the coating solution and block for 1 h with 200 μl of PBS, 0.2% BSA. At the same time, block an equal number of uncoated wells as a negative control.
3. Remove the block solution and wash four times with PT buffer.
4. Add 100 μl of library phage solution in PBT buffer to each of the coated and uncoated wells. Incubate at room temperature for 2 h with gentle shaking.
5. Remove the phage solution and wash 10 times with PT buffer.
6. To elute bound phage, add 100 μl of 100 mM HCl. Incubate 5 min at room temperature. Transfer the HCl solution to a 1.5-ml microfuge tube.
7. Adjust to neutral pH with 1.0 M Tris-HCl, pH 8.0.
8. Add half the eluted phage solution to 10 volumes of actively growing *E. coli* XL1-Blue (OD$_{550}$ <1.0) in 2YT/tet medium. Incubate for 20 min at 37°C with shaking at 200 rpm.
9. Plate serial dilutions on LB/carb plates to determine the number of phage eluted. Determine the enrichment ratio by dividing the number of phage eluted from a well coated with antigen by the number of phage eluted from an uncoated well.

10. Add M13KO7 helper phage to a final concentration of 10^{10} phage/ml. Incubate for 45 min at 37°C with shaking at 200 rpm.
11. Transfer the culture from the antigen-coated wells to 25 volumes of 2YT/carb/kan medium and incubate overnight at 37°C with shaking at 200 rpm.
12. Isolate phage by precipitation with PEG/NaCl solution, resuspend in 1.0 ml of PBT buffer and estimate phage concentration spectrophotometrically.
13. Repeat the selection cycle until the enrichment ratio has reached a maximum. Typically, enrichment is first observed in round 3 or 4, and sorting beyond round 6 is seldom necessary.
14. Pick individual clones for sequence analysis and phage enzyme-linked immunosorbent assay (ELISA).

8.3.3.2 Selection against Biotinylated Antigen

1. Biotinylate the antigen with EZ-Link NHS-SS-biotin.
2. Coat Maxisorp immunoplate wells with 100 µl of Neutravidin solution (5 µg/ml in coating buffer) for 2 h at room temperature or overnight at 4°C. The number of wells required depends on the diversity of the library.
3. Remove the coating solution and block for 1 h with 200 µl of PBS, 0.2% BSA.
4. In a 1.5 ml microfuge tube, combine the biotinylated antigen with 1.0 ml of library phage solution in PBT buffer. The stringency of the selection can be increased by reducing the concentration of the antigen, and the antigen concentration (0.1–10 nM) can be reduced in successive rounds of selection.
5. Incubate at room temperature for 2 h with gentle shaking.
6. Add 100 µl of the phage/antigen mixture to each of the Neutravidin-coated wells.
7. Incubate at room temperature for 15 min with gentle shaking.
8. Remove the phage solution and wash 10 times with PT buffer.
9. To elute bound phage, add 100 µl of 100 mM HCl. Incubate 5 min at room temperature.
10. Transfer the HCl solution to a 1.5 ml microcentrifuge tube and adjust to neutral pH with 1.0 M Tris-HCl, pH 8.0.
11. Amplify phage for further rounds of selection by growth in *E. coli* XL1-blue.

8.3.3.3 Analysis of Binding Clones by Phage ELISA

The selection process described above produces a phage pool that is enriched for antigen-binding clones. However, the population may also contain nonbinding clones and clones that bind nonspecifically. Binding of phage-displayed antibodies to immobilized antigens can be detected by using antibodies raised against the phage particle, and ELISA methods have been adapted for the analysis of specific binding of phage-displayed antibodies to antigen. The culture medium into which phage particles are secreted can be used for direct-binding phage ELISAs to rapidly identify antigen-specific binding clones.

The affinities of specific antigen-binding clones can be estimated by competitive phage ELISAs. In this type of ELISA, binding of phage to immobilized antigen is competed by defined concentrations of antigen in solution. The assay can be done in a relatively high-throughput manner as a single-point competitive phage ELISA in which inhibition of binding is measured at a single antigen concentration. This allows for rank ordering of many unique clones on the basis of percent inhibition of binding, but does not provide an accurate measure of affinity.

Alternatively, a more accurate multipoint competitive phage ELISA can be used to accurately estimate the affinity as an IC_{50} value, which is defined as the concentration of antigen that blocks 50% of antibody-phage binding to immobilized antigen. The multipoint phage ELISA comprises two steps. First, the antibody-phage solution is titered by serial dilution to determine a suitable concentration that provides a detectable but subsaturating signal on antigen-coated wells. Second, the binding of the subsaturating phage concentration to immobilized antigen is competed by serial dilutions of antigen in solution, and this produces a dose-response inhibition curve that is used to determine the IC_{50} value.

8.3.3.3.1   Direct-Binding Phage ELISA

1. Inoculate 450 µl aliquots of 2YT/carb/KO7 medium in 96-well microtubes with single colonies harboring phagemids and grow overnight at 37°C with shaking at 200 rpm.
2. Centrifuge at 4,000 rpm for 10 min and transfer phage supernatants to fresh tubes.
3. Dilute phage supernatant threefold with PBT buffer.
4. Transfer 100 µl of diluted phage supernatant to 96-well Maxisorp immunoplates coated with antigen and blocked with BSA, as described previously. In addition, binding should also be assayed using plates coated with an irrelevant protein or blocked with BSA only, as negative controls.
5. Incubate for 15 min with gentle shaking.
6. Wash eight times with PT buffer.
7. Add 100 µl of horseradish peroxidase/anti-M13 antibody conjugate (diluted 1:3000 in PBT buffer). Incubate 30 min with gentle shaking.
8. Wash six times with PT buffer and two times with PBS.
9. Add 100 µl of freshly prepared TMB substrate. Allow color to develop for 5–10 min.
10. Stop the reaction with 100 µl of 1.0 M H_3PO_4 and read spectrophotometrically at 450 nm in a microtiter plate reader.
11. To assess specific binding, compare the signal strength on plates coated with antigen to that on plates coated with BSA or another protein.

8.3.3.3.2 Single-Point Competitive Phage ELISA

1. Grow and harvest phage as described previously.
2. Dilute phage supernatant fivefold, using PBT buffer alone or PBT buffer containing antigen at a concentration close to the expected affinity.
3. Incubate for 1 h and transfer 100 µl of each solution to Maxisorp immunoplates coated with antigen and blocked with BSA, as described previously.
4. Incubate for 15 min with gentle shaking.
5. Wash, develop and read the plates as described previously.
6. For each clone, calculate the fraction of antibody-phage uncomplexed with solution-phase antigen by dividing the A_{450} in the presence of the solution-phase antigen by the A_{450} in the absence of the solution-phase antigen. This ratio is inversely proportional to the affinity of the interaction and allows for the rank ordering of clones.

8.3.3.3.3 Multipoint Competitive Phage ELISA

1. Inoculate 1 ml 2YT/carb/tet/KO7 medium with a single colony of *E. coli* XL1-blue harboring phagemid. Incubate for 2 h at 37ºC with shaking at 200 rpm.
2. Add kanamycin (25 µg/ml) and continue incubation to mid-log phase ($OD_{550} = 0.6$).
3. Transfer the culture to 25 ml of 2YT/carb/kan medium in a 250-ml baffled flask. Grow overnight at 37°C with shaking at 200 rpm.
4. Isolate phage by precipitation with PEG/NaCl solution, resuspend in 1.0 ml of PBT buffer and estimate phage concentration spectrophotometrically, as described previously.
5. Prepare threefold serial dilutions of phage stock in PBT buffer.
6. Transfer 100 µl of phage solution to Maxisorp immunoplates coated with antigen and blocked as described previously.
7. Incubate for 15 min with gentle shaking.
8. Wash, develop and read the plates as described previously.
9. For the subsequent steps, use a subsaturating concentration of the phage stock which gives ~50% of the ELISA signal at saturation.
10. Aliquot 135 µl of subsaturating phage solution into each of 12 wells of a 96-well plate.
11. Add 15 µl of serially diluted antigen to each well. Use 2-fold dilutions, and ideally, start with an antigen concentration 100-fold greater than the expected IC_{50} value.
12. Incubate for 1.0 h and transfer 100 µl to Maxisorp immunoplates coated with antigen and blocked as described previously.
13. Incubate for 15 min with gentle shaking.
14. Wash, develop, and read the plates as described previously.
15. Plot the OD_{450} reading as a function of the antigen concentration and determine the IC_{50} value by standard curve fitting.

8.3.3.4 DNA Sequencing

The protocols described enable the isolation and functional characterization of anti-body fragments displayed on phage particles. The sequence of the displayed antibody fragment can be deduced by sequencing the encoding DNA, which is encapsulated within the phage particle. Furthermore, the whole DNA sequencing procedure can be performed in a high-throughput 96-well format.

1. Inoculate a single colony–harboring phagemid into 200 μl of 2YT/carb/KO7 medium in a 96-well plate and grow overnight at 37°C with shaking at 200 rpm.
2. Centrifuge at 4000 rpm for 5 min and transfer 100 μl of phage supernatant to a fresh plate.
3. Sterilize the phage supernatant by incubation at 60°C for 1 h.
4. Dilute phage supernatant 1:10 with distilled water and add 2 μl to the following PCR mix: 19.7 μl distilled water, 2.5 μl 10x PCR buffer, 0.25 μl 100 mM dNTPs, 0.25 μl of each PCR primer, and 0.5 units of Amplitaq DNA polymerase. The primers are designed to amplify the DNA fragment that is to be sequenced. The mix can be made up as a cocktail and dispensed into a 96-well PCR plate for high-throughput sequencing.
5. Amplify the DNA fragment with the following PCR program: 5 min at 95°C, 25 cycles of amplification (30 s at 94°C, 30 s at 55°C, 60 s at 72°C), 7 min at 72°C, and storage at 4°C.
6. Analyze representative reactions by electrophoreses of 3.0 μl on a TAE/agarose gel.
7. Dispense 2 μl of clean up mix into each well of a fresh 96-well PCR plate.
8. Transfer 5 μl of PCR product to each well and mix carefully.
9. Incubate the cleanup reactions at 37°C for 15 min, at 80°C for 15 min and store at 4°C.
10. The sample can be used directly as the template in Big-Dye terminator sequencing reactions (PE Biosystems, Foster City, CA).

8.3.3.5 Protein Purification

The phage display phagemid can be converted into an expression vector for Fab protein by inserting a stop codon into the DNA at the fusion point between the heavy chain and the phage coat protein. An amber (TAG) stop codon is most convenient, as the Fab can still be displayed on phage in an amber suppressor host, such as *E. coli* XL1-blue, or it can be expressed as free protein in a nonsuppressor strain. Essentially any convenient *E. coli* expression strain can be used, but we use the protease-deficient strain 34B8. The following protocol is designed for expression and purification of Fab protein from vectors under the control of the alkaline phosphatase promoter, which is activated by phosphate depletion of the medium.

1. Inoculate 25 ml of 2YT/carb medium with a single colony of *E. coli* 34B8 harboring the expression vector.
2. Incubate overnight at 37°C with shaking at 200 rpm.

3. Use 5 ml of the culture to inoculate 500 ml of complete C.R.A.P medium supplemented with 50 µg/ml of carbenicillin.
4. Incubate for 18–24 h at 30°C with shaking at 200 rpm.
5. Centrifuge at 5.5 krpm (5,000 g) and 4°C for 15 min in a Sorvall GS-3 rotor.
6. Discard the supernatant and store the pellet at –70°C.
7. Resuspend the pellet in 50 ml of resuspension buffer.
8. Homogenize the culture with a T-50 Ultra-Turrax Homgenizer at 11,000 rpm for 5 min on ice.
9. Microfluidize the sample twice with an M-110Y Microfluidizer processor (Microfluidics, Newton, MA) at 6000 psi.
10. Eliminate cell debris by centrifugation for 60 min at 15 krpm and 4°C in a Sorvall SS-34 rotor (27,000 g), followed by filtration through a 0.45-micron membrane.
11. Equilibrate protein A-Sepharose resin with PBS.
12. Load the sample twice onto the protein A-Sepharose resin.
13. Wash the resin with PBS until the absorbance at 280 nm is negligible.
14. Elute the Fab protein with 0.1 M acetic acid in a volume equal to half of the resin volume.
15. Neutralize with 1.0 M Tris base, pH 8.0.

REFERENCES

1. Simmons, L.C. and Yansura, D.G., Translational level is a critical factor for the secretion of heterologous proteins in Escherichia coli, *Nat. Biotechnol.*, 14, 629, 1996.
2. Presta, L., Sims, P., Meng, Y.G., Moran, P., Bullens, S., Bunting, S., Schoenfeld, J., Lowe, D., Lai, J., Rancatore, P., Iverson, M., Lim, A., Chisholm, V., Kelley, R.F., Riederer, M., and Kirchhofer, D., Generation of a humanized, high affinity anti-tissue factor antibody for use as a novel antithrombotic therapeutic, *Thromb. Haemost.* 85, 379, 2001.
3. Better, M., Chang, C.P., Robinson, R.R., and Horowitz, A.H., *Escherichia coli* secretion of an active chimeric antibody fragment, *Science*, 240, 1041, 1988.
4. Huston, J.S., Levinson, D., Mudgett-Hunter, M., Tai, M.S., Novotny, J., Margolies, M.N., Ridge, R.J., Bruccoleri, R.E., Haber, E., Crea, R., et al., Protein engineering of antibody binding sites: recovery of specific activity in an anti-digoxin single-chain Fv analogue produced in *Escherichia coli. Proc. Natl. Acad. Sci. U. S. A.* 85, 5879, 1988.
5. Skerra, A. and Pluckthun, A. Assembly of a functional immunoglobulin Fv fragment in *Escherichia coli, Science,* 240, 1038, 1988.
6. Hamers-Casterman, C., Atarhouch, T., Muyldermans, S., Robinson, G., Hamers, C., Songa, E.B., Bendahman, N., and Hamers, R., Naturally occurring antibodies devoid of light chains. *Nature* 363, 446, 1993.
7. Decanniere, K., Desmyter, A., Lauwereys, M., Ghahroudi, M.A., Muyldermans, S., and Wyns, L., A single-domain antibody fragment in complex with RNase A: non-canonical loop structures and nanomolar affinity using two CDR loops. *Structure* 7, 361, 1999.

8. Desmyter, A., Spinelli, S., Payan, F., Lauwereys, M., Wyns, L., Muyldermans, S., and Cambillau, C., Three camelid VHH domains in complex with porcine pancreatic alpha-amylase. Inhibition and versatility of binding topology, *J. Biol. Chem.* 277, 23645, 2002.

9. Simmons, L.C., Reilly, D., Klimowski, L., Raju, T.S., Meng, G., Sims, P., Hong, K., Shields, R.L., Damico, L.A., Rancatore, P., and Yansura, D.G.. Expression of full-length immunoglobulins in *Escherichia coli*: rapid and efficient production of agly-cosylated antibodies, *J. Immunol. Methods,* 263, 133, 2002.

10. Smith, G.P. Filamentous fusion phage: novel expression vectors that display cloned antigens on the virion surface. *Science*, 228, 1315, 1985.

11. Barbas, C.F. III and Burton, D.R., Selection and evolution of high-affinity human anti-viral antibodies, *Trends Biotechnol.,* 14, 230, 1996.

12. Barbas, C.F. III, Bain, J.D., Hoekstra, D.M., and Lerner, R.A., Semisynthetic combinatorial antibody libraries: a chemical solution to the diversity problem, *Proc. Natl. Acad. Sci. U. S. A.*, 89, 4457, 1992.

13. Marks, J.D., Hoogenboom, H.R., Bonnert, T.P., McCafferty, J., Griffiths, A.D., and Winter, G., By-passing immunization: human antibodies from V-gene libraries displayed on phage, *J. Mol. Biol.*, 222, 581, 1991.

14. Griffiths, A.D., Malmqvist, M., Marks, J.D., Bye, J.M., Embleton, M.J., McCafferty, J., Baier, M., Holliger, K.P., Gorick, B. D., Hughes-Jones, N.C., Hoogenboom, H.R., Winter, G., Human anti-self antibodies with high specificity from phage display libraries, *EMBO J.*, 12, 725, 1993.

15. Nissim, A., Hoogenboom, H.R., Tomlinson, I.M., Flynn, G., Midgley, C., Lane, D., and Winter, G., Antibody fragments from a 'single pot' phage display library as immunochemical reagents, *EMBO J.*, 13, 692, 1994.

16. Knappik, A., Ge, L., Honegger, A., Pack, P., Fischer, M., Wellnhofer, G., Hoess, A., Wolle, J., Pluckthun, A., and Virnekas, B., Fully synthetic human combinatorial antibody libraries (HuCAL) based on modular consensus frameworks and CDRs randomized with trinucleotides, *J. Mol. Biol.* 296, 57, 2000.

17. Fellouse, F.A., Wiesmann, C., and Sidhu, S.S., Synthetic antibodies from a four-amino-acid code: a dominant role for tyrosine in antigen recognition, *Proc. Natl. Acad. Sci. U. S. A.* 101, 12467, 2004.

18. Lee, C.V., Liang, W.-C., Dennis, M.S., Eigenbrot, C., Sidhu, S.S., and Fuh, G., High-affinity human antibodies from phage-displayed synthetic Fab libraries with a single framework scaffold, *J. Mol. Biol.* 340, 1073, 2004.

19. Jespers, L., Schon, O., James, L.C., Veprintsev, D., and Winter, G., Crystal structure of HEL4, a soluble refoldable human V_H single domain with a germ-line scaffold, *J. Mol. Biol.* 337, 893, 2004.

20. Frisch, C., Brocks, B., Ostendorp, R., Hoess, A., von Ruden, T., and Kretzschmar, T., From EST to IHC: human antibody pipeline for target research, *J. Immunol. Methods*, 75, 203, 2003.

21. Bond, C.J., Wiesmann, C., Marsters, J.C.J., and Sidhu, S.S., A structure-based database of antibody variable domain diversity, *J. Mol. Biol.* 348, 699, 2005.

22. Fellouse, F.A. and Sidhu, S.S. (2005). Synthetic antibody libraries, In Sidhu, S.S., ed., *Phage Display in Biotechnology and Drug Discovery*, vol. 3, CRC Press, New York, 2005.

23. Marvin, J.S. and Lowman, H.B. (2005). Antibody humanization and affinity maturation using phage display, In Sidhu, S.S., ed., *Phage Display in Biotechnology and Drug Discovery*, vol. 3, CRC Press, New York, 2005.

24. Berry, J.D. and Popkov, M., (2005). Antibody libraries from immunized repertoires, In Sidhu, S.S., ed., *Phage Display in Biotechnology and Drug Discovery*, vol. 3, CRC Press, New York, 2005.

25. Dobson, C.L., Minter, R.R., and Hart-Shorrock, C.P. (2005). Naive antibody libraries from natural repertoires, In Sidhu, S.S., ed., *Phage Display in Biotechnology and Drug Discovery*, vol. 3, CRC Press, New York, 2005.

26. Sidhu, S.S., Fairbrother, W.J., and Deshayes, K., Exploring protein-protein interactions with phage display, *ChemBioChem.* 4, 14, 2003.

27. Sidhu, S.S., Phage display in pharmaceutical biotechnology, *Curr. Opin. Biotechnol.* 11, 610, 2000.

28. Smith, G.P. and Petrenko, V.A., Phage display. *Chem. Rev.* 97, 391, 1997.

29. Hoogenboom, H.R., Griffiths, A.D., Johnson, K.S., Chiswell, D.J., Hudson, P., and Winter, G., Multi-subunit proteins on the surface of filamentous phage: methodologiesfor displaying antibody (Fab) heavy and light chains, *Nucl. Acids. Res.* 19, 4133, 1991.

30. McCafferty, J., Griffiths, A.D., Winter, G., and Chiswell, D.J., Phage antibodies: filamentous phage displaying antibody variable domains, *Nature*, 348, 552, 1990.

31. Knappik, A., Ge, L., Honegger, A., Pack, P., Fischer, M., Wellnhofer, G., Hoess, A., Wolle, J., Pluckthun, A., and Virnekas, B., Fully synthetic human combinatorial antibody libraries (HuCAL) based on modular consensus frameworks and CDRs randomized with trinucleotides, *J. Mol. Biol.* 296, 57, 2000.

32. Tanha, J., Dubuc, G., Hirama, T., Narang, S.A., and MacKenzie, C.R., Selection by phage display of llama conventional V(H) fragments with heavy chain antibody V(H)H properties, *J. Immunol. Methods*, 263, 97, 2002.

33. Burton, D.R., Barbas, C.F. 3rd, Persson, M.A., Koenig, S., Chanock, R.M., and Lerner, R.A., A large array of human monoclonal antibodies to type 1 human immunodeficiency virus from combinatorial libraries of asymptomatic seropositive individuals, *Proc. Natl. Acad. Sci. U. S. A.* 88, 10134, 1991.

34. Griffiths, A.D., Williams, S.C., Hartley, O., Tomlinson, I.M., Waterhouse, P., Crosby, W.L., Kontermann, R.E., Jones, P.T., Low, N.M., Allison, T.J., Prospero, T.D., Hoogenboom, H.R., Nissim, A., Cox, J.P.L., Harrison, J.L., Zaccolo, M., Gherardi, E., and Winter, G., Isolation of high affinity human antibodies directly from large synthetic repertoires, *EMBO J.* 13, 3245, 1994.

35. Clackson, T., Hoogenboom, H.R., Griffiths, A.D., and Winter, G., Making antibody fragments using phage display libraries, *Nature,* 352, 624, 1991.

36. Hoogenboom, H.R. and Winter, G., By-passing immunisation. Human antibodies from synthetic repertoires of germline VH gene segments rearranged *in vitro, J. Mol. Biol.* 227, 381, 1992.

37. Sidhu, S.S., Li, B., Chen, Y., Fellouse, F.A., Eigenbrot, C., and Fuh, G. Phage-displayed antibody libraries of synthetic heavy chain complementarity determining regions, *J. Mol. Biol.* 338, 299, 2004.

38. Sidhu, S.S., Lowman, H.B., Cunningham, B.C., and Wells, J.A. Phage display for selection of novel binding peptides, *Methods Enzymol.* 328, 333, 2000.

39. Kunkel, T.A., Roberts, J.D., and Zakour, R.A., Rapid and efficient site-specific mutagenesis without phenotypic selection, *Methods Enzymol.* 154, 367, 1987.

40. Fellouse, F.A., Wiesmann, C., and Sidhu, S.S., Synthetic antibodies from a four-amino-acid code: a dominant role for tyrosine in antigen recognition, *Proc. Natl. Acad. Sci. U. S. A.* 101, 12467, 2004.

41. Fellouse, F.A., Li, B., Compaan, D.M., Peden, A.A., Hymowitz, S.G., and Sidhu, S.S., Molecular recognition by a binary code, *J. Mol. Biol.,* 348, 1153, 2005.

42. Lechner, R.L., Engler, M.J., and Richardson, C.C., Characterization of strand displacement synthesis catalyzed by bacteriophage T7 DNA polymerase, *J. Biol. Chem.* 258, 11174, 1983.

43. Sidhu, S.S., Lowman, H.B., Cunningham, B.C., and Wells, J.A., Phage display for selection of novel binding peptides. *Methods Enzymol.* 328, 333, 2000.

44. Miller, J.H., *Experiments in Molecular Biology*, Cold Spring Harbor Laboratory Press, New York, 1972.

9 Chemical and Proteolytic Modification of Antibodies

George P. Smith

CONTENTS

9.1 INTRODUCTION

With the emergence of a vigorous biologic research supply industry in the last quarter century, researchers seldom need to modify their own antibodies. Most conspicuously, multitudinous catalogs are thick with lists of second antibodies labeled with fluorescent dyes, biotin, enzymes, and other coupled groups, allowing an unmodified primary antibody to be detected or quantified in a profusion of research contexts. Still, there are circumstances in which an antibody must be directly modified to be of use. For instance, if two primary antibodies from the same species but of differing specificity need to be distinguished in the same cytologic specimen, they must be directly modified with different labels. For example, one antibody could be labeled with a fluorescent dye of one color, whereas the other could be biotinylated, allowing it to be detected with streptavidin labeled with a fluorescent dye of another color. Recognizing this need, commercial suppliers now offer numerous reagents for facile modification of antibodies and other proteins.

This review aims to provide practical guidance to researchers needing to modify their antibodies. It will cover the most common chemical and proteolytic modifications of immunoglobulin (Ig)G antibodies that preserve their antigen-binding function. Harsher modifications that require or result in extensive denaturation or functional inactivation of the protein will not be considered. Radiolabeling of antibodies will also not be included. Throughout the chapter, detailed exemplar protocols for typical modification reactions will be provided, along with explanations of the underlying principles. There will be no attempt to cover the subject of modification exhaustively, however; for a comprehensive review of chemical and proteolytic modification techniques, including radiolabeling, the reader is referred to the book by Hermanson.[1]

The chemical modifications in this review all consist in covalent coupling of a group to antibody; the term *appendage* will be used generically for the group that is thus conjugated to the antibody, regardless of its nature. The focus of this review will not be on the nature of the appendage, but rather on the chemistry of coupling, because it is the latter that the researcher must keep primarily in mind to modify antibody successfully. Subsequent use of the resulting adducts, which depends

TABLE 9.1
Typical Appendages in Modified Antibodies

Appendage or Appendage Type	Typical Applications
Biotin	Affinity tag; binds extremely strongly to avidin and streptavidin
Fluorescent dyes	Label for immunolocalization by conventional fluorescence or scanning laser confocal microscopy; high-throughput fluorescence binding assays; other immunodetection applications
Solid support matrices	Preparation of affinity matrices for affinity purification or depletion of antigens
Enzymes or other proteins	Preparation of enzyme-labeled primary and second antibodies for immunoblots, enzyme-linked immunosorbent assay, and other immunodetection applications; construction of immunotoxins
Reactive groups (e.g., thiols, maleimides)	Activation of antibody for use as intermediate in coupling a protein or other group
Radioactive iodine and other tracer radioisotopes	Radiolabel for autoradiographic localization on immunoblots or microscopic specimens; detection and quantitation of antigens by gamma counting
Chelators	Target-specific radiotherapy or *in vivo* radioimaging with radioactive metal ions
Polyethylene glycol (or other hydrophilic polymers such as dextran)	Modification of antigenic or pharmacokinetic properties of antibodies *in vivo*, in natural fluids, or other circumstances

crucially on the nature of the appendage, lies outside the scope of this chapter. Still, it will be useful to summarize major types of appendage briefly in order to provide a more concrete setting for modification reactions, and because they will serve as illustrations in the many exemplar protocols in the chapter. Accordingly, Table 9.1 lists classes of appendages without regard to the chemistry by which they are coupled to proteins.

After describing some general procedures for handling antibodies (and other proteins), major classes of chemical and proteolytic modifications will be reviewed in turn.

9.2 GENERAL PROCEDURES

9.2.1 STABILITY OF ANTIBODY SOLUTIONS

IgGs are among the most durable of globular proteins, being generally soluble and stable at concentrations up to at least ~20 mg/ml in a wide range of buffers between about pH 5 and 9. Buffered salines are the most common storage buffers; they have NaCl at a concentration of about 0.15 M and a buffer appropriate for the pH at concentrations ranging from about 5 to about 100 mM. Many of the protocols in this chapter use either a weak phosphate-buffered saline (LoPBS; 0.15 M NaCl,

5 mM NaH_2PO_4, pH adjusted to 7.0 with NaOH), a strong phosphate-buffered saline (PBS; 0.15 M NaCl, 0.1 M NaH_2PO_4, pH adjusted to 7.2 with NaOH), or a strong Tris-buffered saline (TBS; 0.15 M NaCl, 50 mM Tris·HCl pH 7.5). An advantage of weak buffering is that the pH can be readily changed by adding a high concentration of another buffer at the desired pH. Antibody solutions can be stored for months or even years at 4°C, especially if they contain an antimicrobial preservative (generally 0.05% NaN_3; this poison is as toxic as cyanide though not volatile, and high concentration stocks [e.g., 5%] must be handled with great care). IgG solutions usually survive multiple freeze-thaw cycles with little damage, and therefore can be stored at –20°C or at –80°C. An increasingly popular storage method that ensures excellent survival with no damage from freeze-thaw cycles is to mix the antibody solution with an equal volume of ultrapure glycerol and store the mixture at –20°C, at which temperature 50% glycerol does not freeze. At high concentrations, glycerol can reduce the pH of the solution; it is therefore a good idea for the antibody solution to be strongly buffered at the desired pH before adding the glycerol.

Small amounts of antibody are lost by adsorption to the surface of the vessels that contain it; polystyrene and glass adsorb protein particularly well, but smaller losses occur on polypropylene and other surfaces. Nevertheless, when necessary, IgG solutions with concentrations as low as 1 µg/ml or less can be processed successfully in polypropylene labware. Whenever feasible, however, dilute antibody solutions (~50 µg/ml or less) should be protected by adding a purified noninterfering carrier protein such as bovine serum albumin (BSA) to a concentration of about 100 µg/ml or higher.

Most antibodies can be lyophilized for long-term storage with little loss of function. For this purpose, the solvent should first be changed to water or to a weak volatile buffer such as 10 mM ammonium bicarbonate or ammonium acetate using one of the buffer exchange methods later in this section. The solution is shell-frozen in a lyophilization flask and freeze-dried; to remove the last traces of volatile salt, the fluffy white powder should be redissolved in pure water and freeze-dried again. The final salt-free powder can be stored desiccated in the dark for decades and reconstituted as needed at any desired concentration in any desired buffer. A simpler alternative is to freeze-dry the antibody directly in a nonvolatile buffer, thus skipping the buffer exchange step; in this case, care should be taken that the vacuum remains sufficiently high to keep the solution frozen as the salt concentration rises. Antibodies that have been freeze dried in this manner should be reconstituted in the correct volume of water to result in the desired buffer concentration.

9.2.2 QUANTITATION OF ANTIBODIES

Table 9.2 lists the most common methods used to quantify IgG; the approximate minimum amount and concentration of IgG that can be accurately quantified for each method are shown in the second and third columns. None of these methods distinguishes IgG from other proteins; they can only be applied to solutions in which the IgG is the dominant protein present. The three methods have different compatibilities, advantages, and disadvantages, as will be summarized as each is discussed in turn.

TABLE 9.2
Summary of Methods for Quantifying Antibodies

Method	Minimum Amount	Minimum Concentration	Major Interfering Solutes	Comments
Spectrophotometry	2 µg[1]	20 µg/ml	Many ultraviolet-absorbing contaminants	A 1 mg/ml solution gives A_{278} of 1.4
Bradford assay	1.5 µg	60 µg/ml[2]	Detergents	Include immunoglobulin G standards from the same species in same buffer
Bicinchoninic acid (BCA) assay	1.3 µg	10 µg/ml	Reducing agents; many other organic compounds, including Tris	Include immunoglobulin G Standards in same buffer; compatible with phosphate buffers

[1]Assumes an ultramicrocuvette with a fill volume of 100 µl.
[2]Can be reduced to about 10 µg/ml by using a homemade concentrated dye solution.

9.2.2.1 Scanning Spectrophotometry

Figure 9.1 shows the ultraviolet (UV) absorption spectrum of an IgG solution in PBS; the instrument was blanked against the same buffer as a reference solution. Essentially, the same spectrum would have been obtained in any near neutral buffer. As for most large proteins containing both tryptophan and tyrosine (the main UV chromophores), peak absorption occurs at ~278 nm, though by tradition protein absorption is reported at 280 nm. A solution of purified, nonaggregated IgG such as the one shown has little absorption at 320 nm, but if there is a significant amount of aggregated or insoluble material there can be significant light scattering at that wavelength. Because light scattering is only relatively weakly dependent on wavelength (inversely with the fourth power for Rayleigh scattering by very small aggregates), the optical density at 320 nm, where essentially no absorption occurs, can be taken as a crude estimate of the light scattering at 278 nm; the difference between the two can therefore be taken as an estimate of the true absorbance at 278 nm. In the example in Figure 9.1, for example, the true absorbance at 278 nm, corrected for light scattering, would be $0.371 - 0.002 = 0.369$ (the light-scattering correction is often much greater than in this case). Mandy and Nisonoff[2] determined the absorbance of a 1 mg/ml solution of rabbit IgG to be 1.4, and this factor (or something close to it) is usually used to convert absorption to concentration for IgGs of any origin. For the sample in Figure 9.1, for example, the IgG concentration would be calculated as $0.369/1.4 = 0.264$ mg/ml, which is equivalent to 1.76 µM, assuming a molecular weight of 150 kDa. UV-absorbing contaminants frequently distort the

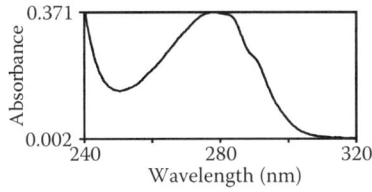

FIGURE 9.1 Typical immunoglobin G ultraviolet absorption spectrum.

spectrum at shorter wavelengths; as long as a slight hump or shoulder at around 278 nm is discernible and the shape of the descending slope as wavelength increases beyond that value is more or less as shown, absorbance at 278 nm can be taken as a reasonably reliable guide to protein concentration. If the IgG solution can be confidently assumed to be sufficiently free of interfering UV-absorbing contaminants and light-scattering aggregates, a single absorbance measurement at 275–280 nm can be used instead of a wavelength scan.

9.2.2.2 Bradford Assay

This color assay exploits the shift in absorption maximum of the dye Coomassie Brilliant Blue G-250 from 470 nm in acidic solution to 595 nm when bound to a protein in that acidic solution. In a typical assay, 250-μl portions of a commercial stabilized acidic dye solution (e.g., Coomassie protein assay reagent from Pierce Chemical Co., Rockford, IL) are mixed with 5–25 μl of the sample solutions with unknown IgG concentrations and of standards (preferably in the same buffer) with known amounts of IgG ranging from 0.625 to 7.5 μg. After an equilibration period of at least 10–15 min to allow dye to bind protein, the absorbance at 595 nm is determined in a spectrophotometer or plate reader; in the former case it is a good idea to use disposable plastic cuvettes as the dye stains glass and quartz surfaces. Figure 9.2 shows an example of a Bradford assay with quadruplicate IgG standards in PBS. The data have been fitted by least squares to an empirical curvilinear equation, which captures the information from the standards in a compact, mathematically useful form. That equation could then be inverted to calculate the IgG concentrations in unknown samples analyzed in parallel from their observed optical densities.

In general the Bradford assay is compatible with a wide range of solutes and contaminants, including almost all commonly used buffer components. However, it is very sensitive to detergents. Different protein standards—even IgGs from different species—can yield very different standard curves, depending on amino acid composition and other traits. For example, rabbit IgG yields about half the absorbance as human or bovine IgGs at comparable concentrations.

9.2.2.3 Bicinchoninic Acid Assay

This color assay measures reduction of Cu^{2+} to Cu^+ by protein amide bonds and some side chains; the Cu^+ combines with bicinchoninic acid (BCA) to form a

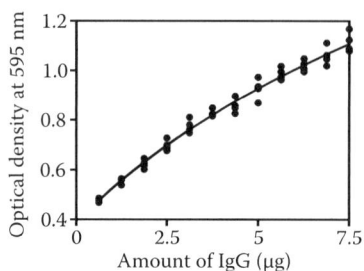

FIGURE 9.2 Typical dependence of optical density on amount of immunoglobin G (IgG) in the Bradford assay. The observed data (filled circles) were fitted to a theoretical curve (solid line) of the form $y = b + \dfrac{m}{1 + (k/c)^n}$, where y is the optical density, c is the amount of IgG, and b, m, k, and n are parameters whose values are adjusted to optimize the fit to the data.

chelate that absorbs light strongly at 562 nm. In a typical assay, 128-µl portions of samples and IgG standards in a compatible buffer like PBS are dispensed into a 96-well enzyme-linked immunosorbent assay (ELISA) dish or microtubes. Meanwhile, a suitable volume of BCA working solution from a commercial kit is mixed in a separate vessel. For example, for a 96-well ELISA dish, and using the Micro BCA protein quantitation assay kit from Pierce, mix 7.5 ml reagent MA, 7.2 ml reagent MB, and 300 µl reagent MC. A 128-µl portion of the working solution is then added to each standard and sample, and the dish or tubes are incubated 1 h at 37°C to allow full development of the color. After cooling to room temperature, the optical density at 562 nm is read in a plate reader or spectrophotometer; in the latter case, it is a good idea to use disposable plastic cuvettes as glass or quartz surfaces become coated with dye. Results for a series of IgG standards are shown in Figure 9.3; the optical density increases linearly with IgG concentration over the range investigated.

The BCA assay is generally much less sensitive to detergents than the Bradford assay, but is incompatible with many buffers, including Tris at very low concentrations. Thiols such as mercaptoethanol (Section 9.4.2) interfere, but Pierce offers a kit that neutralizes thiols by alkylation with iodoacetamide before adding the working solution. Because of the sensitivity of the assay to many solutes, it is highly advisable to dissolve or dilute the IgG standards in exactly the same buffer as the unknown IgG samples.

Because the chief protein reactant in the BCA assay is the amide bond, which is common to all polypeptides, there is much less protein-to-protein variation in standard curves than in the case of the Bradford assay or spectrophotometry. The BCA assay is therefore particularly suited to estimating the overall concentration of protein in complex samples such as natural fluids or tissue extracts containing hundreds or thousands of different protein species.

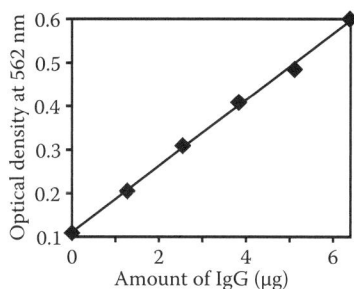

FIGURE 9.3 Dependence of optical density on amount of immunoglobin G in the bicincho-ninic acid assay. The solid straight line is the best linear fit to the data (filled diamonds).

9.2.3 CONCENTRATION AND BUFFER EXCHANGE

A wide selection of centrifugal concentration devices is commercially available. The antibody solution at a relatively low concentration is forced centrifugally through an ultrafiltration membrane that allows small solutes to pass but retains antibodies and other proteins beyond a certain size (e.g., 30 kDa). The protein thus remains in a so-called "retentate" solution of diminishing volume, whereas small solutes pass with the water through the membrane into a filtrate chamber. Table 9.3 lists a few such devices, giving the capacity (the maximum volume of the starting solution) and approximate deadstop (volume of remaining retentate after maximal concentration) for each; these volumes usually differ depending on whether the device is centrifuged in a swinging bucket (90°) or fixed angle rotor. The achievable concentration factor is the capacity divided by the deadstop. In some cases, a starting antibody solution fully concentrated to this maximum factor is a viscous gel that is very difficult to dilute fully to a homogenous solution. It is thus a good idea to limit the concentration factor so that the maximum antibody concentration never exceeds about 20 mg/ml. Unfortunately, this requires tediously tending to the progress of concentration by repeatedly stopping and restarting the centrifugation. To avoid this inconvenience, a few of the available devices have modest maximal concentration factors, allowing the user to let concentration go to completion without overconcentrating the antibody. Although the iCON 7- and 20-ml devices from Pierce have among the highest available concentration factors, their deadstop can be increased to any desired volume by adding buffer to the filtrate chamber, so that the reverse hydrostatic pressure from the rising level of filtrate eventually equals the forward hydrostatic pressure from the falling level of retentate, bringing filtration to a halt. Pierce's instructions give examples of the required total volumes (starting volume plus added filtrate) to reach various final deadstops at various angles of rotation.

An antibody dissolved in one buffer can be dialyzed against a second buffer, using either traditional dialysis tubing or a dialysis cassette (e.g., syringe-injected 0.5-, 3-, 12-, or 20-ml Slide-A-Lyzers from Pierce). To minimize relative losses from hold-up volumes (e.g., in the syringe, needle or cassette itself in the case of Slide-A-Lyzers), it is advantageous to match the capacity to the volume to be dialyzed.

TABLE 9.3
Summary of Centrifugal Concentration Devices

Device	Supplier	Rotor Well Size	Angle	Capacity	Deadstop
Centricon Plus-80	Millipore	250 ml	90°	80 ml	300 μl
Jumbosep	Pall	250 ml	90°	60 ml	3.5–4 ml
Centricon Plus-20	Millipore	50 ml	90°	20 ml	200 μl
iCON 20 ml	Pierce	50 ml	90°	20 ml	20 μl
			23–45°	13.5 ml	65 μl
Macrosep	Pall	50 ml	90°	15 ml	0.5–1 ml
			45°	15 ml	1–1.5 ml
			34°	12.5 ml	1–1.5 ml
			23°	9 ml	1.5–2 ml
Centriprep	Millipore	250 ml	90°	15 ml	3 ml
			23°	12 ml	2.5 ml
Ultra-15	Millipore	15 ml	90°	15 ml	50 μl
			35°	12 ml	50 μl
Centriplus	Millipore	50 ml	90°	15 ml	500 μl
			25°	10 ml	500 μl
iCON 7-ml	Pierce	15-ml	90°	7 ml	10 μl
			23–45°	4.5 ml	10 μl
Ultra-4	Millipore	15 ml	45–90°	4 ml	20 μl
			23°	3.5 ml	20 μl
Microsep	Pall	15 ml	45°	3.5 ml	30–40
			34°	3.5 ml	40–50 μl
Centricon	Millipore	15 ml	23–45°	2 ml	40 μl
Microcon	Millipore	Microcentrifuge	Fixed	500 μl	5–15 μl
Nanosep	Pall	Microcentrifuge	Fixed	500 μl	15 μl

For small volumes, it is often more convenient to exchange buffer by multiple cycles of concentration and redilution on a centrifugal concentration device of appropriate capacity and deadstop.

In both dialysis tubing or cassettes and centrifugal concentration devices, the choice of membrane is important. It should have a molecular weight cutoff (the nominal molecular weight of the largest molecule that can pass through the membrane's pores) well below the molecular weight of the protein to be concentrated (150 kDa for an intact IgG, 50 kDa for an Fab fragment), a low capacity for nonspecifically adsorbing protein, and good resistance to all solutes (including trace contaminants) in the antibody solution. Regenerated cellulose, the most common membrane material, has these desirable properties.

When buffer exchange must be accomplished rapidly—for example, because the IgG adduct is very unstable—gel filtration on a desalting column is preferable to dialysis and centrifugal concentration devices because it can be completed in 1–15 min; the IgG is diluted by a factor of 2 or more in the process. Many suppliers offer prepacked gravity-flow and centrifugal ("spin") columns with a wide range of capacities.

9.3 MODIFICATION OF AMINES

9.3.1 GENERAL CONSIDERATIONS

A typical IgG molecule in its native state has about 30 surface-exposed α- and ε-amino groups (Hermanson[1]); in their unprotonated (nonionized) form, they are strong nucleophiles that can be efficiently modified by a number of reagents under mild, nondenaturing conditions. An ε-amino group in a small, unstructured model compound typically has a pKa of about 9.5, and therefore should be only about 1% unprotonated at pH 7.5. Nevertheless, many exposed ε-amino groups in IgG and other native proteins react readily with amine-reactive reagents at near neutral pH. Free (non–disulfide-bonded) thiols of cysteines and the phenolic hydroxyls of tyrosines in their unprotonated (negatively charged) forms can also react with amine-reactive reagents, and the resulting adducts are sometimes sufficiently stable to be operationally indistinguishable from amine adducts. Other hydroxyls are generally much less nucleophilic than thiols and amines, however, and in IgG free thiols are very rare (Figure 9.14). In any case, groups that can be stably modified by this class of reagents will be referred to operationally as "amines" in this chapter, even if an occasional thiol or phenolic hydroxyl is thereby included.

Although a few modifiable amines lie in or near antigen-binding sites, many more lie elsewhere on the surface of the molecule. Random modification of a few of these groups with small molecular appendages is therefore very unlikely to functionally inactivate more than a small fraction of the IgGs. For these reasons, amine modification has become the most commonly used method of coupling biotin, fluorescent dyes, and other appendages to antibodies, and for coupling antibodies to solid-phase affinity matrices. Controlling the level of modification is key to success in this technology: insufficient modification can fail to confer the desired properties on the IgG, whereas overmodification can substantially impair antigen-binding function. For this reason, a basic understanding of modification kinetics is necessary for designing useful reaction protocols.

The upper part of Figure 9.4 shows a generic amine modification reaction, in which nucleophilic attack of a protein amino group on an electrophilic center in the reagent results in coupling the amino group to the coupled group R, which consists of all or part of the reagent. Because the milieu is aqueous, this modification reaction inevitably competes to a greater or lesser degree with hydrolytic inactivation of the reagent, as depicted in the lower part of Figure 9.4; all amine-reactive reagents must therefore be kept dry during long-term storage, usually in a desiccator. It is obviously desirable that the reaction milieu should be as free as practicable of competing nucleophiles other than water; this includes thiols such as 2-mercaptoethanol (2-ME) or dithiothreitol (DTT), primary amines such as Tris base, and secondary amines such as Bis-Tris-propane. The selectivity of a reagent for amines versus water and other nucleophiles is governed not only by the electrophilicity of the electrophilic center but also by other factors such as steric hindrance by the surrounding atoms or availability of those atoms for participation in coordinated bond rearrangements.

Hydrolysis can be divided into two components: a protein-independent component whose rate is independent of IgG concentration, and a protein-

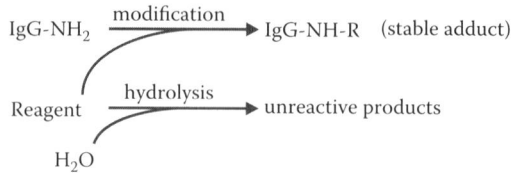

FIGURE 9.4 Generic amine modification reaction.

dependent component whose rate is proportional to IgG concentration. Protein-dependent hydrolysis might include, for example, futile modifications (e.g., of histidine), leading to adducts that are immediately hydrolyzed to restore the original protein group. The rates of both types of hydrolysis depend on temperature, pH, and buffer composition.

In light of the foregoing considerations, and with the aid of some simplifying assumptions, the author developed a mathematical model that describes with reasonable accuracy the degree of modification achieved at completion of the reaction, when all initial reagent has been consumed by hydrolysis and modification:[3]

$$R = \left[m - \frac{Q}{k_m} \ln\left(1 - \frac{m}{M} \right) \right] P - \frac{k_h}{k_m} \ln\left(1 - \frac{m}{M} \right), \qquad (9.1)$$

where

P = IgG concentration (μM)
M = total number of modifiable amines per IgG molecule (dimensionless)
m = modification level (number of modified amines per IgG) (dimensionless)
R = initial concentration of reagent required to achieve a final modification level of m (μM)
k_m = second-order modification rate constant (μM^{-1} s^{-1})
k_h = first-order protein-independent hydrolysis rate constant (s^{-1})
Q = second-order protein-dependent hydrolysis rate constant (μM^{-1} s^{-1})

The equation includes three kinetic parameters—M, k_h/k_m, and Q/k_m—whose values are not specified a priori by the model itself, but must instead be determined empirically for each reagent and set of reaction conditions (pH, buffer composition and temperature). Their values are estimated by measuring the modification level m achieved over a range of reagent concentrations R and IgG concentrations P. After parameter values have been thus established, Eq. 9.1 is a great practical convenience: it tells an experimenter what reagent concentration R will be required to achieve any desired modification level m when the concentration of the IgG to be modified is P.

Quantitative use of Eq. 9.1 will be illustrated in practice below (section 9.3.3). Meanwhile, it is important to emphasize a qualitative principle that is key to practical success. According to the equation, the reagent concentration R required to achieve a given fixed modification level m increases linearly with the IgG concentration P, with slope $\left[m - \dfrac{Q}{k_m} \ln\left(1 - \dfrac{m}{M}\right) \right]$ and intercept $-\dfrac{k_h}{k_m} \ln\left(1 - \dfrac{m}{M}\right)$ (both expressions are always positive, since the logarithm is negative). Consider two extreme cases. On the one hand, if the intercept term is negligible—either because the protein concentration P is high or because the protein-independent hydrolysis rate is low compared to the modification rate (i.e., the ratio k_h/k_m is relatively low)—the reagent concentration R required to achieve a given desired modification level m is directly proportional to the IgG concentration P. In such circumstances, m is not sensitive to R and P separately, but only to their ratio R/P. On the other hand, if the intercept term dominates (P relatively low and/or k_h/k_m relatively high), the reagent concentration required to achieve a given desired modification level m is independent of protein concentration. In that situation, it is not necessary to know the IgG concentration P accurately to modify the IgG to a specified modification level. Between these two extremes, the modification level m depends on both the reagent concentration R and the IgG concentration P individually—not just on their ratio and not just on R alone. Unfortunately, the literature and suppliers' catalogs abound in misleading descriptions and recommendations that encourage the inexperienced user to focus on the R/P ratio even when that ratio is a patently incomplete specification of reaction circumstances.

When only a relatively small fraction of the available amines are modified—when, say, $m/M < 40\%$—the logarithmic terms in Eq. 9.1 approach the value $-m/M$ and Eq. 9.1 can be simplified and rearranged to the approximate form

$$m \approx \frac{1}{1 + \dfrac{Q}{Mk_m} + \dfrac{k_h}{Mk_m P}} \times \frac{R}{P} \qquad (9.2)$$

In these circumstances, therefore, which obtain in almost all amine modification reactions to be discussed in this chapter, the modification level achieved m upon completion of the reaction at a given fixed protein concentration P will be nearly proportional to the ratio of reagent concentration to protein concentration R/P, with the proportionality constant given by the first fraction on the right-hand side of the equation. In those same circumstances, the number of kinetic parameters governing reaction kinetics is effectively reduced from three to two: Q/Mk_m and k_h/Mk_m. When the full three-parameter model Eq. 9.1 is fitted to data to determine optimal kinetic parameters, therefore, the fit will not be very sensitive to the values of $M, Q/k_m$, and k_h/k_m individually, but rather only to the quotients Q/Mk_m and k_h/Mk_m. By the same token, optimal values of M, Q/k_m, and k_h/k_m determined in this manner cannot be taken uncritically as estimates of the physical parameters they nominally gauge.

9.3.2 CLASSES OF AMINE-REACTIVE REAGENTS

Figure 9.5 catalogs classes of commercially available reactive groups that are commonly used to modify IgG amines. In each case, the R groups that are coupled to the IgG—one of the appendages listed in Table 9.1—are highly varied and left unspecified in Figure 9.5.

The most common class of amine-reactive compounds is acylating reagents, in which a carbonyl is activated by linkage to a leaving group. Preeminent among these are the N-hydroxysuccinimide (NHS) esters, but the figure also includes a new family of reagents with a more activating pentafluorophenol-leaving group and a family with long standing in protein chemistry, the acid anhydrides. The anhydride depicted has identical R groups, but asymmetric anhydrides are also sometimes used; in cyclic anhydrides a single molecular structure connects the two carbonyls (e.g., maleic or succinic anhydride). The detailed chemistry of acylation by these reagents differs depending on the nature of the leaving group, but the product of amine modification

Reagent		Modification	
Acylating reagents	N-hydroxy-succinimide (NHS) ester		Amide (very stable)
	Penta-fluoro-phenyl (PFP) ester		Thioester (sometimes stable)
			Ester (some-times stable)
	Acid anhydride		
Isothiocyanate			Thiourea (very stable)
Sulfonyl chloride			Sulfonamide (very stable)
Aldehyde (e.g., periodate-oxidized carbohydrate)			Schiff base (equilibrates with aldehyde)
			Secondary amine (very stable)

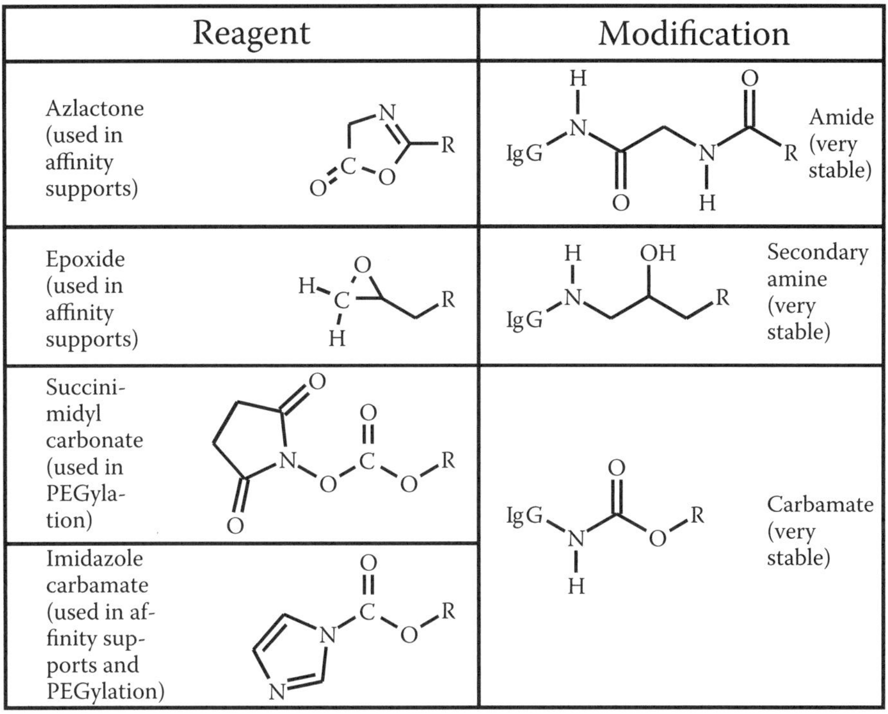

Reagent	Modification
Azlactone (used in affinity supports)	Amide (very stable)
Epoxide (used in affinity supports)	Secondary amine (very stable)
Succinimidyl carbonate (used in PEGylation)	Carbamate (very stable)
Imidazole carbamate (used in affinity supports and PEGylation)	

FIGURE 9.5 Major classes of amine-reactive modifying reagents.

is the same in each case: a very stable amide linkage between R and the protein. Acylation of thiols (if present—free thiols are rare in IgGs) results in thioesters, which can be stable in the absence of nucleophiles—either in solution or nearby on the protein itself. Acylation of tyrosine hydroxyls is generally very slow compared with acylation of amines, but if formed, the resulting phenolic esters can also be quite stable. Acylation of imidazole nitrogens of histidine leads to unstable adducts that are rapidly hydrolyzed to regenerate unmodified histidine; this futile modification cycle is an example of protein-dependent hydrolysis, accelerating reagent consumption without resulting in stable modification.

Aldehydes react with amines reversibly to make a Schiff base, which is largely dissociated at pHs below about 5; stabilization of the adduct requires subsequent treatment with the mild reducing agent sodium cyanoborohydride, whose toxicity is a drawback of this mode of coupling. Although an aldehyde can theoretically react with water, the resulting dihydroxyl would immediately lose water to regenerate the aldehyde; in the absence of reducing or oxidizing agents, therefore, aldehydes can be considered stable.

Isothiocyanates, epoxides, and imidazole carbamates react relatively slowly with protein amines, but also are hydrolyzed only slowly in water. Solid-phase affinity matrices with the latter two groups can be hydrated or washed with aqueous buffers before being added to a solution of the protein to be coupled.

9.3.3 BIOTINYLATION WITH NHS-PEO$_4$-BIOTIN

NHS-PEO$_4$-biotin (Pierce), a typical NHS ester, is shown in Figure 9.6. The hydrophilic four-unit polyethylene oxide (PEO, also known as polyethylene glycol or PEG) linker makes the compound water soluble and increases the hydrophilicity of the adduct. This reagent will be used to illustrate modification of IgG with NHS esters.

For this reagent, optimal values of the kinetic parameters of Eq. 9.1 are available for a particular set of reaction conditions—overnight at room temperature in PBS (0.15 M NaCl, 0.1 M NaH$_2$PO$_4$, pH adjusted to 7.2 with NaOH)—allowing the reagent concentration to be calculated for any desired biotinylation level for any IgG concentration. In Figure 9.7, these parameter values have been substituted into

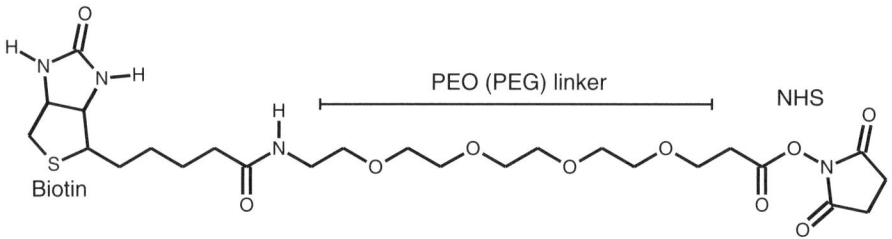

FIGURE 9.6 NHS-PEO$_4$-Biotin, an amine-reactive biotinylating reagent.

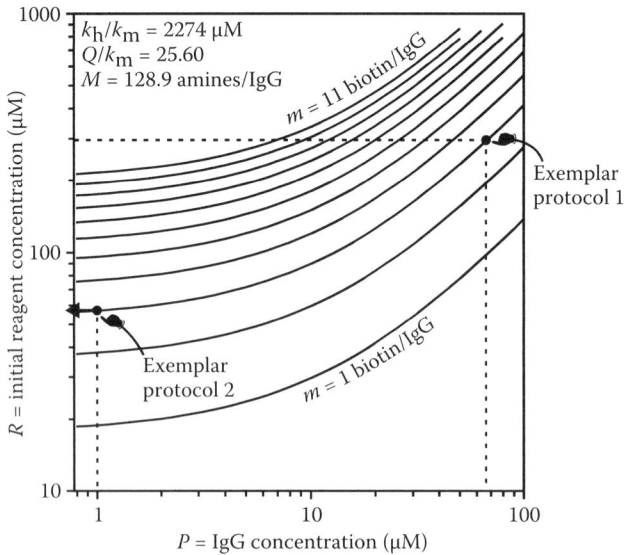

FIGURE 9.7 Iso-modification contours for biotinylation of immunoglobin G (IgG) with NHS-PEO$_4$-biotin. The graph gives the concentration of reagent R required to biotinylate IgG at a given concentration P to a given prespecified modification level m. The contours were calculated from Eq. 9.1 using the optimized parameter values given in the graph.

Eq. 9.1 for modification levels m from 1 to 11 to plot iso-modification contour curves. As explained under Eq. 9.2, the optimal value for the parameter M, 129 amines per IgG, cannot be considered an estimate of the actual number of accessible amines; indeed, the total number of amines in an IgG, whether accessible or not, is only about 90. We will illustrate biotinylation with this reagent to a level of three biotins per IgG molecule (a suitable modification level for almost all applications) under two contrasting conditions: a large amount of IgG at high concentration (exemplar protocol 1; upper right-hand dot in the graph), and a small but unknown amount of IgG at low concentration (exemplar protocol 2; lower left-hand dot).

9.3.3.1 Biotinylation of a Large Amount of IgG at High Concentration

In exemplar protocol 1, 1 mg (6.7 nmol) of normal human IgG in 90 µl PBS (0.15 M NaCl, 0.1 M NaH_2PO_4, pH adjusted to 7.2 with NaOH) is biotinylated to a level of three biotins per molecule in a total reaction volume of 100 µl. If the protein were in some other buffer, it would have to be equilibrated with PBS by dialysis or some other method (see Section 9.2.3). The classification of 1 mg as a "large amount" is somewhat arbitrary, but in this context two key considerations are that this amount is enough to be measured by standard physico-chemical methods, and to manipulate freely without carrier protein to prevent loss by adsorption to vessel surfaces. Because the final IgG concentration P will be 10 mg/ml (67 µM; right-hand vertical dashed line in Figure 9.7), we calculate from Eq. 9.1 that the concentration of reagent R required to achieve a modification level of $m = 3$ is

$$R = \left[m - \frac{Q}{k_m} \ln\left(1 - \frac{m}{M}\right) \right] P - \frac{k_h}{k_m} \ln\left(1 - \frac{m}{M}\right)$$

$$= \left[3 - 25.6 \times \ln\left(1 - \frac{3}{128.9}\right) \right] \times 67 - 2274 \times \ln\left(1 - \frac{3}{128.9}\right)$$

$$= 295 \text{ µM (horizontal dashed line in Figure 9.7)},$$

where the values of the kinetic parameters are those given in Figure 9.7. Accordingly, a 2.95-mM dilution of the reagent—i.e., 10 times the final concentration—will be prepared and 10 µl will be added to the 90-µl protein solution.

9.3.3.1.1 Exemplar Protocol 1: Biotinylation of a Large Amount of IgG at High Concentration

1. Into a 1.5 ml microtube pipette 90 µl of IgG at a concentration of 11.3 mg/ml in PBS.
2. Pipette 100 µl dimethylsulfoxide (DMSO) into a Pierce No-Weigh tube containing 2 mg (3.4 µmol) NHS-PEO$_4$-biotin, using the pipette tip to break through the foil seal; dissolve the oily solid completely by pipetting

up and down and scraping with the tip. (Although the reagent is soluble in water, it is difficult to dissolve the oily solid in water directly, and significant hydrolysis of the reagent can occur in the process; in contrast, the reagent is readily dissolved and stable in dry DMSO, and the tiny concentration of the solvent in the ultimate reaction mixture is completely harmless to IgG.)

3. Pipette the entire 100-μl DMSO solution into 1.05 ml PBS in a 1.5-ml microtube, thus making a 2.95-mM dilution; immediately vortex, pipette 10 μl into the microtube with the IgG, vortex that microtube, microfuge it briefly to drive the solution to the bottom, and allow it to react at least 4 h away from light. The molar ratio of reagent to IgG is 4.4. (The kinetic parameters used to calculate the reagent concentration refer to room temperature incubations; the parameters for incubation at 4°C, say, are unknown, but in general might be somewhat different. A 4-h incubation is undoubtedly more than sufficient for the reaction to reach completion—i.e., for hydrolysis and modification together to have consumed all starting reagent.)

4. Into the reaction microtube pipette 1 ml TBS (0.15 M NaCl, 50 mM Tris·HCl pH 7.5), vortex, and transfer to a Centricon centrifugal concentration device (Millipore) with a 30-kDa molecular weight cutoff; add an additional 700 μl TBS; remove uncoupled biotin by four cycles of concentration to ~50 μl according to the supplier's instructions and redilution to 1.8 ml; concentrate to ~50 μl one more time; add 400 μl TBS to the final retentate, vortex the Centricon device to ensure all the protein is swept off the membrane, and collect the diluted retentate by back-centrifugation according to the supplier's instructions; transfer the retentate into an appropriate vessel for storage. The Centricon device is suitable for up to ~1 mg IgG, corresponding to a maximum concentration of ~20 mg/ml in the ~50-μl deadstop (fully concentrated retentate). For larger amounts of protein, uncoupled biotin can be removed by dialysis (see Section 9.2.3).

5. The theoretical IgG concentration in the final ~450-μl solution from this protocol would be ~2.2 mg/ml, assuming no losses. The concentration can be measured empirically spectrophotometrically or by the Bradford reaction, but not by the BCA assay because of the high concentration of Tris (see Section 9.2.2). A typical yield is ~80% of the starting IgG at the 1-mg scale, but can approach 100% at larger scales.

This protocol can be scaled down at least to 500 μg and scaled up indefinitely, using suitable dialysis and/or concentration devices to remove uncoupled biotin and exchange buffer as needed. IgG concentrations from about 150 μg/ml (roughly the upper limit for exemplar protocol 2), all the way to the solubility limit can be accommodated by adjusting the reagent concentration according to Eq. 9.1.

9.3.3.2 Measuring Biotinylation Levels

Unfortunately, accurate measurement of biotinylation level is difficult. Standard biotin assays rely on competition between the analyte (the biotin to be quantified) and another ligand for binding to avidin or streptavidin. In the 4′-hydroxyazobenzene-2-carboxylic acid (HABA) assay, for example, analyte displaces HABA dye from streptavidin or avidin, thereby decreasing its molar absorbance by about 23,000 at 500 nm[4] (see also Hermanson[1]). This assay is quite insensitive, requiring sufficient volume of ~200-µg/ml Bio-IgG to fill a cuvette or microplate well. Moreover, protein-coupled biotins, unlike free biotins, are unable to block all four biotin-binding sites on a tetrameric avidin or streptavidin molecule as a result of severe steric hindrance; steric hindrance therefore leads to great uncertainty in the quantitative interpretation of the results. Competition ELISA, as exemplified in exemplar protocol 3 below, is much more sensitive than the HABA assay, but still suffers from uncertainties stemming not only from steric hindrance but also from multivalency. These uncertainties can be bypassed by releasing biotins from the protein by acid hydrolysis[5] or by digestion with pronase[1] or proteinase K[3,6] before assay. For a well-characterized biotinylating agent like NHS-PEO$_4$-biotin, however, such quantitation should be unnecessary: the biotinylation level actually achieved should not be far off what is predicted in Figure 9.7, or by Eq. 9.1 for appropriate values of the kinetic parameters.

9.3.3.3 Biotinylation of a Small Unknown Amount of IgG at Low Concentration

In exemplar protocol 2, it is assumed that the concentration of IgG is unknown but no higher than ~1 µM (150 µg/ml) in the final reaction mixture. As can be seen in Figure 9.7, at such low IgG concentrations, the concentration of reagent required to achieve a given biotinylation level m (e.g., three biotins per IgG molecule; dot with the leftward-pointing arrow in the graph) is essentially independent of the IgG concentration. Using Eq. 9.1 with the protein concentration P set to 0, we can calculate the reagent concentration required to achieve a biotinylation level $m = 3$ with very low IgG concentrations:

$$R = -\frac{k_h}{k_m}\ln\left(1 - \frac{m}{M}\right) = -2274 \times \ln\left(1 - \frac{3}{128.9}\right) = 54 \ \mu M.$$

There is no sharply demarcated lower limit on the IgG concentration, but at exceedingly low concentrations (less than ~100 ng/ml) loss of IgG from adsorption to vessel walls might be substantial. After biotinylation is complete, however, a large amount of BSA carrier protein will be added to the Bio-IgG, thus eliminating any further losses from adsorption. The final reaction volume in the exemplar protocol will be 10 µl, consisting of 5 µl IgG and 5 µl 108-µM reagent (giving the desired final reagent concentration of 54 µM); in those circumstances, the upper limit on the amount of IgG would be ~1.5 µg in 5 µl PBS. If the concentration of Bio-IgG

product is to be quantified by ELISA titering as in exemplar protocol 3, some of
the Bio-IgG will have to be sacrificed, ~50 ng being required for an accurate titer.
Thus, for example, if 33% of the final product is sacrificed for titering, the total
yield of Bio-IgG could be as little as 150 ng and still be quantified with reasonable
accuracy.

9.3.3.3.1   Exemplar Protocol 2: Biotinylation of a Small Amount of IgG at Low Concentration

1. Into a 500-µl microtube pipette 5 µl of IgG in PBS. The IgG concentration need not be known, but should be no higher than ~300 µg/ml.
2. Pipette 100 µl DMSO into a Pierce No-Weigh tube containing 2 mg (3.4 µmol) NHS-PEO$_4$-biotin, using the pipette tip to break through the foil seal; dissolve the oily solid completely by pipetting up and down and scraping with the tip (see comment at step 2 of exemplar protocol 1).
3. Pipette 10 µl of the DMSO solution into 3.14 ml PBS in a 4-ml test tube, thus making a 108-µM dilution; immediately vortex, pipette 5 µl into the microtube with the IgG, vortex that microtube, microfuge it briefly to drive the solution to the bottom, and allow it to react at least 4 h at room temperature away from light.
4. Into the reaction microtube pipette 100 µl 1 M ethanolamine, pH adjusted to 9.0 with HCl; vortex, microfuge briefly to drive solution to the bottom, and incubate at room temperature for at least 1 h away from light. The high concentration of primary amine quenches (reacts rapidly with) any residual unhydrolyzed reagent, ensuring that the BSA to be added at the next step will not be biotinylated.
5. Add 400 µl 2.5-mg/ml BSA in TBS (0.15 M NaCl, 50 mM Tris·HCl pH 7.5), vortex, and transfer to a Centricon centrifugal concentration device with a 30-kDa molecular weight cutoff; add an additional 1.3 ml TBS; remove uncoupled biotin by four cycles of concentration to ~50 µl according to the supplier's instructions and redilution to 1.8 ml; concentrate to ~50 µl one more time; add 100 µl TBS to the final retentate, vortex the Centricon device to ensure all the protein is swept off the membrane, and collect the diluted retentate by back-centrifugation according to the supplier's instructions; transfer the retentate into a 500-µl microtube for storage. The BSA acts as a carrier protein, blocking any further loss of Bio-IgG by adsorption to vessel walls; it is added only after all residual unconsumed reagent is quenched in the previous step. The total amount of BSA added, 1 mg, is fully soluble in 50 µl—the approximate minimal volume achieved in each cycle of concentration on the Centricon device.
6. If desired, the concentration of Bio-IgG in the final preparation can be determined by ELISA titering as in the next subsection.

9.3.3.4   Estimating Bio-IgG Concentration by ELISA Titering

In this procedure, serial dilutions of Bio-IgG samples at unknown concentrations and
of a Bio-IgG standard at known concentration are reacted with streptavidin-coated

wells in an ELISA dish; the standard and sample Bio-IgGs must match with regard to species of origin, isotype distribution (if applicable) and number of biotins per molecule. The least dilute standard should have a Bio-IgG concentration around 100 nM; if the least dilute unknown sample turns out to have a Bio-IgG concentration less than ~3 nM, accuracy may be compromised. After unbound Bio-IgG is washed away, the Bio-IgG that remains bound is detected with an enzyme-linked second antibody. This assay requires a plate reader capable of reading ELISA wells at around 405 nm (a kinetic reader is preferable, but a single-read instrument is acceptable), and is facilitated if a plate washer is also available.

9.3.3.4.1 Exemplar Protocol 3: Estimating Bio-IgG Concentration by ELISA Titering

1. In a 50-ml tube, make 8.5 ml 10-µg/ml streptavidin (from a refrigerated 1-mg/ml stock in water) in 0.1 M $NaHCO_3$, pH adjusted to 8.5 with NaOH; pipette 80 µl portions into each well of a modified flat-bottom ELISA dish; cover the dish with a lid and allow the protein to coat the plastic surface of the wells overnight in the cold in a humid box (small sealed plastic box with a damp paper towel on the bottom to maintain humidity).

2. Mix 15 ml Sea Block blocking solution (Pierce; almost any other blocking solution will work; in this case, bovine protein was avoided in the blocking solution because the Bio-IgG to be quantified was from that species) and 15 ml water; pipette 285 µl into each well, filling the well to brimming. Allow the dish (not covered with a lid) to block for at least 2 h in the cold in a humid box.

3. Meanwhile, using red TTB (TBS containing 0.5% Tween 20, 1 mg/ml BSA, and 100 µM phenol red) as diluent, make serial dilutions (typically twofold) of the unknown Bio-IgG sample(s) and of a 120-nM Bio-IgG standard; 50 µl will be needed for each well at the next step, plus ~10 µl extra to compensate for pipetting errors. In this example, the samples and standards were polyclonal bovine IgGs biotinylated to the same level; the standard dilutions had volumes of 250 µl (more than enough for four 50-µl repetitions), the least dilute having a Bio-IgG concentration of 120 nM; the sample dilutions had volumes of 120 µl (more than enough for two 50-µl repetitions).

4. When the blocking step 2 is finished, wash the ELISA wells with TBS/Tween (TBS containing 0.5% Tween 20) by hand or with a plate washer and fill the wells rapidly (to avoid excessive drying) with 150 µl TTB (same as red TTB but without phenol red; this step is greatly facilitated if a multichannel pipetter is available). Into the wells (which already contain the 150 µl of uncolored diluent) pipette 50-µl portions of the red sample and standard Bio-IgG dilutions previous step (this results in a further fourfold dilution of the Bio-IgGs); the red color helps avoid missing wells or loading them twice. Cover the dish with a lid and incubate it at least 2 h in the cold in a humid box; during this time, the Bio-IgG molecules are captured by the immobilized streptavidin.

5. Mix 12.5 ml TTB and the appropriate amount (see last sentence of this step) of alkaline phosphatase (AP) conjugated second antibody; wash the ELISA dish as previously described; pipette 100 µl of the second antibody dilution into each well (again, this step is greatly facilitated if a multichannel pipetter is available); cover the dish with a lid, and incubate it 2.5 h in the cold in a humid box. The specificity of the AP-conjugated second antibody will obviously depend on the origin of the Bio-IgGs. In this exemplar, in which the IgGs were polyclonal and of bovine origin, 3.5 µl of AP-conjugated goat anti-bovine IgG (H+L) from Southern Biotechnology (Cat#6030-04; Birmingham, AL) was used; the 3.5-µl volume had been determined in a preliminary experiment to be just sufficient to give a strong ELISA signal (next step) when the immobilized streptavidin is saturated with Bio-IgG.

6. As the 2.5-h incubation previous step is nearing completion, prepare substrate solution by mixing 10 ml 1 M diethanolamine (pH adjusted to 9.8 with HCl), 10 µl 1 M MgCl$_2$, and 100 µl 50-mg/ml p-nitrophenylphosphate (stored frozen). Wash the ELISA dish as previously described; working rapidly to avoid excessive drying, pipette 90 µl of substrate solution into the wells (again, this step is facilitated if a multichannel pipetter is available). The difference between the optical density (OD) at 405 and 490 nm is read at 3-min intervals over a 57-min period on a kinetic plate reader in order to obtain a slope (mOD/min) for each well (1 mOD corresponds to an OD of 1/1000), which was taken as the ELISA signal. (If a kinetic plate reader is not available, a single OD measurement at 405 nm after 30–60 min can be substituted for the slope.) Results for the exemplar experiment are shown in Figure 9.8. The smallest dilution factor is 4, reflecting the fourfold dilution at step 4. Comparison of the two titer curves indicates that the sample Bio-IgG in this exemplar was 36.9 times less concentrated than the Bio-IgG standard, which had a concentration of 120 nM; thus, the estimated sample Bio-IgG concentration is 3.25 nM. In fact, the "unknown sample" in this experiment actually had a known Bio-IgG concentration of 3.75 nM; the assay thus

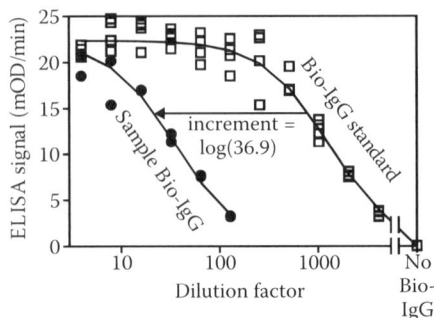

FIGURE 9.8 Quantifying Bio-IgG by enzyme-linked immunosorbent assay titering. See exemplar protocol 3 for details.

underestimated the true Bio-IgG concentration by ~13%—an entirely acceptable error in most circumstances. It is evident from the graph that if the "unknown sample" had had a Bio-IgG concentration less than ~1 nM, accuracy would have been somewhat compromised.

9.3.3.5 Solid-Phase Biotinylation

Most IgGs have a cluster of three highly conserved, solvent-exposed histidines near the C terminus of the H chain. This cluster is present in human IgG1, IgG2, and IgG4 (but not IgG3); mouse IgG1, IgG2a, and IgG3 (but not IgG2b); rabbit IgG; and bovine IgGs. As a consequence, most purified IgGs can be captured on nickel-charged immobilized metal affinity chromatography (IMAC) matrices (IMAC is not effective at capturing IgG from whole serum). It is not known whether IgGs such as human IgG3 and mouse IgG2b that have only partial histidine clusters also bind nickel. This binding can be exploited to biotinylate IgG with NHS-PEO$_4$-biotin while the protein is immobilized on a nickel column.[7] The advantage of solid-phase biotinylation is that the reagent can be added and uncoupled biotin removed simply by flowing reagent or wash buffer through the column, after which the biotinylated IgG can be released from the column with buffered 0.2 M imidazole. This advantage disappears, however, if the concentrated imidazole must be removed before the biotinylated antibody can be used. The kinetics of solid-phase biotinylation are complicated and not entirely understood. The concentration of reagent required to achieve a given level of modification seems to be higher than in the conventional liquid-phase reaction. It is not clear whether reagent in the solution that lies upstream or downstream of the IMAC column itself can gain significant access to the immobilized IgG during the reaction.

9.3.4 LABELING WITH A FLUORESCENT DYE

AlexaFluor 647 carboxylic acid succinimidyl ester (AF647NHS; Molecular Probes, Eugene, OR) is a typical amine-reactive fluorescent dye; its structure is proprietary, but its molecular weight is given as ~1250 Da. When coupled to a protein, it has an absorption maximum at 650 nm with a molar extinction coefficient of 239,000 and an emission maximum at 665 nm; its absorption at 280 nm (the approximate absorption maximum for the protein it is coupled to) is 3% of its absorption at 650 nm. The foregoing spectral parameters will differ for different dyes. In this exemplar protocol, AF647NHS is reacted with bovine IgG. After removal of uncoupled dye by dialysis, the fluorescent conjugate is analyzed spectrophotometrically to determine the IgG concentration and the number of dye molecules conjugated per IgG molecule. Because fluorescent dyes are light sensitive, they should be handled in subdued light and kept out of the light altogether when feasible.

9.3.4.1 Exemplar Protocol 4: Labeling IgG
with AlexaFluor 647

1. In a 1.5-ml microtube pipette 10 mg IgG in 606 μl LoPBS (0.15 M NaCl, 5 mM NaH$_2$PO$_4$, pH adjusted to 7.0 with NaOH) and 67 μl 1 M NaHCO$_3$,

pH adjusted to 8.5 with NaOH. The weak buffering capacity of LoPBS
allows the pH to be readjusted as desired (in this case to 8.5) by adding
small volumes of concentrated buffer.

2. Dissolve the entire 1-mg content of a tube of AF647NHS in 100 µl DMSO
    by vigorous vortexing (the concentration is 10 mg/ml = 8 mM); pipette
    50 µl into the microtube previous step, vortexing immediately; microfuge
    briefly to drive solution to the bottom; allow the tube to react overnight
    at 4°C away from light. The final concentrations of IgG and reagent are
    92 and 553 µM, respectively.

3. Add 100 µl 3 M ethanolamine, pH adjusted to 9 with HCl; allow to react
    1 h at room temperature away from the light to quench unreacted reagent
    (see exemplar protocol 2 step 4).

4. Transfer the reaction mixture to a 4-ml vessel, dilute to a total volume of
    3 ml with LoPBS, and dialyze against three changes of 1 l LoPBS.

5. Remove the dialyzed IgG from the dialysis bag or cassette and concentrate
    it to 610 µl by two cycles of concentration on a Centricon centrifugal
    concentrator (this concentration step is unnecessary if the labeled IgG is
    not needed at high concentration).

6. Make 500 µl of a 1/100 dilution, using LoPBS as diluent; scan spectro-
    photometrically from 220 to 320 nm and from 610 to 690 nm using LoPBS
    as reference solution. The spectra are shown in Figure 9.9. From the peak
    absorbance of 0.778 at 650 nm, the dilution factor of 100, and the given
    dye molar extinction coefficient of 239,000, the dye concentration was
    calculated as 325.5 µM. Correcting the peak absorbance of 0.214 at 280
    nm for the dye contribution (3% of 0.778), the absorbance at 280 that is
    attributable to IgG was calculated as 0.19066; from the dilution factor of
    100 and the IgG molar extinction coefficient of 210,000 (corresponding
    to an absorbance of 1.4 for a 1-mg/ml solution), the IgG concentration

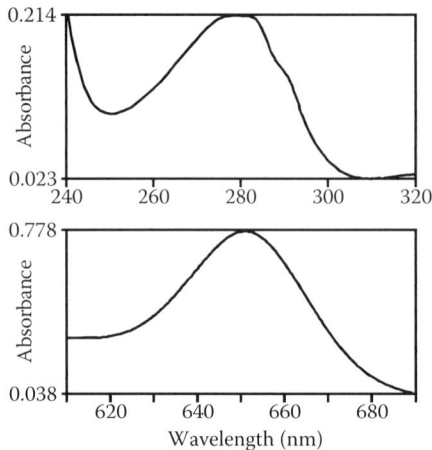

FIGURE 9.9 Ultraviolet and visible absorption spectrum of AF647-labeled immunoglobin G.

was calculated as 90.8 µM (13.6 mg/ml). Therefore, the number of dye molecules coupled per IgG molecule was 325.5/90.8 = 3.6; this level of modification is sufficient to give bright immunocytologic images by laser scanning confocal microscopy with little risk of blocking a significant fraction of the antigen binding sites. The final IgG yield was 0.61 ml × 13.6 mg/ml = 8.3 mg, or 83% of the starting IgG—a typical and entirely acceptable yield for multistep processes. It should be noted that no attempt to correct for light scattering was made (see Section 9.2.2.1), because there was significant residual dye absorbance at both 320 and 690 nm.

The foregoing protocol can be adapted to a wide range of IgG amounts and concentrations. However, because the kinetics of modification with AF647NHS have not been studied in the same depth as the kinetics of modification with NHS-PEO$_4$-biotin (Figure 9.7), it may be necessary to explore a range of reagent concentrations if the IgG concentration, temperature, pH, or buffer composition differ substantially from those in the exemplar.

9.3.5 INTRODUCTION OF THIOLS

IgGs themselves have few or no free thiols (Figure 9.14), but thiols can be introduced artificially by means of a heterobifunctional cross-linker with an amine-reactive group on one end and a thiol on the other. In most cases, the purpose of thiolation is to allow the antibody to be coupled to an enzyme or other protein derivatized with thiol-reactive maleimide groups (see Section 9.8). SATA (*N*-succinimidyl *S*-acetylthioacetate) is a typical cross-linker for this purpose, having an amine-reactive NHS group linked to a thioester, which can be considered a protected form of thiol. Reaction of SATA with IgG acylates a few of its amino groups to yield the thioester-containing adduct shown schematically in the upper part of Figure 9.10. In the absence of nucleophilic solutes, this thioester is much more stable than free thiols. Just before use, the protected thiols are deprotected under mild conditions by treatment with hydroxylamine, yielding the bottom adduct in Figure 9.10; the resulting thiol-containing IgG is used immediately to prevent gradual loss of thiol to oxidation or other side-reactions (Section 9.4.3). Exemplar protocol 5 is based on the SATA instructions from Pierce.

9.3.5.1 Exemplar Protocol 5: Thiolation with SATA

1. In a 1.5-ml microtube pipette 1 ml 9-mg/ml (60-µM) IgG in PBS (0.15 M NaCl, 0.1 M NaH$_2$PO$_4$, pH adjusted to 7.2 with NaOH).
2. Dissolve SATA in DMSO to a concentration of 5.36 mg/ml (55 mM); pipette 10 µl into the microtube previous step, vortex immediately, and allow to react at room temperature for 30 min. The final concentrations of IgG and SATA are 59.4 and 545 µM, respectively; the resulting modification level is said to be 3–3.6 protected sulfhydryls per IgG molecule. It is not a good idea to quench unreacted reagent with ethanolamine after

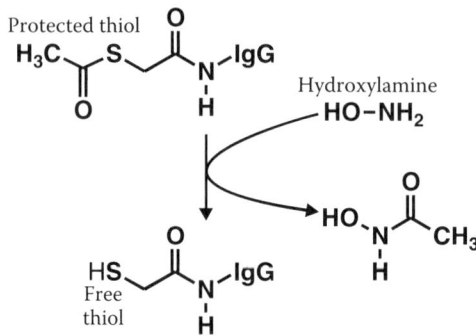

FIGURE 9.10 Introduction of protected thiols into immunoglobin G with SATA (*N*-succinimidyl *S*-acetylthioacetate) and deprotection of the thiols with hydroxyl amine.

reaction with SATA (see exemplar protocol 2 step 4) because the ethanolamine at high concentration may deprotect thiols prematurely.

3. Remove unreacted and hydrolyzed SATA by dialysis against PBS or by multiple cycles of dilution in PBS and concentration on a centrifugal concentrator; determine the IgG concentration spectrophotometrically or by the Bradford or BCA assays (Section 9.2.2). The exact final concentration of IgG is not important, but it is assumed in the next step that that concentration is not far from the starting concentration of 9 mg/ml. The derivatized IgG can be stored in the refrigerator or freezer for months or years.

4. When deprotected IgG is needed, pipette the required volume of derivatized IgG into a microtube and add 0.1 volume of freshly-prepared deacylation solution (dissolve 1.74 g hydroxylamine · HCl and 0.365 g EDTA disodium salt in 40 ml PBS; add water to a final volume of 50 ml; adjust pH to 7.2 with NaOH); vortex, microfuge to drive solution to the bottom of the tube, and allow to react 2 h at room temperature.

5. Remove reactants by dialysis against PBS/EDTA (PBS supplemented with 10 mM EDTA) or by multiple cycles of dilution in PBS/EDTA and concentration on a centrifugal concentrator; determine the IgG concentration spectrophotometrically or by the Bradford or BCA assays.

6. Quantify thiols with Ellman's reagent (exemplar protocol 8).

The EDTA helps protect thiols from metal-catalyzed oxidation; nevertheless, the deprotected IgG should be used as soon as possible because EDTA does not block all modes of thiol loss.

9.3.6 PEGYLATION

Attaching PEG chains to therapeutic monoclonal antibodies can dramatically alter their pharmacokinetic properties.[8] The polymer chains are very hydrophilic and organize extensive shells of hydration around themselves, thus tending to exclude macromolecules from close approach. This excluded volume effect can

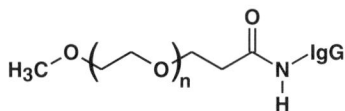

FIGURE 9.11 Schematic diagram of immunoglobin G after PEGylation with an amine-reactive PEGylating reagent.

extend the half-life of IgG in the serum and block its binding to anti-IgG antibodies and other macromolecules. Larson and colleagues[9] studied the kinetics of PEGylation of IgG with the NHS ester mPEG-SPA, which has a PEG chain with an average molecular weight of 5000 and yields an adduct with the structure shown in Figure 9.11.

IgG at a fixed concentration of 3 mg/ml (20 μM) was reacted with mPEG-SPA at various concentrations and the resulting level of modification was determined using nuclear magnetic resonance because there is no simple, sensitive color reaction for PEG. In accord with Eq. 9.2, it was found that the modification level m was proportional to the ratio of reagent to protein R/P, the proportionality constant being 0.73 under their reaction conditions. Exemplar protocol 6 is based on the article by Larson et al.[9]

9.3.6.1 Exemplar Protocol 6: PEGylation with mPEG-SPA

1. Dissolve mPEG-SPA with a molecular weight of 5000 (Shearwater Polymers Inc, Huntsville, AL) in DMSO to a concentration at least 20 times higher than the final concentration at the end of the next step (the final DMSO concentration will thus not exceed 5%).
2. Add the required volume of mPEG-SPA solution dropwise to a stirring or vortexing solution of IgG at a concentration of 3 mg/ml in borate buffer pH 9.2. The molar ratio of mPEG-SPA to IgG R/P should be $1/0.73 = 1.37$ times higher than the desired modification level m (number of PEG chains per IgG). Allow the reaction to continue with stirring for 1 h at room temperature, then overnight at 4°C.
3. Remove uncoupled PEG and exchange into PBS by multiple rounds of concentration and redilution on a centrifugal concentration device with a molecular weight cutoff of between 50 and 100 kDa.
4. Measure the protein concentration spectrophotometrically.
5. If a nuclear magnetic resonance instrument is available, the number of PEG chains coupled per IgG can be determined as described.[9]

9.3.7 COUPLING TO A CHELATOR

Amine-reactive chelators serve as bifunctional linkers for attaching radioactive metals to antibodies (or other proteins) for radiotherapy or *in vivo* imaging.[10,11] Labeling occurs in two stages. In the first, empty chelator (i.e., with no metal ligand) is covalently coupled to antibody (or other biomolecule) to yield a stable adduct that can be stored indefinitely. In the second, a portion of the chelator-coupled adduct is reacted with radionuclide and freed of unbound metal; because of the short half-life of the radionuclides used in these applications, the second stage must be carried out rapidly and the resulting adduct used without delay.

Amine-reactive derivatives of diethylene triamine pentaacetic acid (DTPA) are particularly suitable for labeling antibodies because chelation occurs rapidly at moderate temperatures. DTPA effectively chelates ^{90}Y (a β emitter used for radiotherapy), ^{86}Y (for imaging by positron emission tomography), ^{111}In (γ emitter for imaging by single photon emission computed tomography), and other radionuclides. Two amine-reactive isothiocyanate derivatives of DTPA are available commercially from Macrocyclics (Dallas, TX), and the reaction kinetics of a third, 1M3B-DTPA, has been extensively investigated by Mirzadeh and colleagues.[12] Their article serves as the basis of exemplar protocol 7; the thiourea adduct formed by reaction with 1M3B-DTPA is diagrammed in Figure 9.12.

Isothiocyanates react much more slowly with amines (and with water) than do the NHS esters used in exemplar protocols 1, 2, and 4–6; even after a 35-h reaction with 1M3B-DTPA coupling is not complete.[12] Equations 9.1 and 9.2 therefore cannot be used confidently to extrapolate from reaction conditions that have been directly explored to conditions that have not been so analyzed. Using the reaction conditions of exemplar protocol 7 with IgG at a concentration of 46 μM, the authors determined the modification level m for several different molar ratios of reagent to IgG R/P, with the results shown in Figure 9.13. The data fit well to the empirical equation

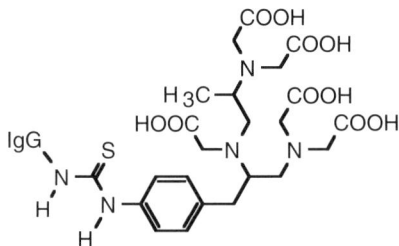

FIGURE 9.12 Schematic diagram of immunoglobin G modified with 1M3B-DTPA, an amine-reactive isothiocyanate derivative of the metal chelator DTPA.

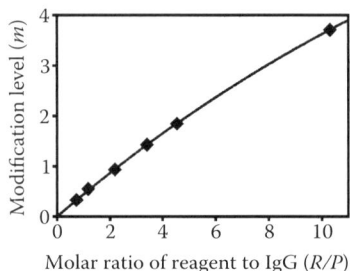

FIGURE 9.13 Modification of immunoglobin G (IgG) with 1M3B-DTPA; *see* text for details. The IgG concentration is 46 μM. The solid line is a theoretical curve, as explained in the text.

(not based on any kinetic model) $m = 9.939 \times [1 - \exp(-0.04533\ R/P)]$ (continuous curve in the graph), which can therefore be used to interpolate the reagent concentration required to modify 46-µM IgG to any desired modification level m. It is very likely (but not proven empirically) that the commercially available DTPA isothiocyanate derivatives will modify IgG with very similar kinetics.

9.3.7.1 Exemplar Protocol 7: Modification with an Isothiocyanate Derivative of DTPA

1. To 46-µM IgG in 0.15 M NaCl, 50 mM N-(2-hydroxyethyl)piperazine-N'-ethanesulfonic acid (HEPES), pH adjusted to 8.6 with NaOH, add the desired volume of 10-mM reagent (isothiocyanate derivative of DTPA freshly dissolved in water) as calculated previously; allow the reaction to proceed 17.3 h at room temperature (27°C).
2. Remove uncoupled DTPA by dialysis against PBS or by multiple cycles of concentration and dilution with PBS on a centrifugal concentration device.
3. Determine protein concentration by spectrophotometry or Bradford or BCA assays.

In the foregoing studies,[12] the 1M3B-DTPA reagent was available in ^{14}C-labeled form, which allowed the modification level m to be determined by scintillation counting. Since then, a simple colorimetric method has been developed for titrating DTPA when the reagent is not available in labeled form.[13]

9.4 MODIFICATION OF THIOLS AND DISULFIDES

9.4.1 DISULFIDES AND THIOLS IN IgG

Immunoglobulins are made up of multiple ancestral homology units having a conserved overall conformation called the *immunoglobulin fold*. Within each immunoglobulin fold is a single highly conserved intrachain disulfide bond; in the schematic diagram of a generic IgG in Figure 9.14, these disulfide bonds are represented by square brackets. Breaking these bonds requires denaturation of the antibody; reactions involving these cysteines therefore lie outside the scope of this chapter.

Interchain disulfides bond the two heavy chains to each other in and near the hinge region, and each heavy chain to one of the light chains, as shown with dashed lines in Figure 9.14. The number and exact positions of these bonds differ from species to species and from subclass to subclass within a species, but are highly conserved in IgGs of a given subclass in a given species, regardless of their variable region sequences and antigen-binding specificities. Because these interchain disulfides are usually accessible to solvent and can be cleaved without denaturing the IgG or destroying its antigen-binding ability, reactions involving them do lie within the scope of this chapter.

In some IgGs, there are additional intrachain disulfides in the H-chain constant region; for example, rabbit IgG has such a bond, which is accessible without

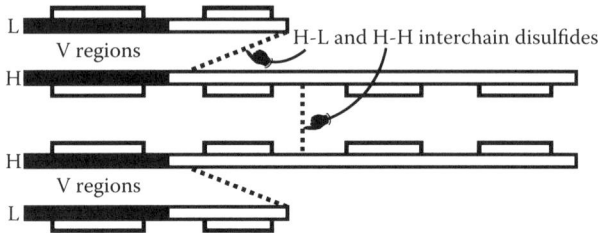

FIGURE 9.14 Disulfide bonds in a generic immunoglobin G molecule. The interchain di-sulfide bonds shown as dashed lines can be reduced by thiols or tri-carboxyethylphosphine without denaturing the protein. The intrachain disulfide bonds shown with solid brackets are highly conserved in the basic immunoglobulin fold, and are not accessible in nondenatured protein.

denaturing the protein (Figure 9.28). In addition to the constant-region intrachain and interchain disulfide bonds, individual molecular species of IgG (e.g., monoclonal antibodies) can have idiotypic thiols or disulfides in their variable regions. Free (non–disulfide-bonded) thiols are generally quite rare in IgGs.

9.4.2 Disulfide Interchange Reaction

Fundamental to thiol/disulfide transactions in the cell and to manipulation of thiols and disulfides *in vitro* is the disulfide interchange reaction. The minimal disulfide interchange, involving only two distinguishable sulfur-coupled groups R and R′, is illustrated in Figure 9.15. Even this minimal reaction encompasses four simultaneous equilibria: two acid-base equilibria between the uncharged sulfhydryl and negatively charged thiolate forms of the thiols, and the two disulfide interchange reactions themselves. In practice, a typical disulfide interchange reaction might involve a dozen or so distinguishable thiols and many hundreds of simultaneous equilibria. No matter how complex the system, however, there is no net gain or loss of disulfide, though the distribution of disulfides can change dramatically depending on the equilibria constants and initial reactant concentrations.

It is easy to see that if a small amount of disulfide-containing protein, represented in the diagram as R′SSR′, is treated with a vast molar excess of a sulfhydryl-containing reagent like mercaptoethanol, represented in the diagram as RSH, the two disulfide interchange equilibria are driven overwhelmingly in the upward direction. The net effect is transfer nearly all the disulfide bonds that were originally in the protein to disulfide bonded reagent, represented in the diagram as RSSR, concomitantly releasing non–disulfide-bonded protein, represented in the diagram as the thiol pair R′S⁻/R′SH. In these circumstances the added RSH thiol is said to have "cleaved" or "reduced" all the protein disulfides. Conversely, if a small amount of non–disulfide-bonded protein R′SH is treated with a vast excess of a disulfide-containing reagent RSSR (e.g., cystamine), the upper disulfide interchange equilibrium is driven overwhelmingly in the downward direction, with the result that almost all the protein is converted to the mixed disulfide R′SSR, in which R is disulfide-bonded to nearly all the protein thiols.

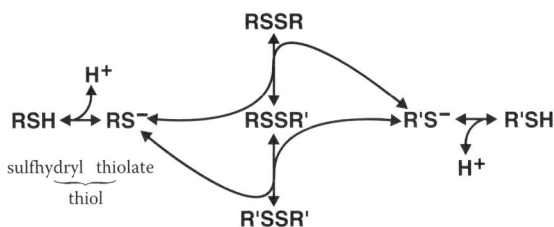

FIGURE 9.15 A minimal disulfide interchange reaction involving only two distinct sulfur-bonded groups R and R′.

Because free thiols are subject to acid-base equilibrium, as shown in Figure 9.15, the percent of thiol available as thiolate depends on pH. Cysteine residue thiols in small model compounds have pKs around 8–9, but the pKs of thiols in native proteins can diverge markedly from that range as a result of their immediate microenvironment. Small alkyl thiols like mercaptoethanol have pKs around 9.5. Accordingly, disulfide interchange reactions are typically carried out at a slightly alkaline pH (about 8). Conversely, acidifying a sample to about pH 5 or lower generally greatly reduces disulfide interchange.

Conformational constraints may greatly skew the equilibrium in disulfide interchange reactions, as exemplified by Cleland's reagents DTT and dithioerythritol (DTE).[14] The two thiols of this reagent react with a disulfide in two steps (Figure 9.16), the second of which is essentially irreversible. The irreversibility stems from the facts that the thiol in the mixed disulfide is positioned to attack the DTE- or DTT-derived sulfur far more readily than the protein-derived sulfur, and that the resulting intramolecular disulfide is a very stable six-membered ring. DTE or DTT therefore reduce accessible protein disulfide bonds quantitatively even if the DTE or DTT is only slightly in molar excess.

Thiols and disulfides in native proteins, including IgG, often behave very differently from those in small model compounds. Some are not accessible under nondenaturing conditions, as has already been emphasized for the conserved intrachain disulfide bond in each immunoglobulin fold (solid brackets in Figure 9.14). Others, although perhaps accessible (to small molecules at least), may be so highly constrained conformationally as to drive the equilibrium overwhelmingly in one direction or the other—just as in the case of DTE and DTT in the previous paragraph. For example, 10 mM 2-mercaptoethylamine (2-MEA) at pH 5 preferentially reduces the H–H interchain bond of rabbit IgG under nondenaturing conditions without cleaving other disulfides, including the H–L interchain disulfide (Figure 9.28).[15] A study of preferential cleavage of various monoclonal antibody interchain disulfide bonds under various conditions has recently been published.[16]

The symmetrical aryl disulfides in Ellman's reagent 5,5′-dithio-*bis*-(2-nitrobenzoic acid) (DTNB; Figure 9.17)[17] and in 2,2′-dithiopyridine (2,2′-DTP) and 4,4′-dithiopyridine (4,4′-DTP; Figure 9.18)[18] are irreversibly reduced by ordinary alkyl thiols (including protein thiols) to produce a mixed disulfide and a highly colored leaving group. The leaving group is 5-thio-(2-nitrobenzoic acid) (TNB) in the case of DTNB (Figure 9.17), 4-thiopyridone (4-TP) in the case of 4,4′-DTP (Figure 9.18).

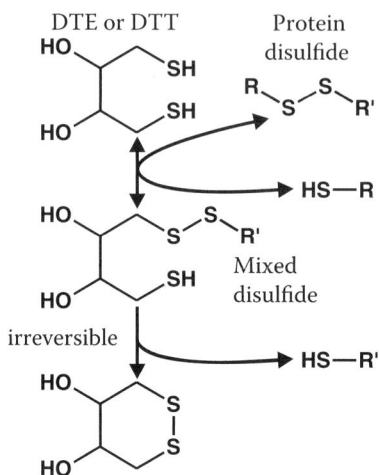

FIGURE 9.16 Complete cleavage of protein disulfide bonds with dithioerythritol (DTE) and dithiothreitol (DTT).

The increase in color can be used to quantify free thiols in an IgG or other protein, as will be illustrated in exemplar protocol 8. If a second alkyl thiol attacks the mixed disulfide in a second-stage reaction, it almost always attacks the alkyl sulfur rather than the aryl (reagent) sulfur, thus releasing the same colored leaving group (TNB or 4-TP) as in the first-stage reaction. Figure 9.18 illustrates how this feature can be exploited to couple two thiols RSH and R'SH (for example, two thiol-bearing proteins) to create a disulfide-bonded adduct RSSR'.[19] In the first step, RSH is reacted with excess 4,4'-DTP to create the mixed disulfide, which is freed of the 4-TP leaving group and excess 4,4'-DTP reagent. In the second step, the mixed disulfide is reacted with a roughly equivalent amount of R'SH to create the final RSSR' adduct. Both steps can be quantitatively monitored spectrophotometrically.

9.4.2.1 Exemplar Protocol 8: Determination of Free Thiols with Ellman's Reagent (DTNB)

1. Make a fresh 100-mM (15.4-mg/ml) solution of DTE (or DTT) in weak acid buffer (5 mM acetic acid, pH adjusted to 5 with NaOH); using the same buffer as diluent, make a linear dilution series covering the range 0–90 μM (0–180 μM thiol, because each DTE molecule has two thiols); these are the DTE standards.

2. Again using the weak acid buffer as diluent, dilute the unknown protein (or other sample) so that its anticipated thiol concentration is in the range 60–180 μM (if anticipated thiol concentration has more than threefold uncertainty, prepare an appropriate dilution series).

3. Into a tared glass vial (wipe off with damp towel and dry to dissipate static electricity) weigh 1–3 mg DTNB; reweigh and note exact net weight; dissolve in sufficient 1 M Tris·HCl pH 8 to give a concentration of 0.8

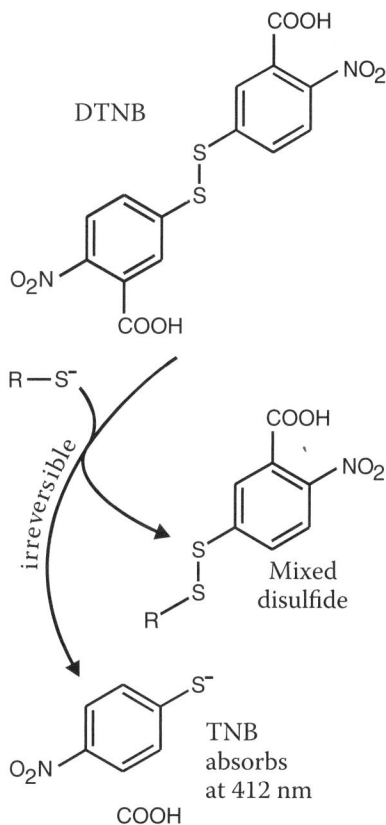

FIGURE 9.17 Reduction of Ellman's reagent 5,5'-dithio-*bis*-(2-nitrobenzoic acid) to release a mixed disulfide and the intensely colored thiol 2-nitrobenzoic acid.

mg/ml (2 mM). In a 1.5-ml microtube make a 1/7 dilution with water to give 286 μM DTNB in 143 mM Tris buffer.

4. Carefully pipette 150 μl of this mix into 500-μl microtubes or wells of an ELISA dish. Into each well, pipette 50 μl of one of the DTE standards step 1 or the protein sample(s) step 2, stirring with the pipette tip (the 143-mM pH 8 Tris in the 150 μl DTNB dominates over the 5-mM pH 5 acetate in the DTE standard or protein sample). Allow the color to develop at least 15 min.

5. Determine the absorbance at 412 nm (the absorption maximum of the TNB leaving group; see Figure 9.17) spectrophotometrically, or the optical density at a nearby wavelength on an ELISA plate reader (a plate reader at 405 nm was used to obtain the results in Figure 9.19). Plot the DTE standard and calculate the best-fitting straight line (see Figure 9.19); use the line's slope and intercept to convert the observed absorbance or optical density of the unknown protein sample to the equivalent thiol concentration.

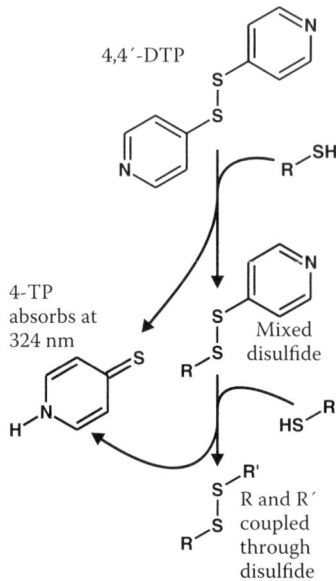

FIGURE 9.18 Reduction of a large excess of 4,4′-DTP with a thiol RSH to yield a mixed disulfide, and subsequent reaction of the mixed disulfide with a second thiol R′SH to couple R and R′ through a disulfide linkage. Both stages release the intensely colored 4-TP.

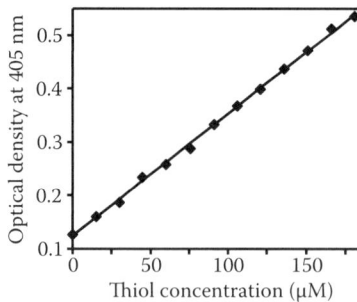

FIGURE 9.19 Thiol titration with Ellman's reagent 5,5′-dithio-*bis*-(2-nitrobenzoic acid); *see* exemplar protocol 8 for details. The filled diamonds are the observed optical densities for the dithioerythritol standards; the solid line is the best linear fit to the data.

Using an ELISA plate reader introduces a small error because the protein sample has a curved meniscus in the plastic wells whereas the DTE standards, having no protein to wet the plastic, give a flat meniscus. To avoid this optical distortion, determine the absorbance of the standards and samples in a spectrophotometer with a cuvette. If the thiol content of the protein is very low, so that high concentrations of protein are required for reasonable accuracy, there may be some background absorbance at 412 nm from light scattering or contaminants. Under these circumstances, it is advisable to scan the standards and protein samples spectrophotometrically from

365 to 465 nm so that the true contribution of the TNB leaving group to the absorption profile can be determined with less uncertainty.

9.4.3 OTHER OXIDATION-REDUCTION REACTIONS INVOLVING THIOLS AND DISULFIDES

Thiols can be oxidized to disulfides by mild oxidants like iodosobenzoic acid,[20,21] provided the participating thiols are able to approach each other with appropriate geometry. Even in the absence of added oxidant, however, dissolved oxygen from the air can slowly oxidize thiols to disulfides at neutral and alkaline pH; this air oxidation is accelerated by trace heavy metal contaminants. Three measures are commonly taken to avoid unwanted air oxidation: reducing the pH to 5 or less if feasible; adding chelators (e.g., EDTA at a concentration of at least 1 mM) to sequester heavy metal ions; and purging dissolved oxygen with nitrogen or other inert gas. Small alkyl thiols such as mercaptoethanol are very susceptible to air oxidation and should be made up within a few hours of use. DTE and DTT, on the other hand, are quite resistant to air oxidation despite the great stability of the disulfide-bonded six-membered ring (Figure 9.16); stock solutions can be stored in the freezer for months with little loss of potency. TCEP, described in the next paragraph, is even more resistant to air oxidation than DTE and DTT.

Tri-carboxyethylphosphine (TCEP) reduces disulfide bonds as shown in Figure 9.20, a single TCEP molecule releasing two thiols and thereby becoming oxidized to TCEP oxide. TCEP has at least three important advantages over DTE and DTT as a disulfide reductant. First, the reaction is essentially irreversible and takes place over a very wide pH range. Second, as mentioned at the end of the previous paragraph, it is resistant to air oxidation so that stock solutions are stable for an extended period of time. Third, it is much less reactive with alkylating agents than thiols.

Sulfitolysis[22] cleaves disulfide bonds oxidatively rather than reductively, as shown in Figure 9.21. The resulting S-sulfonated cysteine residues are stable at neutral pH, but the cysteine thiols are readily regenerated by treatment with excess thiol such as DTE or mercaptoethanol. Despite these potential advantages, no specific studies of sulfitolysis of IgG disulfides under nondenaturing conditions could be found.

9.4.4 ALKYLATION OF THIOLS WITH ALKYL HALIDES AND N-ALKYL MALEIMIDES

Alkyl halides and N-alkyl maleimides are moderate electrophiles that react preferentially with thiols in their thiolate form to create stable thioether (i.e., alkylated thiol) adducts, as shown in Figure 9.22. The optimal reaction pH for alkyl halides and alkyl maleimides are about 8.5 and 7.0, respectively; under those optimal conditions the reagent is both reactive with and selective for thiols. At a pH far below the optimum the fraction of thiol in the thiolate form is so low that the reaction is very slow; at a pH far above the optimum other protein nucleophiles, particularly amines, react. To ensure selectivity, the alkylating reagent is generally

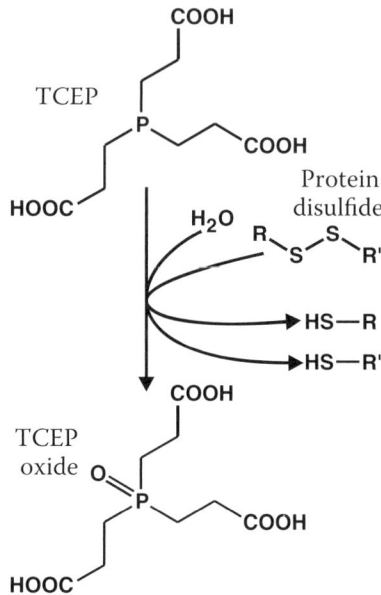

FIGURE 9.20 Reduction of disulfide bonds by tri-carboxyethylphosphine.

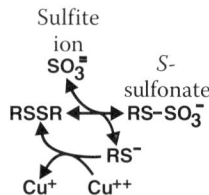

FIGURE 9.21 Oxidative cleavage of disulfide bonds by sulfitolysis.

added in only slight molar excess over the total concentration of reactive thiols. If the protein thiols are generated by addition of a thiol reductant like DTE or DTT, the alkylating reagent is added in slight molar excess of the reductant thiols (e.g., twice the DTE or DTT concentration). If the protein thiols are generated by addition of TCEP, however, the alkylating reagent need only be in molar excess of the protein thiols themselves, since TCEP reacts poorly with such reagents. At a pH above about 7, N-alkyl maleimides are hydrolyzed to the corresponding N-alkyl maleamic acids; nevertheless, even they are much more stable in aqueous solution than are NHS esters. Heterobifunctional cross-linking reagents with both an amine-reactive NHS ester and a thiol-reactive N-alkyl maleimide exploit this relative stability. In a first step, the cross-linker is coupled to a protein's amino groups as in exemplar protocol 1, yielding a modified protein with a number of thiol-reactive maleimide appendages; the maleimide group is sufficiently stable to survive these manipulations. Then, in a second step, the maleimide-modified protein is reacted with a thiol-containing molecule, which can be another protein.

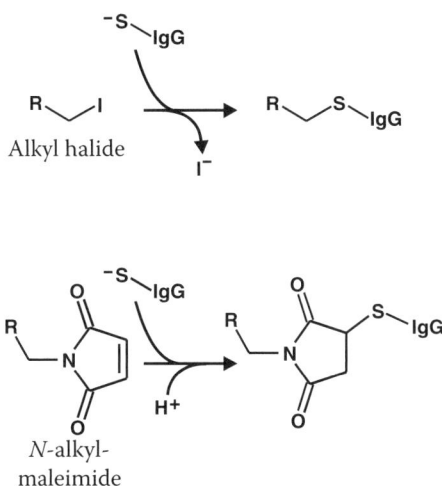

FIGURE 9.22 Alkylation of protein thiols with alkyl halides and *N*-alkyl malemides to create stable thioether adducts.

Alternatively, the maleimide-modified protein can be freeze-dried and stored indefinitely in a desiccated state without losing its reactive maleimides. A number of such maleimide-modified proteins are commercially available. (Cross-linking will be discussed more fully in Section 9.8.)

In exemplar protocol 9, based on Pierce instructions, the cysteines from the interchain disulfides of IgG are biotinylated with maleimide-PEO$_2$-biotin (Figure 9.23), which, as with NHS-PEO$_4$-biotin (Figure 9.6), has a hydrophilic PEO linker.

9.4.4.1 Exemplar Protocol 9: Biotinylation of IgG with Maleimide-PEO$_2$-Biotin

1. Dissolve 6 mg of 2-mercaptoethylamine (2-MEA) in 1 ml of 2.5-mg/ml IgG in reducing buffer (0.1 M NaH$_2$PO$_4$, 2.5 mM Na$_2$EDTA, pH adjusted to 6.0 with NaOH); the final concentration of 2-MEA is 53 mM; incubate the solution at 37°C for 90 min. These conditions are intended to cleave the H-H interchain disulfide bonds while leaving the other disulfides (including the H-L interchain disulfide bonds) intact.
2. Use a desalting gel filtration column (for example, D-Salt from Pierce) to rapidly remove 2-MEA and reequilibrate the protein with conjugation buffer (1 mM Na$_2$EDTA in PBS); dilute the IgG to 2.5 ml (theoretical concentration 1 mg/ml). Buffer exchange is much faster on a desalting column than in a dialysis bag or cassette or centrifugal concentration device, thus minimizing reoxidation of the disulfide bonds.
3. Dissolve 2 mg malemide-PEO$_2$-biotin (Pierce No-Weigh tube) in 100 μl DMSO; add 25 μl to the 2.5 ml reduced IgG solution; mix and allow to react at least 2 h at room temperature.

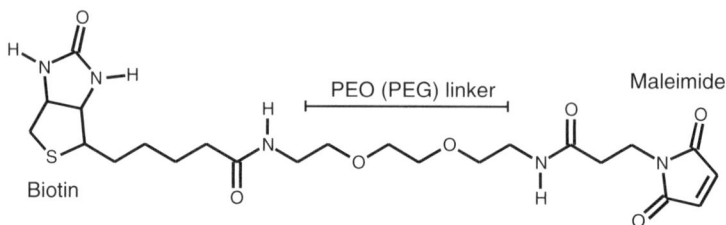

FIGURE 9.23 Maleimide-PEO$_2$-biotin, a thiol-reactive biotinylating reagent.

4. Remove excess reagent and equilibrate the biotinylated IgG in PBS by dialysis or with a centrifugal concentration device.
5. Determine the IgG concentration spectrophotometrically or by the BCA or Bradford assay (Section 9.2.2).
6. If, as intended, H-H interchain bonds have been reduced and the resulting free thiols biotinylated while H-L interchain remain intact (Figure 9.14), electrophoresis without reduction of disulfide bonds (steps 6–7 of exemplar protocol 17) should reveal predominantly a 75-kDa heterodimer consisting of one H chain and one L chain disulfide bonded to each other.

Biotinylation of goat IgG after reduction of interchain disulfide bonds with TCEP results in cleavage of both H-H and H-L disulfide bonds[7]; yet, the IgG remains intact and retains full antigen-binding potency. This shows that preservation of the H-L interchain disulfide bonds is not required for the conformational integrity and functional activity of IgG under nondenaturing conditions. In some applications, however, performance of a thio conjugate can be significantly affected depending on which interchain disulfide bonds are cleaved.[16]

9.5 MODIFICATION OF CARBOHYDRATES

Naturally produced antibodies are glycoproteins with N-linked oligosaccharides,[23] though some monoclonal antibodies are underglycosylated or not glycosylated at all. Because the carbohydrates occur almost exclusively in the Fc domain, far away from the antigen-binding sites in the Fab domains, modification of IgG carbohydrate very seldom affects antigen binding. When antibodies are to be modified with small appendages, the practical effect of this theoretical advantage over random modification of amines is not normally discernible. When the appendage is bulky, however— for example, when the IgG is to be coupled to an enzyme or other macromolecule, or to a solid surface—coupling via carbohydrate may be noticeably superior to coupling via amines. Moreover, a rare individual monoclonal antibody may happen to have a particularly reactive amine near its antigen-binding site, so that amine modification inactivates a substantial fraction of the IgG molecules.

The N-linked oligosaccharides are typical complex carbohydrates whose reducing ends are capped with sialic acid residues, so that no aldehydes are exposed. However, aldehydes can be generated artificially in one of three ways: by removing sialic acids with neuraminidase (however, the aldehyde form of the exposed galactose sugar

IgG | neuraminidase or NaIO$_4$

Hydrazide Hydrazone

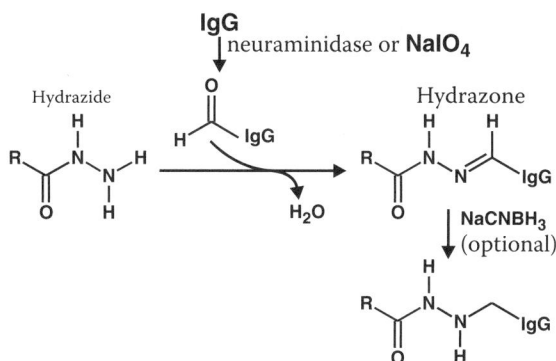

FIGURE 9.24 Reaction of a hydrazide with oxidized immunoglobulin G to form a hydrazone adduct, and further optional reduction of the hydrazone to form an even more stable substituted hydrazide adduct.

residues is in equilibrium with the hemiacetal form, the latter greatly predominating); by oxidizing vicinal hydroxyls of sialic acids with 1 mM sodium periodate; and by oxidizing vicinal hydroxyls of other sugar residues with 10 mM sodium periodate. As already indicated in Figure 9.5, the resulting aldehydes react reversibly with amino groups under slightly alkaline conditions to form Schiff bases, which in turn can be stabilized by mild reduction with NaCNBH$_3$. This is not a very attractive way of modifying IgG, however, because the IgG molecule itself has many amino groups that can form Schiff bases with aldehydes on other IgG molecules. Much more useful is reaction with hydrazides under slightly acidic conditions in which Schiff base formation is minimal (Figure 9.24); the resulting hydrazone adducts are sufficiently stable at neutral pHs for most applications, but if necessary can be reduced with NaCNBH$_3$ to the even more stable substituted hydrazide form without reducing disulfides or otherwise modifying the protein (Figure 9.24). An important advantage of this mode of coupling is that the aldehyde and hydrazide reactants are stable, and therefore need not be prepared immediately before use.

In exemplar protocol 10, a tumor-binding mouse monoclonal IgG is oxidized with NaIO$_4$ to create aldehydes in its carbohydrate moiety. In exemplar protocol 11 the resulting aldehyde-activated IgG is reacted with 4-desacetylvinblastine-3-carbohydrazide (DAVLB hydrazide), a hydrazide derivative of a cytotoxic vinca alkaloid (Figure 9.25). The resulting conjugated IgGs, with about 5 DAVLB appendages per molecule, are immunotoxins with anti-tumor properties. Both protocols are adapted from an extensive study of 10 anti-tumor mouse monoclonal IgGs representing all four subclasses by Laguzza and colleagues.[24]

9.5.1 Exemplar Protocol 10: Oxidation of IgG Carbohydrates with NaIO$_4$

1. To an ice-cold 10 mg/ml solution of IgG in 0.1 M sodium acetate buffer pH 5.6, add sufficient granular sodium *meta*periodate to bring the periodate concentration to 34 mg/ml (160 mM), stirring, rotating or gently vortexing the slurry continuously for 21 min on ice.

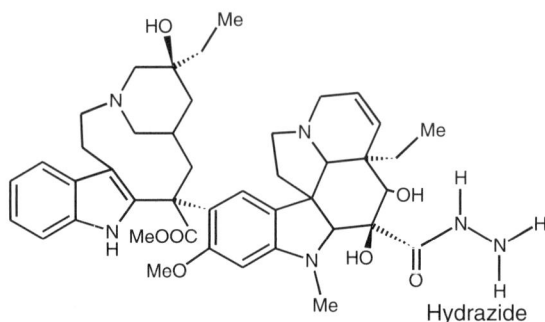

FIGURE 9.25 DAVLB-hydrazide, a hydrazide derivative of a cytotoxic vinca alkaloid.

2. Rapidly, and with constant stirring, rotating or vortexing at 0°–2°C, add 1/15.7 vol 12.5 M ethylene glycol (final concentration 800 mM) and continue the incubation for 1 h to quench the oxidation reaction.
3. Centrifuge at 2000g for 20 min at room temperature or 4°C to pellet a small amount of insoluble material.
4. Dialyze the clear, colorless supernatant against several changes of 0.1 M sodium acetate buffer pH 5.6.
5. Quantify the antibody spectrophotometrically.

The periodate concentration used in this protocol, 160 mM, is 16 times higher than the standard concentration; the investigators found that IgG oxidized with 10 mM periodate under the same conditions yielded conjugates with about two DAVLB appendages per molecule. One of the mouse IgGs, an IgG1, entirely lost its ability to bind tumor cells after oxidation; the other nine, including another IgG1, retained full cell-binding potency.

9.5.2 EXEMPLAR PROTOCOL 11: COUPLING DAVLB-HYDRAZIDE TO OXIDIZED IgG

1. To an ice-cold 2.76-mg/ml (18.4-μM) solution of oxidized IgG in 0.1 M sodium acetate buffer pH 5.6 add 1/12.4 vol 53.7-mg/ml (70-mM) DAVLB hydrazide in dimethylformamide dropwise (or in small increments) with constant stirring or gentle vortexing; continue stirring or end-over-end rotation at 4°C for 24 h.
2. Centrifuge at 2,000g for 20 min at room temperature or 4°C to pellet a small amount of insoluble material.
3. Dialyze the supernatant against PBS in the cold.
4. Quantify the conjugated antibody and determine the modification level (number of DAVLB residues per IgG molecule) spectrophotometrically. The modification level was calculated from the ratio of absorbances at 280 and 270 nm, assuming the following molar extinction coefficients: for IgG, 180,000 and 214,000 at 270 nm and 280 nm, respectively; for DAVLB, 12,300 and 10,800 at 270 nm and 280 nm, respectively.

The investigators[24] found that the conjugates, which were not stabilized by NaCNBH$_3$ reduction, were fairly stable at pH 7.4 at 4°C (~2.5% loss of DAVLB residues after 7 days) and at 37°C (~10% loss in 7 days), but unstable at pH 5.6 at 4°C (~18% loss in 7 days) and 37°C (27–30% loss in 7 days, mostly in the first 24 h).

9.6 IMMOBILIZATION OF ANTIBODIES

9.6.1 IMMOBILIZATION ON POLYSTYRENE DISHES AND ELISA WELLS

Many common procedures in cell and molecular biology, such as ELISA (for example, exemplar protocol 3) and affinity selection from phage display libraries,[25] require immobilization of antibodies or other proteins on a polystyrene surface. In this subsection, four modes of immobilization will be considered: direct nonspecific adsorption onto the plastic surface, immobilization through a biotin-streptavidin bond, immobilization through Fc-binding proteins (e.g., second antibodies or protein A or protein G), and covalent attachment.

Antibodies, like other proteins, can be nonspecifically adsorbed onto the surface of a polystyrene dish or ELISA well, as illustrated previously for streptavidin in steps 1 and 2 of exemplar protocol 3. The protein is dissolved at a concentration of ~1–10 μg/ml in a detergent-free nondenaturing buffer, such as PBS, TBS or 0.1 M NaHCO$_3$. This coating solution is reacted with the solid surface for at least 4 h (preferably overnight), after which the surface is blocked with a high concentration of noninterfering protein (e.g., 5 mg/ml BSA) in detergent-free buffer for at least 1 h. The coating solution should fill the dish or well only partially, but the blocking solution should fill it to brimming to ensure that no unblocked plastic remains accessible to subsequent reactants. After blocking, the dish or ELISA well is thoroughly washed in preparation for subsequent solid-phase reactions. The amount of protein adsorbed depends on the conditions and the particular polystyrene preparation; a typical amount would be ~10 ng in an ELISA well—far less than the amount of protein in the coating solution. During prolonged storage or reactions the plastic is held in a sealed humid container to guard against desiccation. Nonspecific adsorption is thought to require local denaturation of the protein in order to expose hydrophobic side chains to the hydrophobic polystyrene. In most cases, only a small fraction of adsorbed antibody molecules lose their antigen-binding potency as a result of this localized denaturation. Similarly, other adsorbed proteins typically retain their functional activity to a large extent. Nondenaturing detergents such as Tween 20 block adsorption, but do not desorb the IgG (or the blocking protein) after adsorption is complete; such detergents are frequently included at high concentration (e.g., Tween 20 at 0.5%) during subsequent reactions with the coated surface in order to discourage nonspecific interactions.

Indirect immobilization of biotinylated IgG (Bio-IgG) onto streptavidin-coated polystyrene was illustrated in steps 1–4 of exemplar protocol 3. In the first step, the dish or wells are coated with streptavidin as in the previous paragraph. Then, in a second step, the streptavidin-coated surface is reacted with Bio-IgG (or other biotinylated protein) under nondenaturing conditions. The buffer can be supplemented with a noninterfering carrier protein like BSA and a nondenaturing detergent such as 0.5% Tween 20 to discourage nonspecific adsorption of the biotinylated protein. Indirect immobilization of the IgG has several advantages. Because neither biotinylation nor the reaction with the streptavidin-coated surface denatures the IgG, the protein remains in its native form throughout. Because of the extremely high affinity of the biotin-streptavidin bond, immobilization is nearly quantitative as long as the capacity of the immobilized streptavidin is not exceeded, thus increasing predictability and greatly reducing the amount of IgG required compared with direct nonspecific adsorption. Because of the exceedingly slow dissociation rate of the biotin-streptavidin interaction, little of the immobilized Bio-IgG is released by subsequent exposure to biotin at high concentration (e.g., 10 µM). The little release that does occur is probably from partial denaturation of some of the adsorbed streptavidin molecules, because no comparable dissociation of the biotin-streptavidin complex occurs at high biotin concentrations in solution. In some applications, biotin is included during subsequent reactions with the coated plastic surface to block all empty biotin-binding sites.

Several suppliers offer polystyrene microplates precoated with protein A, protein G, protein A/G, or protein L. IgGs that bind these proteins can be captured on these surfaces in native form without occluding their antigen-binding sites. It is important to choose an appropriate microplate, because different IgG subclasses from different species have different affinities for the four IgG-binding proteins.

Microplates that have been pre-activated with maleic anhydride (amine-reactive), maleimide (thiol-reactive), and hydrazide (reactive with oxidized IgG; see Figure 9.24 and exemplar protocols 10 and 11) can be obtained from several suppliers. Covalent capture of the antibody on these surfaces does not denature the antibody, and seldom blocks the antibody's antigen-binding site.

9.6.2 Covalent Coupling to an Affinity Matrix

Affinity chromatography has become a core technology in cell and molecular biology in the last three decades. Affinity purification of antibodies using antigen as the immobilized bait is the most frequent application in immunology; but the converse procedure, in which antigen is affinity purified using immobilized antibody as bait, is also common. For many years, the bait was usually coupled via its amines to CNBr-activated agarose beads, but more recently a number of alternative amine-reactive solid supports with important advantages have become commercially available. These include not only beaded matrices like the original CNBr-activated agarose, but also flat membrane sheets (e.g., Immunodyne ABC and UltraBind modified nylon membranes from Pall Life Sciences, Ann Arbor, MI). The affinity matrix in exemplar protocol 12 is UltraLink Biosupport Medium (Pierce), which consists of hydrophilic cross-linked bis-acrylamide/azlactone copolymer beads with a diameter of 50–80 µm, a pore size of ~100 nm, and a size exclusion limit of ~2 MDa. The matrix is more rigid than beaded agarose, being able to withstand a maximum pressure drop of 100 psi; the swelling ratio is 8–10 ml/g of dry weight, and is essentially independent of buffer ionic strength. Although the amine-reactive azlactone groups (Figure 9.5) are in vast molar excess over the protein amino groups, their immobility and steric hindrance prevent a single protein molecule from reacting with more than a few of them. Immobilization of an antibody is therefore very unlikely to block the antigen-binding sites on a significant fraction of the coupled molecules.

9.6.2.1 Exemplar Protocol 12: Coupling Protein to UltraLink Biosupport Medium

1. Mix 5 ml of protein at a concentration of 1.724 mg/ml (exact concentration not important; 1–2 mg/ml is a good range) in LoPBS (or any other weakly buffered solution with no nucleophilic solutes) with an equal volume of coupling buffer (1 M trisodium citrate, 0.2 M $NaHCO_3$, pH adjusted to 8.5 with NaOH).

2. Remove a 40-μl sample of the mixture into a microtube and store in freezer for use at step 6; this is the precoupling protein solution.

3. Weigh out 0.625 g UltraLink Biosupport Medium (Pierce; swelling ratio for this batch was 8 ml per gram dry weight) into a 15-ml polypropylene conical centrifuge tube; without delay add the remainder of the mixture step 1 and vortex to suspend the dried beads; rotate the tube slowly end-over-end overnight at 4°C to allow the protein to couple to the beads via their amino groups.

4. Centrifuge the 15-ml tube 5 min at low speed (~1000 rev/min); after removing the 40-μl supernatant sample at the next step, store the centrifuge tube in the refrigerator while awaiting the results of the Bradford assay at step 6.

5. Remove a 40-μl sample of the supernatant to a microtube for use in the next step.

6. Make Bradford diluent in a 1.5-ml microtube by mixing 500 μl LoPBS (or whatever buffer the protein was in at step 1) and 500 μl coupling buffer (step 1). Using this diluent, make 11 serial twofold dilutions of both the precoupling protein solution step 2 and the postcoupling supernatant previous step by passing 20 μl into 20 μl diluent. Mix 10 μl portions of all 24 samples (including the undiluted precoupling protein solution and postcoupling supernatant) with 250-μl portions of Coomassie protein assay reagent (Pierce), allow color to develop 15–45 min, and read the optical density at 595 nm on a plate reader or in a spectrophotometer. The exemplar results are shown in Figure 9.26; it is evident that in this case most of the input protein was coupled to the matrix and thus absent from the postcoupling supernatant. (Coupling cannot be assessed spectrophotometrically because of UV-absorbing material in the dried matrix; the supplier's instructions suggests the BCA assay as an alternative way of quantifying coupling.)

7. Centrifuge the 15-ml reaction tube step 4 for 3 min at ~1000 rev/min; carefully draw off the supernatant without disturbing the pellet; into the tube pipette 6 ml 2 M ethanolamine, pH adjusted to 9.0 with HCl; vortex to resuspend the matrix; rotate the tube slowly end-over-end overnight at 4°C to allow the ethanolamine to couple to and thus quench all remaining azlactone reactive groups.

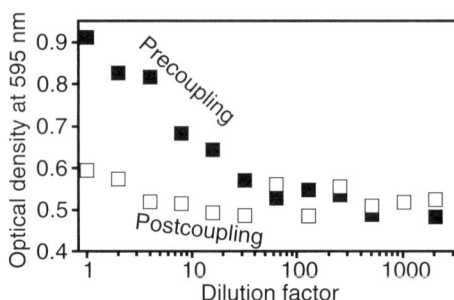

FIGURE 9.26 Bradford assays of the input protein solution and the postcoupling supernatant from a reaction of a protein with the amine-reactive matrix UltraLink Biosupport Medium. See exemplar protocol 12 for details. The data show that very little protein remains in the supernatant after coupling (open squares), implying that most of the input protein was successfully coupled to the matrix.

8. Transfer the suspension to a 50-ml centrifuge tube and fill the tube with TBS; wash the coupled matrix ten times with TBS, centrifuging for ~5 min at ~1000 rev/min and carefully removing the supernatant each time; after the final wash resuspend the matrix in a total volume of 45 ml and store at 4°C (do not freeze). As expected from the swelling ratio of 8 given in the specifications for this batch of support, the packed volume is ~5 ml. If desired, NaN_3 can be added to 0.02% as a preservative (note that concentrated NaN_3 is very toxic and should be handled with care).

9.6.3 Biotin-Mediated Immobilization on an Affinity Matrix

Just as the extremely high-affinity biotin-streptavidin bond can be exploited to immobilize a Bio-IgG on a polystyrene surface, so also it can be used to immobilize a Bio-IgG on an affinity matrix. Suitable affinity matrices with streptavidin already coupled are available commercially in many formats. This mode of immobilization is extremely reliable and efficient, essentially all the input biotinylated protein being captured on the matrix until the binding capacity of the coupled streptavidin is exceeded. The experimenter thus has control over the density of immobilized protein. To fully exploit this benefit the Bio-IgG (or other biotinylated protein) should be bound to the matrix batchwise with mixing, so that it is distributed uniformly in the matrix material. If, contrary to this advice, the Bio-IgG is loaded in limiting amounts onto a matrix that is already packed in a chromatography column, the upper parts of the column bed will have much higher densities of the protein than the lower parts.

Surface plasmon resonance (SPR; reviewed by Karlsson and Larsson[26,27]) is a common application of this type of affinity matrix, and illustrates well its advantages. An SPR instrument continuously records the refractive index in a ~100-nm layer on the surface of a chip exposed to a stream of fluid. Typically, the surface of the chip is coated with carboxymethylated dextran polymer chains to which proteins or other biomolecules can be covalently coupled. SPR chips with streptavidin precoupled at high density are commercially available from several sources. The SPR chip is mounted in an integrated microfluidics cartridge, whose channels and computer-controlled valves allow selected solutions to be pumped over small patches of the chip surface called flow cells. In the Biacore 2000 SPR instrument (Biacore Inc., Piscataway, NJ), for example, each chip has four independently addressable flow cells with a volume of 60 nl each, and the instrument can record from all four flow cells independently and simultaneously as small volumes of solution flow through them individually or series. One member of a binding pair—the "ligand"—is immobilized on the chip surface. In exemplar protocol 13 below, the ligand is streptavidin that has been precoupled to the chip surface. A solution containing the other member of the binding pair—the "analyte"—is then pumped through ("injected" into) the flow cell. In the exemplar protocol, the analyte is a Bio-IgG. As analyte molecules bind to immobilized ligand molecules, the refractive index at the chip surface increases by minute amounts, which are accurately measured and continuously recorded by the instrument in standardized resonance units (RU) that are proportional to the mass of analyte bound. Because analyte solution flows continuously over the chip surface, this binding does not result in any significant decrease in analyte concentration in the bulk solution.

Streptavidin that has been coupled to an SPR chip is extraordinarily stable. Multiple 1-min pulses of a strongly alkaline stripping solution (0.05 N NaOH, 1 M NaCl) are routinely used to clean a streptavidin-coupled flow cell of all adventitiously bound material before injection of a biotinylated protein. After a biotinylated protein is bound, it can be used in turn as the ligand for a second mobile analyte in multiple second-stage binding experiments. Harsh stripping solutions can be used to strip the flow cell of residual analyte after each such second-stage experiment without desorbing the biotinylated protein, provided the latter survives in those stripping solutions. Examples of such stripping solutions are: 0.025 N HCl; pH 2.2 elution buffer (0.1 N HCl, pH adjusted to 2.2 with glycine); ImmunoPure IgG elution buffer from Pierce (pH 2.75); and ImmunoPure gentle elution buffer from Pierce (pH 6.55; very high ionic strength).

In exemplar protocol 13, four different Bio-IgGs were immobilized to a level of ~100 RU in the four flow cells of a streptavidin-coupled chip. Three of the Bio-IgGs were antipeptide antibodies that had been affinity-purified from the total Bio-IgG population in a complex anti-pathogen antiserum, which had subspecificities against each of the three peptides and against hundreds or thousands of other antigens besides. The fourth Bio-IgG was the unbound Bio-IgG population from the affinity purifications; since this peptide-absorbed Bio-IgG showed virtually no binding activity against any of the three peptides, it served as a negative control flow cell in the second-stage binding experiments to be described after the exemplar protocol. It was important in this application to control the level of Bio-IgG immobilization. If the immobilization level had been too low, analyte binding couldn't have been measured in the subsequent second-stage binding experiments; if the immobilization level had been too high, those second-stage binding experiments would have been compromised by artifacts. The total consumption of each Bio-IgG was less than 25 ng.

9.6.3.1 Exemplar Protocol 13: Immobilization of Bio-IgGs on a Streptavidin-Coupled Chip

1. Equilibrate all four flow cells with TBS/Tween (0.15 M NaCl, 0.1 M Tris·HCl pH 7.5, 0.5% Tween 20) at a flow rate of 10 µl/min.

2. Prepare 150 µl of a 1-nM dilution of each Bio-IgG in TBS/Tween; inject each into a single flow cell at a flow rate of 10 µl/min, continuously monitoring the increase in RU; when that increase reaches 100 RU, stop the injection and pump TBS/Tween through the flow cell. Results for one of the Bio-IgG injections are shown in Figure 9.27A. Immediately after injection is begun there is a sharp rise in RU; this rise, called the bulk shift, reflects a slight difference in the bulk refractive index of the TBS/Tween buffer and the Bio-IgG dilution (if the Bio-IgG dilution had a lower refractive index than TBS/Tween, the bulk shift would have been negative). After the bulk shift, there is a gradual increase in RU, reflecting binding of the Bio-IgG analyte to the streptavidin ligand. It is the simplicity of the biotin-streptavidin interaction that allows immediate, direct readout of immobilization level; more complex immobilization methods, such as covalent coupling, do not allow direct readout.

3. After the cumulative increase (after the bulk shift) reaches about ~100 RU, immediately stop the injection and resume the flow of TBS/Tween through the flow cell. After a sharp negative bulk shift—the mirror image of the positive bulk shift at the start of injection—the RU settles to a level ~100 RU above the original baseline level, indicating successful immobilization of Bio-IgG to a level of ~100 RU.

4. Repeat the previous two steps for the other three Bio-IgGs.

The chip generated in the foregoing exemplar protocol was used in numerous second-stage binding studies in which the immobilized Bio-IgGs served as ligands and peptide-bearing fusion proteins served as analytes. After each such binding experiment, the chip was regenerated with two 1-min pulses of one of the stripping buffers mentioned above: either pH 2.2 elution buffer, ImmunoPure IgG elution buffer, or ImmunoPure gentle elution buffer. Figure 9.27B shows the results of a series of eight binding experiments in which a peptide-bearing fusion protein at four concentrations was injected through all four flow cells for either 2.5 or 10 min, the resonance response being continuously recorded both during the injection (in which analyte associated with ligand) and during the subsequent 10 min of dissociation. After each experiment, the flow cells were stripped with two 1-min pulses of ImmunoPure gentle elution buffer. The resonance response was recorded from all four flow cells, the background response from the negative control flow cell being subtracted from that from each of the other three flow cells. Figure 9.27B shows the background-subtracted response from the flow cell in which the cognate anti-peptide Bio-IgG was immobilized (the other flow cells showed essentially no response during these injections). Because of background subtraction, the bulk shifts are evidenced only by small transients. The superimposibility of the 2.5- and 10-min association phase responses at all four fusion protein concentrations shows

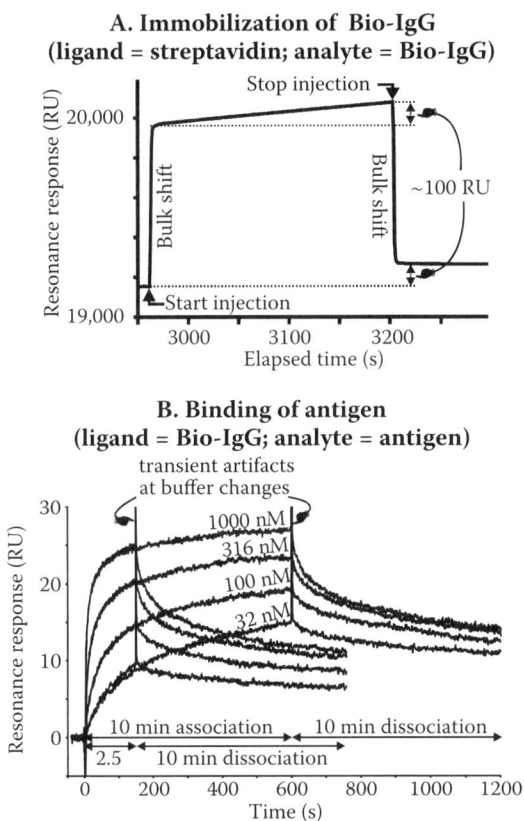

FIGURE 9.27 Two stages of an SPR analysis. *See* the text for details. (A) Capture of Bio-IgG on a streptavidin-coupled SPR chip. (B) Second-stage experiments analyzing the kinetics of binding of the captured Bio-IgG with its cognate antigen.

that the antigen-binding capacity of the immobilized Bio-IgG was fully recovered after each binding experiment. These SPR experiments dramatically illustrate the effectiveness of biotin-mediated immobilization on an affinity matrix.

9.7 PROTEOLYTIC MODIFICATION

Porter's finding that the protease papain cleaves native rabbit IgG into two antigen-binding Fab domains and a single crystallizable Fc domain[28] was key in delineating the basic four-chain/three-domain structure of immunoglobulins. The cleavage that liberates the Fab and Fc domains occurs in the hinge region of the H chain, not because of high specificity of the protease for that region of the amino acid sequence, but rather because the hinge region is flexible and extended, and for that reason much more accessible to protease than the L chain and the remainder of the H chain. Other proteases were similarly shown to cleave native IgG preferentially in the hinge region, as shown in Figure 9.28. Because papain cleaves N-terminal to the H–H interchain disulfide bonds, it liberates Fab and Fc fragments (about 50 kDa each)

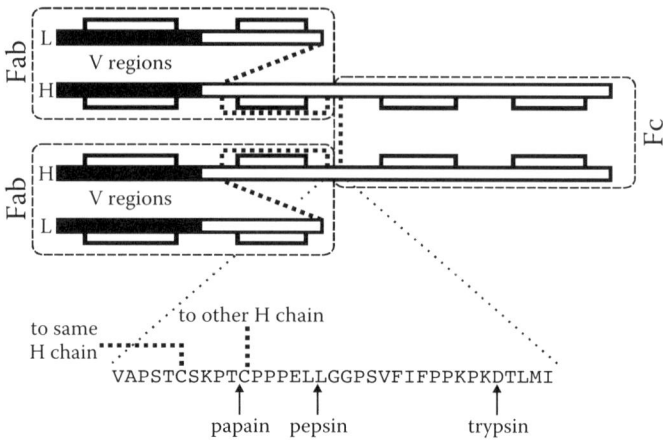

FIGURE 9.28 Schematic diagram of rabbit immunoglobin G (IgG),[45] with the conserved intra-chain disulfide bonds of the ancestral immunoglobulin fold (accessible only under denaturing conditions) shown with solid square brackets, and the other disulfides (accessible under non-denaturing conditions) shown with dashed brackets and lines (as in the diagram of a generic IgG in Figure 9.14). Rabbit IgG is unusual in having an extra intrachain disulfide bond in the constant region of the H chain (dashed bracket). The detail shows the amino acid sequence of the H-chain hinge region (GenBank accession P01870), with dashed lines indicating disulfide bonds and arrows indicating sites of cleavage of native IgG with papain, pepsin, and trypsin.[46] Pepsin also cleaves at multiple sites within the Fc domain (not shown).

that are not disulfide bonded to one another. The N-terminal half of the cleaved H chain, which is part of the Fab fragment, is called the Fd fragment. Pepsin, in contrast, cleaves C-terminal to the H–H bond and at multiple sites within the Fc domain; the Fab domains therefore remain disulfide bonded to each other, forming a ~100-kDa F(ab′)$_2$ fragment. The Fab and F(ab′)$_2$ proteolytic fragments are readily isolated in good yield from polyclonal rabbit IgG, and have a number of practical uses. Fab binds antigen monovalently, thus avoiding the avidity effect of multivalent interaction, while F(ab′)$_2$ fragment binds bivalently and thus retains the avidity effect observed for intact IgG. Both types of fragment are missing most of the specific binding sites involved in antibody effector functions, such as complement fixation. In particular, they do not bind the various Fc receptors on many mammalian cells, and for that reason are often preferable to intact IgG as immunocytochemical or *in vivo* imaging probes. The inter–H-chain disulfide bond of F(ab′)$_2$ fragments can be reduced under non-denaturing conditions to yield monovalent F(ab′) fragments, which in turn can be modified with thiol-reactive modifying reagents (Section 9.4).

Rabbits are unusual among mammals in having a single IgG isotype; papain and pepsin therefore cleave the great majority of IgG molecules in normal or immune serum with much the same kinetics. The multiple IgG isotypes in most other species are cleaved by various proteases with widely differing kinetics, depending on the positions of the interchain disulfide bonds and the conformation and exact amino

TABLE 9.4
Studies of Proteolytic Cleavage of Nondenatured IgG from Various Species

IgG	Proteases	Authors
Human IgG1–IgG4	Pepsin, papain, trypsin, plasmin	Gorevic et al.[30]
Mouse IgG1, IgG2a, IgG2b	Clostripain, lysyl endopeptidase, metalloendopeptidase, V8 protease	Yamaguchi et al.[31]
Mouse mAb OC859	Pepsin, papain, bromelain, ficin	Zou et al.[32]
Mouse IgG subclasses	Papain	Adamczyk et al.[33]
Mouse IgG1, IgG2a, IgG2b	Pepsin, papain, elastase, trypsin, chymotrypsin	Parham[34]
Mouse IgG1, IgG2a, IgG2b, IgG3	Pepsin	Lamoyi[35]
Mouse IgG1, IgG2a, IgG2b, IgG3	Pepsin, papain	Smith-Gill et al.[36]
Mouse IgG isotypes	Various	Smith[37]
Mouse IgG1	Pepsin, papain	Kurkela et al.[38]
Mouse monoclonal B72.3	Pepsin, papain, bromelain	Milenic et al.[39]
Rat IgG isotypes	Pepsin, papain, V8 protease	Rousseaux et al.[44,41]
Bovine IgG1, IgG2a, IgG2b	Pepsin, papain	Butler et al.[42,43]
Chicken IgY	Papain, pepsin	Akita and Nakai[44]

acid sequence in the hinge region. Table 9.4 lists some of the studies on proteolytic fragmentation of IgG isotypes from a number of species. For neither mouse, human, nor rat IgG is there are a single set of conditions that reliably yields Fab or F(ab')$_2$ in good yield from all isotypes, though suitable conditions for each individual isotype have been defined. Preparation of proteolytic fragments will be illustrated with a single set of exemplar protocols.

9.7.1 PREPARATION OF FAB FROM RABBIT IGG BY PAPAIN DIGESTION

This protocol uses immobilized papain, whose key advantage over nonimmobilized enzyme is that the cleaved IgG substrate can be completely removed from the protease after digestion. It also allows complete removal of the disulfide reducing agent that is required to activate the enzyme (a cysteine protease) before digestion (though this advantage was not exploited in the exemplar protocols below). Removal of the reducing agent can be particularly important in the case of rabbit IgG, whose Fab domain contains an intra–H-chain disulfide bond that can be reduced under nondenaturing conditions (Figure 9.28). The papain-digested IgG is passed through a protein A affinity column to remove the Fc fragment and any residual uncleaved or partially cleaved IgG that retains the Fc fragment. Immobilized papain and other enzymes (including pepsin) are available from several sources. In the exemplar protocols below, the immobilized papain was part of the ImmunoPure Fab Preparation Kit from Pierce.

9.7.1.1 *Exemplar Protocol 14: Pilot Digestions to Determine*
         the Time Course of Papain Digestion of Rabbit IgG

1. Dissolve 17.5 mg cysteine·HCl in 5 ml phosphate buffer pH 10 from Pierce
   kit. The pH was 6.6; it was adjusted to 6.8 by small additions of 2 N NaOH.
   This is the digestion buffer; it was kept on ice.
2. Into a 1.5-ml microtube pipette 800 µl of digestion buffer. Cut off the
   end of a pipette tip to make a larger opening. Thoroughly but gently
   resuspend the immobilized papain (6% cross-linked beaded agarose supplied as a 50% slurry in 50% glycerol, 0.1 M sodium acetate, 0.05%
   NaN$_3$, pH 4.4; 250 µg enzyme per ml of settled gel; 28 BAEE units per
   mg enzyme) by inverting the bottle many times. Using the cut-off tip,
   pipette 50 µl of the immobilized papain slurry into the 1.5-ml microtube;
   rotate the tube 5 min. During this incubation the cysteine in the digestion
   buffer reduces any active-site cysteines that may have become oxidized
   during storage, thus activating the immobilized papain.
3. Microfuge the tube at 6000 rev/min for 4 min. Gently pipette off the
   supernatant (being careful to avoid losing any of the pellet), and resuspend the pellet in 800 µl digestion buffer. Mix by inversion, remicrofuge
   as before and again pipette off the supernatant. Resuspend, microfuge,
   and pipette off supernatant once more as before. Add 100 µl digestion
   buffer, giving a nominal total volume of ~150 µl. This is the activated
   papain; the microtube containing it is called the digestion tube in steps
   5 and 6.
4. Into six 1.5-ml microtubes pipette 40 µl 2× sample buffer (125 mM
   Tris·HCl pH 6.8, 20% glycerol, 4% sodium dodecyl sulfate, 0.4 mg/ml
   bromphenol blue); these will be the time-point tubes.
5. To the activated papain in the digestion tube (step 3) add 53.7 µl (962
   µg) 17.9 mg/ml rabbit IgG in LoPBS; the final volume is ~200 µl and
   the final protein concentration is 4.7 mg/ml. Mix the contents of the
   digestion tube by inversion and immediate remove a 4-µl (18.9-µg)
   sample to one of the time-point tubes previous step; this is the 0 h time
   point. Warm the digestion tube to 37°C in a water bath and tape it
   horizontally to the platform of a shaker-incubator at 37°C. Meanwhile,
   heat the 0 h time-point tube to 95°C for a few minutes and store it in
   the deep freeze.
6. At 1, 2, 4, 8, and 20 h, pipette 4-µl samples from the digestion tube into
   additional time-point tubes from step 4. Return the main digestion tube
   to the shaker incubator as fast as possible each time to keep the main
   digestion temperature at 37°C. Heat each time-point tube to 95°C for ~5
   min and store in freezer with the 0 h sample (previous step).
7. In a 1.5-ml microtube mix the standard (10 µl 2x sample buffer, 8 µl
   water, 2 µl Promega mid-range protein standards [Promega, Madison,
   WI]), and 1 µl 2-ME.
8. Thaw the six time-point tubes (steps 5 and 6); to each add 40 µl of a 1/10
   dilution of 2-ME.

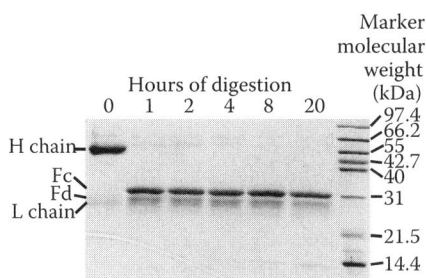

FIGURE 9.29 Time course of digestion of rabbit immunoglobin G (IgG) with immobilized papain. IgG samples after various times of digestion were analyzed by acrylamide gel electrophoresis after reduction of disulfide bonds. See exemplar protocol 14 for details.

9. Heat the standard and time-point tubes (previous two steps) at 95°C for ~5 min, then microfuge briefly to drive solution and the papain resin to bottom.

10. Electrophorese 21-μl (9-μg) portions of the supernatants in the time-point tubes and 5-μl samples of the standard in a standard denaturing 12.5% polyacrylamide Tris·HCl gel and stain with Coomassie Blue as usual.[29] The stained gel is shown in Figure 9.29.

Before digestion (0 h) the intact IgG yields an intensely-stained H chain (apparent molecular weight ~50 kDa) and a faintly-stained L chain (apparent molecular weight ~25 kDa). Already after 1 h of digestion the H chain is almost completely cleaved into halves: an intensely-stained Fc portion (apparent molecular weight ~31 kDa; the native Fc is a homodimer of these chains) and a faintly stained Fd fragment (apparent molecular weight ~27 kDa) just behind the L chain (which remains intact with an apparent molecular weight of ~25 kDa; the native Fab is a heterodimer of Fd with the L chain). In light of these results the large-scale papain digestion (exemplar protocol 15) will be allowed to progress for 1 h.

9.7.1.2 Exemplar Protocol 15: Large-Scale Papain Digestion of Rabbit IgG

1. Dissolve 0.25 g cysteine·HCl in 71.4 ml phosphate buffer pH 10 from Pierce kit. Adjust pH to 6.84 by small additions of 2 N NaOH. This is the digestion buffer, which is kept on ice.

2. Set up a 125-ml vacuum flask with a 15-ml sintered glass filter attached. Resuspend the immobilized papain by inverting the bottle many times; quickly, before the gel has a chance to settle again, pipette 4 ml of the suspension (~2 ml settled bed volume) into the sintered glass filter.

3. Connect vacuum and suction off the fluid. Disconnect the vacuum, resuspend the gel in ~10 ml digestion buffer by stirring with a small spatula, reattach vacuum and suction out the buffer. Repeat the cycle in the previous sentence two more times. Resuspend the gel in 5 ml digestion buffer; use a transfer pipette to transfer all of it into an ice-cold 50-ml conical

centrifuge tube and wash the sintered glass filter into the 50-ml tube with additional digestion buffer as necessary to give a total gel suspension volume of 12 ml (volume estimated from graduations on the 50-ml tube). This is the activated papain.

Note: As explained previously, the interchain and one of the intrachain disulfide bonds of rabbit IgG can be reduced under nondenaturing conditions (dashed brackets and lines in Figure 9.28). Because cysteine is present in the digestion buffer, disulfide interchange is possible during the papain digestion. To eliminate this possibility, the immobilized activated papain could have been washed a few times with a thiol-free digestion buffer at pH 6.8 before being transferred from the sintered glass filter.

4. To the activated papain add 4.3 ml (77 mg) 17.9-mg/ml rabbit IgG in LoPBS. Close cap tightly, vortex gently, warm briefly in a 37°C water bath, and shake gently in a 37°C incubator for 1 h.
5. During the 1-h digestion, rinse out the 15-ml sintered glass filter and mount it on the 125-ml vacuum flask. Run 5 ml 1-mg/ml BSA through the filter to block nonspecific adsorption sites on the glass, followed by many washes of distilled water to thoroughly clean the filter. Mount the filter on a clean 125-ml vacuum flask.
6. When the 1-h digestion is finished, immediately transfer the contents of the 50-ml tube into the 15-ml sintered glass filter connected to vacuum, collecting the filtrate in the vacuum flask. Centrifuge the 50-ml tube to drive residual suspension to the bottom, and pipette this too into the sintered glass filter. Disconnect vacuum, resuspend the gel in 8 ml ImmunoPure IgG (A) binding buffer from the Pierce kit, and immediately suction the wash into the vacuum flask, too. (The used immobilized papain on the sintered glass filter will be regenerated in the next step.) Transfer the filtrate from the vacuum flask to a 50-ml conical centrifuge tube; rinse out the vacuum flask with an additional 5 ml of binding buffer and add that, too, to the filtrate in the 50-ml conical centrifuge tube. This is the papain digest that will be processed in exemplar protocol 16; its volume was ~28 ml, and the theoretical IgG concentration 2.75 mg/ml.
7. To regenerate the used immobilized papain, wash the matrix five times with LoPBS on the sintered glass filter. Transfer the suspended gel into a 15-ml bottle in a total volume of ~10 ml (need not be accurately determined) and add 100 μl 5% sodium azide. This can be stored in the refrigerator for reuse.

9.7.1.3 Exemplar Protocol 16: Removal of Fc and Incomplete Digestion Products by Protein A Affinity Chromatography

1. Remove top stopper and then the bottom stoppers from two 2.5-ml protein A columns from the Pierce kit; pour off the storage solution (caution: contains

azide); mount columns on a ring stand; fill tops of columns with 7 ml IgG binding buffer (see exemplar protocol 15) and allow columns to drain (flow will stop when liquid gets to fritted column top); wash columns with an additional 7-ml portion of IgG binding buffer.

2. Mount each column over a 50-ml conical centrifuge tube. Using a transfer pipette, load half (14 ml) of the papain digest (step 6 of exemplar protocol 15) into each column; centrifuge the 50-ml tube to drive residual digest to the bottom, and load half of this residual solution into each column. When all the papain digest has flowed into a column, add an additional 5 ml IgG binding buffer and allow that to flow in too; when that 5 ml has flowed in, add yet another 5 ml of IgG binding buffer and allow that to flow in too. All this time, the column effluents should be collected into the 50-ml conical centrifuge tubes. The final volume in each tube should be ~24 ml. Transfer the contents of one of the conical centrifuge tubes into the other. This is Fab; the total volume is 45 ml and the Fab concentration is theoretically 1.14 mg/ml; store in the refrigerator until it is processed at step 6 below.

3. As each protein A column is finished, mount it over a waste beaker. Wash the column with two 3-ml portions of IgG binding buffer, allowing the effluent to collect into the waste beaker. Then mount the column over a 15-ml polypropylene tube containing 300 μl 1 M Tris·HCl pH 9.1. Elute each column with six 1-ml portions of ImmunoPure IgG elution buffer from the Pierce kit. Pool the two eluates (one from each column) into a single tared 15-ml bottle: the net weight turned out to be 13.6 g = 13.6 ml. The Fc concentration is theoretically 1.89 mg/ml; incompletely cleaved contaminants may be present in this eluate as well. The pH was estimated to be ~8.4 with an indicator strip. This bottle was stored in the deep freeze.

4. To regenerate the columns, secure each over a large waste beaker and wash it with two 5-ml portions of 0.1 M citric acid, pH adjusted to 3.0 with NaOH, two 5-ml portions of distilled water, and one 7-ml portion of 0.02% NaN_3 in water. Stop the column when there is ~2 ml of the final solution remaining on top of the bed. Put on bottom stopper first, then top stopper. Store the columns in the refrigerator.

5. In a 1.5-ml microtube mix the standard as in step 7 of exemplar protocol 14. Premix 585 μl water, 650 μl 2x sample buffer, and 65 μl 2-ME; pipette 60 μl of this premix into two 1.5-ml microtubes; into one microtube (Fab) pipette 3 μl (theoretically 3.42 μg) of the Fab preparation step 2; into the other microtube (Fc) pipette 1 μl (theoretically 1.89 μg) of the Fc preparation step 3. Heat the samples in a boiling water bath for ~5 min; cool; microfuge briefly to drive samples to bottom of tubes. Electrophorese 21-μl portions of the Fab and Fc samples and a 5-μl portion of the standard as in step 10 of exemplar protocol 14. The results in Figure 9.30 indicate essentially complete digestion by papain (as in the time-course experiment in exemplar protocol 14) and essentially complete separation of Fab from Fc.

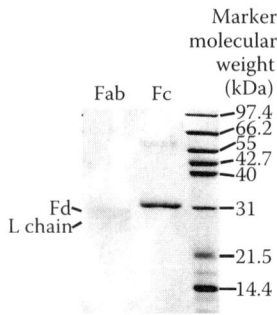

FIGURE 9.30 Fab and Fc samples from protein A chromatography were analyzed by acrylamide gel electrophoresis after reduction of disulfide bonds. See exemplar protocol 16 for details. The data show no evidence for Fc contamination of the Fab fraction.

6. Concentrate the bulk Fab preparation step 2 and wash it twice with LoPBS on a Centriprep 30-kDa centrifugal concentrator; the final volume was 1.935 ml. The protein concentration was determined to be 24 mg/ml by scanning spectrophotometry of a 1/40 dilution, assuming the absorbance of a 1-mg/ml solution is 1.48^2; the total yield was therefore 46 mg, which is 90% of theoretical from 77 mg of input IgG.

Because of the possibility of disulfide interchange during papain digestion (note at step 3 of exemplar protocol 15), the Fab may have some free sulfhydryls. In exemplar protocol 17, any such sulfhydryls will be blocked by alkylation with N-ethylmaleimide (NEM; Figure 9.22). The Fab preparation will then be fractionated by gel filtration chromatography on Sephadex G150 to separate authentic Fab molecules (50 kDa) from contaminants such as Fab dimers (100 kDa), partially cleaved IgG (100 kDa) or uncleaved IgG (150 kDa). As it turned out, the chromatography step was probably unnecessary in this case because no evidence of contaminants was seen.

9.7.1.4 Exemplar Protocol 17: Blocking Sulfhydryls with N-ethylmaleimide and Sephadex G150 Chromatography

1. Swell 10 g of Sephadex G150 in water at 80°C for 5 h, cool, remove fines by three cycles of settling, equilibrate in LoPBS, pour it into a chromatography column with an inner diameter of 2.5 cm, and pack with a pump at a flow rate of 1 ml/min. The final bed volume height was 45 cm, corresponding to a volume of 220 ml.
2. Add 65 μl LoPBS and 9.2 μl 250 mM EDTA, pH adjusted to 8 with NaOH to the Fab preparation (step 6 of exemplar protocol 16); the purpose of the EDTA is to chelate heavy metal ions that otherwise might catalyze oxidation of sulfhydryls.
3. Dissolve a small amount of NEM in sufficient HEPES buffer (100 mM HEPES, pH adjusted to 7.0 with NaOH) to make a 2.5-mg/ml (40-mM)

solution; add 670 µl of 40-mM NEM to the Fab preparation previous step, bringing the final NEM concentration to 10 mM; incubate 40 min at room temperature to allow NEM to alkylate any free sulfhydryls. Note that NEM is toxic and should be disposed of properly.

4. Immediately load the NEM-treated Fab into the G150 column and elute the column with LoPBS at a flow rate of 1 ml/min, collecting 3.9- and 1.9-ml fractions.

5. Determine the absorbance of the fractions at 280 nm. The results are shown in the lower part of Figure 9.31. A single symmetrical protein peak was observed at a position expected for a 50-kDa Fab fragment; no evidence for 150-kDa intact IgG or 100-kDa partial digestion products is apparent in the trace.

6. Premix 585 µl water and 650 µl 2× sample buffer (step 4 of exemplar protocol 14), and pipette 19 µl of the premix into 16 1.5-ml microtubes; into these tubes pipette 2 µl of selected fractions step 4. Pipette 385 and 44 µl of the premix into two additional 1.5-ml microtubes; to these tubes add 1 µl (17.9 µg) untreated IgG and 1.1 µl (theoretically 2.1 µg) Fc fragment (step 3 of exemplar protocol 16), respectively. Heat all 18 microtubes to 95°C for ~5 min; microfuge them briefly to drive the solution to the bottom. These electrophoresis samples contain no mercaptoethanol or other reducing agent; any disulfide bonds should therefore remain intact.

7. Electrophorese the samples from the previous step on two Tris·HCl acrylamide gels and stain the gels as in step 10 of exemplar protocol 14. The results are shown at the top of Figure 9.31. The main species in all fractions has an apparent molecular weight of ~50 kDa, as expected for a Fab whose L-chain and Fd fragment remain disulfide-bonded to each other (see Figure 9.28).

Note: The Fc fragment seems to be a mixture of species with apparent molecular weights of ~25, ~50, and ~75 kDa, corresponding ostensibly to monomers, homodimers, and homotrimers of the Fc portion of the H chain (see Figure 9.28). Such heterogeneity of multimeric state can arise from disulfide interchange during the papain digestion (see note at step 3 of exemplar protocol 15). However, since the Fc portion of the H chain theoretically has only one cysteine accessible under non-denaturing conditions (see Figure 9.28), it should not be able to form a trimer. Perhaps disulfide exchange occurred after denaturation at step 6; alternatively, trimeric molecules may have arisen from papain cleavage of some of the H chains at a position N-terminal to the first cysteine in the detail in Figure 9.28, rather than between that cysteine and the cysteine that is disulfide-bonded to the other H-chain.

8. Pool peak fractions from step 4 (large filled square symbols in the lower part of Figure 9.31); concentrate them and wash them twice with LoPBS on a Centriprep 30-kDa centrifugal concentration device (see Section 9.2.3); the final volume was 1.445 ml. This preparation can be stored in the refrigerator or the deep freeze.

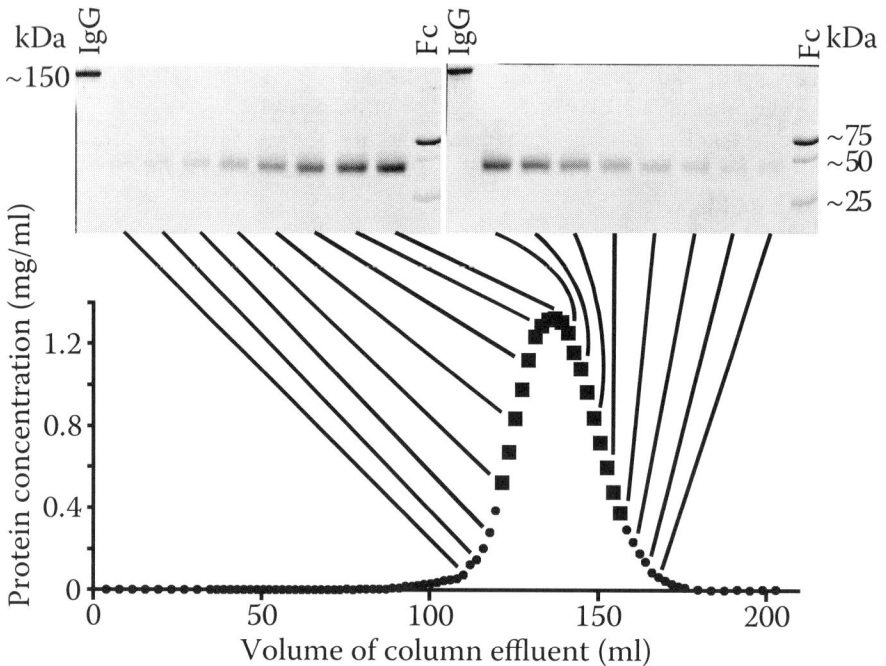

FIGURE 9.31 Analysis of fractions from G150 gel filtration. See exemplar protocol 17 for details. The lower part of the figure shows the protein concentrations in the G150 fractions; large filled squares indicate fractions that were ultimately pooled as purified Fab; small filled circles indicate fractions that were discarded. The upper part of the figure shows analysis of G150 fractions by acrylamide gel electrophoresis without reduction of disulfide bonds. Nearly all the protein has a disulfide-bonded size of ~50 kDa.

9. Scan a 1/40 dilution spectrophotometrically. The concentration was calculated as in step 6 of the previous exemplar protocol to be 24.4 mg/ml. The total yield is thus $1.445 \times 24.4 = 32.3$ mg, which is 69% of the theoretical yield from 77 mg of starting IgG.

9.8 CROSS-LINKING ANTIBODIES TO OTHER PROTEINS

Immunoconjugates in which antibodies are covalently cross-linked to other proteins are ubiquitous in biotechnology today. This class of conjugates includes most prominently antibody-enzyme conjugates for ELISA, immunoblotting, and immunocytochemistry applications; and immunotoxins, in which an antibody specific for a tumor or other therapeutic target is coupled to a cytotoxic "payload" to kill the target. Hermanson[1] has reviewed preparation of enzyme conjugates and immunotoxins; only the most general principles will be discussed and illustrated here.

Almost all antibody-protein conjugates today are prepared by a two-step process using heterobifunctional cross-linkers. Each partner (the antibody and the other protein) is "activated" by modification with a relatively stable reactive group. The two activated proteins are then mixed and allowed to cross-link to each other. Two-step cross-linking greatly favors cross-conjugation of the two partners with each other, which creates functional complexes, over self-conjugation of the proteins, which creates nonfunctional and potentially interfering aggregates. Three general characteristics are key to a useful two-step cross-linking method: (1) The cross-linking chemistry should be "orthogonal," in the sense that the activating groups conjugated to each protein partner should not react with any groups on that protein, but should react readily with the activating groups conjugated to the other protein; (2) The activating groups should be stable, allowing time for the two activated proteins to be purified and quantified before being cross-linked to each other; (3) The activating groups should be quantifiable, so that the cross-linking reaction can be reproducibly controlled.

In some cases the maleimide/thiol reactive system (bottom part of Figure 9.22) meets these criteria reasonably well. For instance, protected thiols can be conjugated to an antibody (having few or no endogenous free thiols) with the amine-reactive NHS ester SATA (exemplar protocol 5), while maleimides can be conjugated to the other protein with the amine-reactive NHS ester succinimidyl-4-(N-maleimidomethyl)cyclohexane-1-carboxylate. Alternatively, the maleimides can be conjugated to antibody amines, whereas the other protein's amines can be thiolated (if it does not have endogenous free thiols). Free thiols, though subject to air oxidation, are sufficiently stable for a brief purification process, and are easily quantified with Ellman's reagent (exemplar protocol 8). Maleimides, though subject to slow hydrolysis, are also sufficiently stable to allow purification (see Section 9.4.4); they have the disadvantage of not being easily quantified.

Solulink (San Diego, CA) has recently introduced an aldehyde/hydrazine cross-linking technology called HydraLinK that meets the orthogonality, stability and quantifiability requirements more fully than does the maleimide/thiol system. As shown in Figure 9.32, the IgG's amines are reacted with SANH to make an MEHN-IgG adduct with acetone-blocked aromatic hydrazines; while the other protein's amines are reacted with SFB to make an FB-protein adduct with aromatic aldehydes. The amine-modified proteins are stable for weeks in aqueous solution—plenty of time for purification and quantitation. When they are mixed, the aromatic aldehydes on one protein rapidly displace the acetone from the aromatic hydrazines on the other protein, thereby cross-linking the two proteins through stable bis-aromatic hydrazone (HydraLinK) cross-links. Unlike nonaromatic hydrazones (Figure 9.24), these linkages are stable in a wide pH range without reduction. After the two proteins have been satisfactorily cross-linked, any unreacted aldehydes can be capped with excess aromatic hydrazine and any unreacted hydrazine can be capped with excess aromatic aldehyde (Figure 9.33). These capping reactions also create near-UV chromophores by means of which the number of hydrazines or aldehydes on small portions of the modified proteins can be quantified before cross-linking (Figure 9.33). In exemplar protocols 18–20, which are based on instructions from Novabiochem (San Diego, CA), it is the IgG that is modified with hydrazines and the other protein that is modified with aldehydes; the order could equally well be reversed.

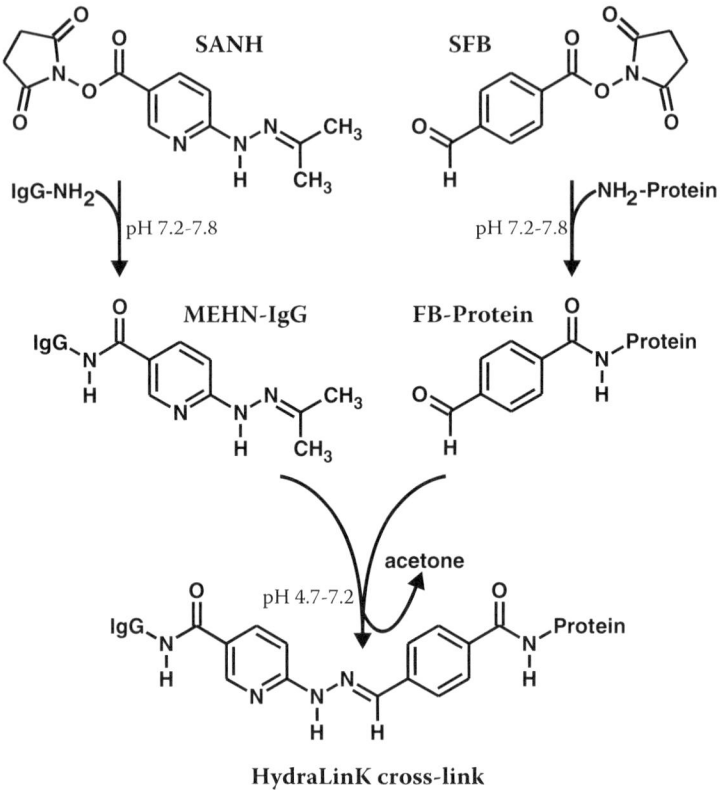

FIGURE 9.32 HydraLinK cross-linking of IgG to another protein. One of the two proteins (in this case, the immunoglobin G [IgG] is amine-modified at neutral pH with the NHS ester acetone 5-(succinimidyloxycarbonyl)-pyridine-2-yl hydrazone (SANH), yielding a 6-[2-(1-methyeth-ylidene)hydrazino]nicotinyl- (MEHN-) IgG adduct with several acetone-blocked aromatic hydrazines. In parallel, the other protein is amine-modified (also at neutral pH) with another NHS ester, succinimidyl 4-formylbenzoate (SFB), yielding a formylbenzoyl- (FB-) protein adduct with several aromatic aldehydes. When the two modified proteins are mixed, the aromatic aldehydes on one protein displace the acetone blocking groups from the aromatic hydrazines on the other protein, cross-linking the proteins through stable bis-aromatic hydrazone linkages.

9.8.1 Exemplar Protocol 18: Conjugation of Hydrazine to IgG Amines

1. Bring IgG to a concentration of 50 μM in PBS (0.15 M NaCl, 0.1 M NaH$_2$PO$_4$, pH adjusted to 7.2 with NaOH). Dissolve solid SANH (Figure 9.32) in sufficient dimethylformamide to give a concentration 1.45 mg/ml (5 mM). Add 1 volume of 5-mM SANH to 10 volumes of 50-μM IgG solution; the final concentrations of IgG and SANH are 45.5 and 455 μM, respectively. Allow the reaction to proceed at room temperature for at least 2–3 h.

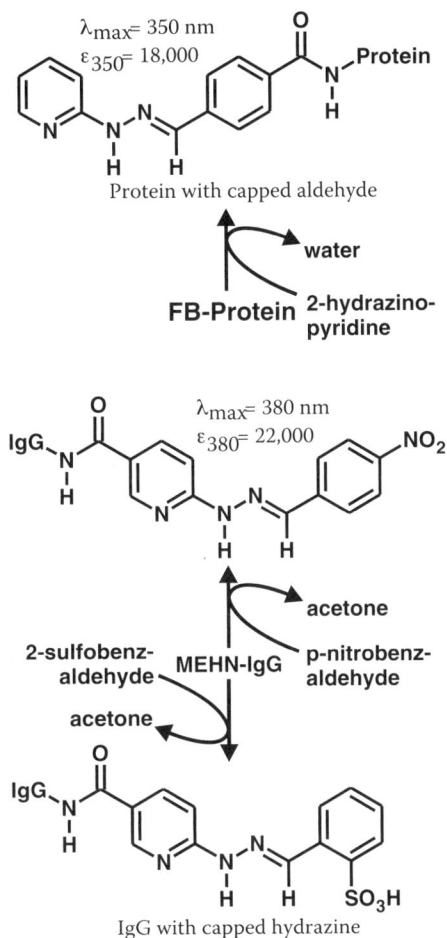

FIGURE 9.33 Quantitation and capping of modified proteins. In order to quantify the number of aldehydes on an FB-protein adduct, a small amount of the modified protein is reacted with excess 2-hydrazinopyridine, yielding an adduct with a capped aldehyde that absorbs light at 350 nm. Similarly, to quantify the number of hydrazines on an MEHN-IgG adduct, a small amount of the modified IgG is reacted with excess of p-nitrobenzaldehyde, yielding an adduct with a capped hydrazide that absorbs light at 390 nm. After FB-protein and MEHN-IgG adducts have been mixed to allow protein-protein cross-linking, unreacted aldehydes can be capped (rendered unreactive) with excess 2-hydrazinopyridine, after which unreacted hydrazines can be capped with excess 2-sulfobenzaldehyde.

Note: The Novabiochem instructions recommend a modification level *m* of 2.5–3 hydrazines per MEHN-IgG molecule (this exemplar protocol) and 2.5–3 aldehydes per FB-protein molecule (exemplar protocol 19), but do not give specific guidelines comparable to Figure 9.7 for reliably achieving those modification levels. Until careful studies have been carried out, some trial and error may be required to achieve a suitable modification level.

2. Free the MEHN-IgG conjugate of unconjugated hydrazine and reequilibrate it with acetate cross-linking buffer by dialysis or a centrifugal concentration device (see Section 9.2.2).

Note on cross-linking buffer: Cross-linking (exemplar protocol 20) is optimal at
pH 4.7, and in exemplar protocol 20 will be carried out in acetate cross-linking
buffer (0.1 M sodium acetate, pH adjusted to 4.75 with HCl). Although antibodies
generally are not irreversibly inactivated at pH 4.75, it is possible that the other
protein to which an antibody is to be coupled cannot withstand that pH. In that
case, a less acidic cross-linking buffer can be chosen and the cross-linking time
in exemplar protocol 20 extended accordingly. Examples of alternative cross-
linking buffers are: 0.15 M NaCl, 0.1 M citric acid, pH adjusted to 6.0 with NaOH;
and PBS pH 7.2.

3. Determine the concentration of MEHN-IgG conjugate using the Bradford
or BCA assays (see Quantitation of Antibodies; the aromatic hydrazide
appendages interfere with spectrophotometric determination).
4. The modification level m (number of hydrazines per IgG molecule) is
determined colorimetrically by reacting to a small portion of the MEHN-
IgG conjugate with excess p-nitrobenzaldehyde (PNBA), as diagrammed
in Figure 9.33. Dissolve a few mg of PNBA in DMSO to give a 7.56-
mg/ml (50-mM) solution. Dilute 1 volume of 50-mM PNBA in 100 vol
acetate cross-linking buffer (see note after step 2; use the acetate buffer
for quantitation even if the cross-linking in exemplar protocol 20 is to
be carried out in a less acidic buffer) to make a 500-µM dilution of
PNBA. Mix 1 volume of 500-µM PNBA with up to 0.25 vol of the
MEHN-IgG conjugate to yield an IgG sample with a final IgG concentration of 7 µM and a final PNBA concentration of at least 400 µM; as
a control, mix 1 vol 500-µM PNBA with an identical volume of blank
buffer (the same buffer as used to equilibrate the IgG at step 2). Both
IgG sample and control should have final volumes sufficient to fill a
spectrophotometric cuvette, and should have a pH no higher than 5.
Incubate both mixtures for 1 h at 37°C or 2 h at room temperature, then
determine their absorbances at 380 nm. The modification level is cal-

culated as $m = \dfrac{(A_{380}\text{ of IgG sample}) - (A_{380}\text{ of control})}{0.022 \times 7}$ As noted after

step 1, a modification level of 2.5–3 hydrazines per protein molecule is
recommended.

9.8.2 Exemplar Protocol 19: Conjugation of Aldehydes to
Amines on the Other Protein

1. Bring the other protein (the protein to be cross-linked to IgG) to a concentration of 50 µM in PBS. Dissolve solid SFB (see Figure 9.32) in sufficient
dimethylformamide to give a concentration of 1.24 mg/ml (5 mM). Add 1
vol 5-mM SFB to 10 vol 50-µM protein solution; the final concentrations
of protein and SFB are 45.5 and 455 µM, respectively. Allow the reaction

to proceed at room temperature for at least 2–3 h (see note after step 1 of Exemplar Protocol 18).

2. Free the FB-protein conjugate of unconjugated aldehyde and reequilibrate it with acetate cross-linking buffer (see note after step 2 of exemplar protocol 18) by dialysis or a centrifugal concentration device.

3. Determine the concentration of FB-protein conjugate using the Bradford or BCA assays (the aromatic aldehyde appendages interfere with spectro-photometric determination).

4. The modification level m (number of aldehydes per protein molecule) is determined colorimetrically by reacting a small portion of the FB-protein conjugate with excess 2-hydrazinopyridine (HP), as diagrammed in Figure 9.33. Dissolve a few mg of 2-hydrazinopyridine·2HCl in water to give a 9.1-mg/ml (50-mM) HP solution. Dilute 1 vol of 50-mM HP in 100 volume of acetate cross-linking buffer (see note after step 2 of Exemplar Protocol 18 use the acetate buffer for quantitation even if the cross-linking in exemplar protocol 20 is to be carried out in a less acidic buffer) to make a 500-μM dilution of HP. Mix 1 volume of 500-μM HP with up to 0.25 volume of the FB-protein conjugate to yield a protein sample with a final protein concentration of 7 μM and a final HP concentration of at least 400 μM; as a control, mix 1 volume of 500-μM HP with an identical volume of blank buffer (the same buffer as used to equilibrate the protein at step 2). Both IgG sample and control should have final volumes suffi-cient to fill a spectrophotometric cuvette, and should have a pH no higher than 5. Incubate both mixtures for 1 h at 37°C or 2 h at room temperature, then measure their absorbances at 350 nm. The modification level is calculated as $m = \dfrac{(A_{350}\ \text{of protein sample}) - (A_{350}\ \text{of control})}{0.018 \times 7}$. As noted after step 1 of exemplar protocol 18, a modification level of 2.5–3 aldehydes per protein molecule is recommended.

Obviously, protocol 19 can only be carried out if the protein has sufficient accessible amines. The enzyme that is most commonly cross-linked to antibodies for ELISA, immunocytochemistry and immunoblotting applications is horseradish peroxidase, which is highly unusual in having only two available ε-amino groups per molecule. On the other hand, horseradish peroxidase has abundant carbohydrate, which can be readily oxidized to aldehydes with periodate (see Section 9.5); periodate-oxidized horseradish peroxidase is also available commercially from several sources. Periodate-oxidized horseradish peroxidase can be substituted for HP-protein conjugate in the cross-linking reaction in exemplar protocol 20. Although the resulting cross-links are not the bis-aromatic HydraLinK linkages, they are nevertheless sufficiently stable that they need not be stabilized by subsequent reduc-tion with sodium cyanoborohydride.

In exemplar protocol 20, the MEHN-IgG and FB-protein conjugates are mixed to allow them to be cross-linked with HydraLinK linkages (Figure 9.32). No gen-erally applicable guidelines for the cross-linking reaction have been developed, but the Novabiochem instructions report results of a pilot demonstration (not fully described) in which FB-IgG (molecular weight 150,000) and MEHN-BSA (molecular weight 66,000) were cross-linked to each other at various ratios. At a BSA:IgG molar ratio of 3:1 all the IgG and most of the BSA were incorporated into cross-linked complexes. It is likely that cross-linking slows dramatically or comes to a halt altogether because of steric hindrance well before all available hydrazines and aldehydes are reacted. By the same token, the degree of cross-linking may depend largely on the ratio of the two reactants and only weakly on their individual concentrations. In any case, it is a good idea to cap unreacted aldehydes on the FB-protein conjugate with excess HP and unreacted hydrazines on the MEHN-IgG conjugate with excess 2-sulfobenzaldehyde (SBA), as diagrammed in Figure 9.33 and illustrated in exemplar protocol 20. Capping prevents any further cross-linking reactions.

9.8.3   Exemplar Protocol 20: Cross-Linking MEHN-IgG and FB-Protein

1. Mix the MEHN-IgG and FB-protein conjugates in the desired molar ratio in cross-linking buffer (*see* note after step 2 in exemplar protocol 18) and allow them to react at room temperature for 1–3 h (or at least 6 h if the coupling buffer has a pH of 7.2).

2. Add 1 vol of 50-mM HP (step 4 of exemplar protocol 19) to 100 vol cross-linking reaction mixture from the previous step, giving a final HP concentration of 500 µM. Allow capping to continue for 1 h at room temperature.

3. Dissolve 2-sulfobenzaldehyde sodium salt in water to give a 10.4-mg/ml (50-mM) SBA solution. Add 1 volume of 50-mM SBA to 50-volume reaction mixture from the previous step, giving a final SBA concentration of ~1 mM (twofold molar excess over the HP added in the previous step). Allow capping to continue for 1 h at room temperature; during this incubation, SBA will not only cap unreacted hydrazines on the MEHN-IgG conjugate, but also react with excess unreacted HP from the previous step.

Note: The two capping reactions must be done sequentially, not simultaneously; otherwise, one of the capping reactants (the one in excess) will inactivate the other before capping is complete.

4. Free the cross-linked proteins of excess capping reactants and equilibrate them with the desired final buffer by dialysis or a centrifugal concentration device.

5. Characterize the cross-linked complex by sodium dodecyl sulfate polyacrylamide gel electrophoresis and by functional tests as needed. If desired, the cross-linked complex can be freed of protein that has not been cross-linked to IgG by nickel immobilized metal affinity chromatography, provided the IgG binds effectively to the nickel chelate (see Section 9.3.3.5).

ACKNOWLEDGMENTS

Supported by U.S. National Institutes of Health grant GM41478 to the author and National Cancer Institute Center Grant P50-CA-10313-01 to Wynn A. Volkert. Robert Davis provided excellent technical assistance in much of the experimental work reported here.

REFERENCES

1. Hermanson, G.T., *Bioconjugate Techniques,* Academic Press, San Diego, CA, 1996.
2. Mandy, W.J. and Nisonoff, A., Effect of reduction of several disulfide bonds on the properties and recombination of univalent fragments of rabbit antibody, *J. Biol. Chem.,* 238, 206, 1963.
3. Smith, G.P., Kinetics of amine modification of proteins, *Bioconjug. Chem.* 17, 501, 2006.
4. Green, N.M., A spectrophotometric assay for avidin and biotin based on binding of dyes by avidin, *Biochem. J.,* 94, 23C, 1965.
5. Rao, S.V., Anderson, K.W., and Bachas, L.G., Determination of the extent of protein biotinylation by fluorescence binding assay, *Bioconjug. Chem.,* 8, 94, 1997.
6. Bayer, E.A. and Wilchek, M., Protein biotinylation, *Methods Enzymol.,* 184, 138, 1990.
7. Strachan, E., Mallia, A.K., Cox, J.M., Antharavally, B., Desai, S., Sykaluk, L., O'Sullivan, V., and Bell, P.A., Solid-phase biotinylation of antibodies, *J. Molec. Recognit.,* 17, 268, 2004.
8. Chapman, A.P., PEGylated antibodies and antibody fragments for improved therapy: a review, *Adv. Drug Deliv. Rev.,* 54, 531, 2002.
9. Larson, R.S., Menard, V., Jacobs, H., and Kim, S.W., Physicochemical characterization of poly(ethylene glycol)-modified anti-GAD antibodies, *Bioconjug. Chem.,* 12, 861, 2001.
10. Liu, S., The role of coordination chemistry in the development of target-specific radiopharmaceuticals, *Chem. Soc. Rev.,* 33, 445, 2004.
11. Liu, S. and Edwards, D.S., Bifunctional chelators for therapeutic lanthanide radiopharmaceuticals, *Bioconjug. Chem.,* 12, 7, 2001.
12. Mirzadeh, S., Brechbiel, M.W., Atcher, R.W., and Gansow, O.A., Radiometal labeling of immunoproteins: covalent linkage of 2-(4-isothiocyanatobenzyl)diethylenetriaminepentaacetic acid ligands to immunoglobulin, *Bioconjug. Chem.,* 1, 59, 1990.
13. Pippin, C.G., Parker, T.A., McMurry, T.J., and Brechbiel, M.W., Spectrophotometric method for the determination of a bifunctional DTPA ligand in DTPA-monoclonal antibody conjugates, *Bioconjug. Chem.,* 3, 342, 1992.
14. Cleland, W.W., Dithiothreitol, a new protective reagent for SH groups, *Biochemistry,* 3, 480, 1964.
15. Nisonoff, A. and Dixon, D.J., Evidence for linkage of univalent fragments or half-molecules of rabbit gamma-globulin by the same disulfide bond, *Biochemistry,* 3, 1338, 1964.
16. Sun, M.M., Beam, K.S., Cerveny, C.G., Hamblett, K.J., Blackmore, R.S., Torgov, M.Y., Handley, F.G., Ihle, N.C., Senter, P.D., and Alley, S.C., Reduction-alkylation strategies for the modification of specific monoclonal antibody disulfides, *Bioconjug. Chem.,* 16, 1282, 2005.
17. Ellman, G.L., Tissue sulfhydryl groups, *Arch. Biochem. Biophys.,* 82, 70, 1959.

18. Grassetti, D.R. and Murray, J.F., Jr., Determination of sulfhydryl groups with 2,2'- or 4,4'-dithiodipyridine, *Arch. Biochem. Biophys.,* 119, 41, 1967.

19. King, T.P., Li, Y., and Kochoumian, L., Preparation of protein conjugates via intermolecular disulfide bond formation, *Biochemistry,* 17, 1499, 1978.

20. Kim, D.S. and Churchich, J.E., The reversible oxidation of vicinal SH groups in 4-aminobutyrate aminotransferase. Probes of conformational changes, *J. Biol. Chem.,* 262, 14250, 1987.

21. Webb, J.L., *Enzyme and Metabolic Inhibitors,* Academic Press, New York, 1966.

22. Cole, R., Sulfitolysis, *Methods Enzymol.,* 11, 206, 1967.

23. O'Shannessy, D.J. and Quarles, R.H., Labeling of the oligosaccharide moieties of immunoglobulins, *J. Immunol. Methods,* 99, 153, 1987.

24. Laguzza, B.C., Nichols, C.L., Briggs, S.L., Cullinan, G J., Johnson, D.A., Starling, J.J., Baker, A.L., Bumol, T.F., and Corvalan, J.R., New antitumor monoclonal antibody-vinca conjugates LY203725 and related compounds: design, preparation, and representative *in vivo* activity, *J. Med. Chem.,* 32, 548, 1989.

25. Scott, J.K. and Smith, G.P., Searching for peptide ligands with an epitope library, *Science,* 249, 386, 1990.

26. Karlsson, R., SPR for molecular interaction analysis: a review of emerging application areas, *J. Mol. Recognit.,* 17, 151, 2004.

27. Karlsson, R. and Larsson, A., Affinity measurement using surface plasmon resonance, *Methods Mol. Biol.,* 248, 389, 2004.

28. Porter, R.R., The hydrolysis of rabbit γ-globulin and antibodies with crystalline papain, *Biochem. J.,* 73, 119, 1959.

29. Harlow, E. and Lane, D., *Antibodies: A Laboratory Manual,* Cold Spring Harbor Laboratory, Cold Spring Harbor, NY, 1988.

30. Gorevic, P.D., Prelli, F.C., and Frangione, B., Immunoglobulin G (IgG), *Methods Enzymol.,* 116, 3, 1985.

31. Yamaguchi, Y., Kim, H., Kato, K., Masuda, K., Shimada, I., and Arata, Y., Proteolytic fragmentation with high specificity of mouse immunoglobulin G. Mapping of proteolytic cleavage sites in the hinge region, *J. Immunol. Methods,* 181, 259, 1995.

32. Zou, Y., Bian, M., Yiang, Z., Lian, L., Liu, W., and Xu, X., Comparison of four methods to generate immunoreactive fragments of a murine monoclonal antibody OC859 against human ovarian epithelial cancer antigen, *Chin. Med. Sci. J.,* 10, 78, 1995.

33. Adamczyk, M., Gebler, J.C., and Wu, J., Papain digestion of different mouse IgG subclasses as studied by electrospray mass spectrometry, *J. Immunol. Methods,* 237, 95, 2000.

34. Parham, P., On the fragmentation of monoclonal IgG1, IgG2a, and IgG2b from BALB/c mice, *J. Immunol.,* 131, 2895, 1983.

35. Lamoyi, E., Preparation of F(ab')$_2$ fragments from mouse IgG of various subclasses, *Methods Enzymol.,* 121, 652, 1986.

36. Smith-Gill, S.J., Finkelman, F.D., and Potter, M., Plasmacytomas and murine immunoglobulins, *Methods Enzymol.,* 116, 121, 1985.

37. Smith, T.J., Purification of mouse antibodies and Fab fragments, *Methods Cell. Biol.,* 37, 75, 1993.

38. Kurkela, R., Vuolas, L., and Vihko, P., Preparation of F(ab')$_2$ fragments from monoclonal mouse IgG1 suitable for use in radioimaging, *J Immunol. Methods,* 110, 229, 1988.

39. Milenic, D.E., Esteban, J.M., and Colcher, D., Comparison of methods for the generation of immunoreactive fragments of a monoclonal antibody (B72.3) reactive with human carcinomas, *J. Immunol. Methods,* 120, 71, 1989.

40. Rousseaux, J., Rousseaux-Prevost, R., and Bazin, H., Optimal conditions for the preparation of proteolytic fragments from monoclonal IgG of different rat IgG subclasses, *Methods Enzymol.,* 121, 663, 1986.

41. Rousseaux, J., Rousseaux-Prevost, R., Bazin, H., and Biserte, G., Tryptic cleavage of rat IgG: a comparative study between subclasses, *Immunol. Lett.,* 3, 93, 1981.

42. Butler, J.E. and Kennedy, N., The differential enzyme susceptibility of bovine immunoglobulin G1 and immunoglobulin G2 to pepsin and papain, *Biochim. Biophys. Acta* 535, 125, 1978.

43. Heyermann, H. and Butler, J.E., The heterogeneity of bovine IgG2—IV. Structural differences between IgG2a molecules of the A1 and A2 allotypes, *Mol. Immunol.,* 24, 1327, 1987.

44. Akita, E.M. and Nakai, S., Production and purification of Fab′ fragments from chicken egg yolk immunoglobulin Y (IgY), *J. Immunol. Methods,* 162, 155, 1993.

45. O'Donnell, I.J., Frangione, B., and Porter, R.R., The disulphide bonds of the heavy chain of rabbit immunoglobulin G, *Biochem. J.,* 116, 261, 1970.

46. Givol, D. and De Lorenzo, F., The position of various cleavages of rabbit immunoglobulin G, *J. Biol. Chem.,* 243, 1886, 1968.

10 Applications

Lee Bendickson and Marit Nilsen-Hamilton

CONTENTS

10.1 INTRODUCTION

Several powerful methods use the specificity of the antigen-antibody interaction to indicate the presence of particular proteins in samples.

Immunoprecipitation (IP) uses antibodies to preferentially precipitate the protein or protein complex of interest from a mixture of proteins. Precipitated proteins are then resolved using polyacrylamide gel electrophoresis (PAGE). If the original mixture of proteins was radiolabeled, then autoradiography is used to visualize the radioactive signal. Otherwise, biotinylated or phosphorylated proteins can be detected by these modifications, and enzymes can be detected by their enzymatic activity. Detection of biotinylated and phosphorylated proteins often also involves using Western blotting.

The closely related methods of immunohistochemistry and immunocytochemistry detect protein *in situ* in tissue sections or intact cells, respectively, and usually employ a chromogenic substrate that leaves a stain in the region of the fixed tissue or cell where the protein of interest is located. These *in situ* methods are useful for locating the protein but are not quantitative and do not provide further characterization of the protein, such as its molecular mass. An interesting extension of these *in situ* methods is the tissue-print technique[2] where, for example, plant tissue sections are blotted directly on nitrocellulose leaving a surface image of the section that can be probed using specific antibodies.

The enzyme-linked immunosorbent assay (ELISA) allows large numbers of samples to be screened quickly. This method relies on the adsorption of protein to plastic. Adsorbed protein is detected by its binding to a specific primary antibody that, in turn, is bound by a secondary antibody linked to an enzyme that catalyzes the color change of a chromogenic substrate. Using a spectrophotometer, the soluble, colored product can be measured.

Since its initial description in 1981,[1] the Western blot method, with its simplicity, specificity, and sensitivity, has found wide application for studying specific proteins in complex mixtures. The number of applications that incorporate Western blotting for visualizing specific proteins continues to increase: by early 2005, almost 1% of the articles annotated in the database of the National Center for Biotechnology Information and accessed by PubMed included the words Western and blot in the title or abstract.

Western blotting is characterized by the transfer and immobilization of complex mixtures of proteins that have been separated by electrophoresis, typically PAGE, where sodium dodecylsulfate is included in the gel and buffers (e.g., SDS-PAGE). Conceptually, the process is similar to Southern and Northern blotting, in which DNA or RNA, respectively, is separated by electrophoresis, transferred to a membrane, and probed to detect the molecule of interest.

For the Western blot, proteins are transferred from a gel to a membrane, such as nitrocellulose, nylon, or polyvinylidene difluoride (PVDF). The membrane is then probed with antibodies that specifically recognize the protein of interest. Variations of the basic Western blot procedure involve the use of ligands other than the antibody for its antigen. For example, aptamers and streptavidin have been used in place of antibodies to recognize specific proteins on membranes. Here we discuss various aspects of the Western blot procedure, its variations, benefits, and pitfalls and describe a step-by-step description of a typical Western blot protocol that we use successfully in our laboratory.

The methods just described are subject to misinterpretation if there is cross-reactivity between the antibody and proteins other than the one being studied. In this regard, the Western blot procedure has the advantage that the proteins are resolved by molecular weight, and therefore the protein of interest is readily distinguished from most cross-reacting proteins. This is probably the main reason that Western blotting has become a standard technique in the biochemistry and molecular biology technology repertoire.

10.2 TRANSFERRING PROTEINS

10.2.1 TRANSFER DEVICES

After electrophoresis, protein transfer from gel to membrane can be accomplished by using an electrical field,[3] by capillary transfer,[4,5] or by passive diffusion. Transfer devices employing an electrical field fall into one of two types: wet transfer or semidry transfer.

Wet transfer devices are typically composed of a tank that holds the transfer buffer as well as one or more cassettes that slightly compress the fragile polyacrylamide gel against the transfer membrane and hold them perpendicular to the electrical field generated by the unit. Platinum wire electrodes are submerged in the tank and connected to an external power supply by insulated leads or connected directly to an integral power supply located in the lid of the unit. To dissipate heat generated during transfer, these devices usually employ some type of cooling method that may consist of (1) a buffer recirculation device connected to an external cooling coil or (2) an aluminum heat sink in the floor of the tank through which cool water is circulated. Heat can also be controlled by doing the transfer in a cold room with chilled buffer. In addition, Burnette recommends that applied voltage not exceed 10 V/cm between electrodes to avoid this heating effect.[1] We have seen buffer temperatures exceed 80°C if these precautions are not taken. Although most electrotransfer procedures include precautions to prevent temperature elevation, Kurien and Scofield[6] have reported using Towbin buffer, without methanol, heated to 70°C to improve transfer efficiency and speed as compared with electrotransfer done at 4°C. They found that transfer of large proteins (between ~70 and 200 kDa) was much improved, which they ascribed to increased permeability of the gel due to the heat.

Semidry transfer devices are composed of two large electrode panels positioned horizontally parallel to each other and are connected to an external power supply via insulated leads. One or more gel-transfer membrane sandwiches are prepared as for wet transfer and placed between the two electrode panels. Buffer-saturated filter paper (e.g., Whatman 3MM Chromatography Paper) cut to the same size as the gel-transfer membrane sandwich separates it from the electrode panels. When properly assembled, the gel-transfer membrane sandwich completes the circuit between the electrodes resulting in protein transfer out of the gel. Because the electrodes serve as large heat sinks, heat can be dissipated rapidly and transfer can usually be completed in a short time.

A comparison of semidry and wet transfer of proteins to nitrocellulose has been conducted by Tovey where the semidry system described by Kyhse-Andersen[7] was compared with the wet transfer system with Towbin's buffer.[8] In this comparison, Tovey verified Burnette's[1] observation that small proteins transfer out of the gel faster than large (>100 kDa) proteins with both transfer systems. For both systems, 0.1-μm nitrocellulose membranes retained more protein after transfer than 0.45-μm membranes, and transfer efficiencies were comparable. The authors concluded that although the amount of protein transferred was similar between the two methods, semidry blotting is advantageous due to its steeper voltage gradient and shorter transfer times.

Capillary transfer is based on the DNA blotting technique developed by Southern in which DNA fragments are drawn from the agarose resolving gel by upward capillary action and deposited onto a suitable membrane.[9] Nagy et al.[4] modified the downward capillary technique of Lichtenstein et al.,[10] applied it to protein transfer out of SDS-agarose gels, and found that downward capillary transfer proceeds about twice as fast as upward. Zeng et al.[5] have described an upward capillary transfer method, similar to Southern blotting, that is applied to protein transfer from SDS-PAGE gels. Compared with electrotransfer, this method requires a longer blotting period but is more efficient at retaining proteins that would otherwise pass through the membrane.

A method called "modi-blotting," a combination of capillary transfer and diffusion blotting, was developed to transfer proteins from gels bound to plastic backing without removing the plastic.[11] In this transfer method, a PVDF membrane is placed between the gel (top) and a Whatman 3MM filter paper wick (bottom), and the stack is clamped together with glass plates and binder clips. One end of the wick is immersed in a buffer reservoir placed slightly higher than the transfer stack, and the other end hangs free, allowing buffer to evaporate. Transfer buffer is therefore constantly being drawn from the reservoir, under the membrane and down to the free end of the wick. This arrangement develops a constant diffusion gradient that helps draw protein out of the gel.

Diffusion blotting, although similar to capillary transfer, differs because there is no buffer flow and molecules are captured by the membrane when they diffuse from the gel. Thus diffusion blotting allows the production of two identical blots at the same time; one on each side of the gel. With repeated blotting, up to 12 blots can be made from one gel.[12] Using ^{14}C-labeled protein markers run on 0.5-mm precast gels immobilized on plastic, Olsen and Wiker showed that to transfer approximately the same amount of protein to the membrane during sequential blots, the transfer time must be doubled for each subsequent transfer. They also show that band spreading (loss of resolution) is not a problem when diffusion blotting for less that 3 h.[13]

Typically, transfer of proteins from the resolving gel is done immediately after the gel run is complete. However, Coomassie blue stained proteins can be transferred from destained gels[14] or destained, dried gels.[15] Because of the irreversible binding of Coomassie blue to protein, the subsequent immunodetection must be done with a nonchromogenic method, such as the ^{125}I-protein A method or an enhanced chemiluminescence method. By contrast, proteins transferred from silver-stained gels can be detected using chromogenic methods because the proteins transfer without the stain.[14,15] The transfer time required for rehydrated gels is much less than for destained wet gels, which may be due to the removal of volatile substances during drying.[15] The ability to produce an acceptable protein transfer from stained and rehydrated gels allows one to gather total protein data and Western blot data from the same gel and stained and dehydrated gels, that are easily stored, provide a resource for later Western blot analysis if new antibodies become available.[15]

10.2.2 TRANSFER BUFFERS

Most Western blot electrotransfer buffers are based on the electrophoresis buffer: 25 mM Tris, 192 mM glycine, and pH 8.3[16], which was later modified for transfer of protein to nitrocellulose by adding 20% methanol.[3] Since then, many different buffer compositions have been described and their effects on protein transfer and retention during Western blotting and subsequent processing have been studied.[17,18] These Tris/glycine/methanol buffers are contrasted with the discontinuous buffer used in semidry blotting in which the cathodic and anodic buffers are designed to set up an isotachophoretic system.[7]

Jacobson and Karsnas[18] showed that the transfer buffer most significantly affects the binding capacity of the immobilizing membrane and that the effect of buffer composition on the transfer of protein from the gel is small. They further suggested that immobilizing transferred proteins is the most difficult problem to overcome when doing Western blotting. Furthermore, they and Tovey[17] concluded that the optimal electrotransfer conditions are assay-dependent. That is, different buffers, membranes, and transfer times need to be tested and carefully optimized for a given assay to achieve maximum sensitivity.

10.2.3 OPTIMIZATION

To find the optimum transfer procedure, various times of transfer and constituents of the transfer buffer must be tried. As well, it is useful to place one or more pieces of nitrocellulose behind the piece to which the proteins are being transferred. These backup membranes will catch protein that has moved through the first membrane and will help establish the maximum time for transfer before proteins move out of the membrane and into the buffer on the other side. If a large amount of protein is observed on the backup membranes, we recommend that several different membrane types be tested. The gel from which the protein has been transferred should also be stained (Coomassie blue, silver stain, or a combination) to determine the conditions required for the majority of the protein to be removed from the gel.

As in protein electrophoresis, protein migration during electrotransfer depends on the protein having a net positive or net negative charge; a protein at its isoelectric point will not be affected by the electrical field and will move only by diffusion. To ensure complete protein transfer, it may be tempting to include SDS in the transfer buffer. Although SDS works well in PAGE and encourages the transfer of protein from the gel during electrotransfer, it tends to interfere with protein binding to some transfer membranes (most notably nylon) when included in transfer buffers.[17] With

SDS included in the transfer buffer, protein binding to nylon membranes was nearly completely blocked, whereas binding of protein to PVDF or nitrocellulose was low, variable, and depended the membrane source. This effect was reduced by lowering the pH from 8.3 to 8.0 and by adding NaCl to the buffer. The reduction was presumably the result of the shielding of ionized groups. However, SDS is critical to the success of large protein transfer with Fairbanks gels[19] in combination with a transfer buffer based on that system (40 mM Tris, 20 mM sodium acetate, 2 mM EDTA, pH 7.4, 20% methanol, 0.05% SDS).[20] Therefore, careful consideration of the advantages and disadvantages should be undertaken before deciding to include SDS in the transfer buffer.

10.2.4 TRANSFER MEMBRANES

A study of the performance of nitrocellulose, mixed ester, nylon, and covalent-binding PVDF membranes after passive protein adsorption and also after electrotransfer was done with several different proteins labeled with [125]Iodine.[17] The membranes exhibited different binding capacities in passive adsorption tests with labeled bovine serum albumin. The PVDF showed the least, and the regenerated cellulose and nylon membranes showed the most protein binding. Nitrocellulose and mixed ester membranes were midway between. In tests measuring protein retention, PVDF retained the most bound protein when washed with detergents or 5% skimmed milk.[17] All the membranes showed virtually the same binding capacity as measured by autoradiography when tested under electrotransfer conditions with Towbin's buffer. In passive adsorption tests, the membranes exhibited a broad range of capacities but gave similar results in electrotransfer tests. These differences were ascribed to active migration of protein into the membrane matrix instead of simple diffusion and the increased hydrophobicity of Towbin's transfer buffer because of the inclusion of methanol.[17] Because of the variability in protein binding and retention of the membranes, Tovey concluded that assay design should include an assessment of membrane performance as it relates to a given application.

The choice of membrane used for Western blotting is more critical if the blotted protein must maintain its native conformation for detection by the antibody. For example, a comparison using a guanosine triphosphate (GTP)-overlay assay showed that the activity of a bovine GTP-binding protein was barely detectable after transfer to hydrophobic PVDF membranes but was clearly detected after transfer to nitrocellulose.[21] Western blot analysis showed the GTP-binding protein to be present on both PVDF and nitrocellulose membranes, with slightly more detected on the PVDF membranes. The authors speculated that the poor performance of the PVDF in the GTP-overlay assay may have been due to an inability of GTP-binding protein, thus immobilized, to renature correctly. Therefore, nitrocellulose might be preferred for a Western blot procedure, in which detection requires that the transferred protein regain its native conformation after transfer, such as when the blotting agent recognizes three dimensional structure; for example, an antigenic epitope consisting of noncontiguous residues.

Materials, such as nitrocellulose, PVDF, nylon, and diazo papers, have been used for many years as solid phase supports for blotted protein. Xerographic (photocopy) paper was also reported to be an alternative to these more expensive transfer membranes.[22] Citing an inability to reliably detect bovine αS1-casein from the milk of transgenic mice by conventional transfer methods with PVDF, Yom and Bremel turned to photocopy paper and found it to be nearly three times as sensitive. Wetting the paper in methanol before use was critical for protein binding as was the exclusion of methanol from the Towbin transfer buffer. They showed that log[protein] was linearly related to the signal over a range of 5 ng to 15 μg protein as measured by direct densitometry of the stained blot. Using such types of paper provides a dramatic cost savings over nitrocellulose at current prices.

Aware that smaller (<10 kDa) proteins are frequently difficult to detect using Western blot analysis, Karey and Sirbasku extended the low end of the LMM range of proteins studied by Tovey and Baldo from 14 kDa to 5.6 kDa by studying the binding of the growth factors basic fibroblast growth factor (15–17 kDa), insulin-like growth factor (7.6 kDa), epidermal growth factor (EGF, 6.1 kDa) and transforming growth factor alpha (5.6 kDa).[23] Although these proteins transferred to nylon (Zeta Probe) and nitrocellulose membranes with high efficiency using Towbin's buffer without methanol, they were not retained well on either membrane during the subsequent processing steps. Retention was further improved 1.5- to 12-fold without loss of antibody detectability by treating proteins blotted to Zeta Probe membranes with 0.5% (v/v) glutaraldehyde.[23] However, the effect of glutaraldehyde fixation on antibody binding should be tested with each antibody because the recognition of epitopes by some antibodies is sensitive to certain fixation methods.

Even proteins as small as 400 Da, such as neuropeptides, can be bound by coating the nitrocellulose with 0.5% gelatin before electrotransfer and fixing with paraformaldehyde after transfer.[24] This method was applied[25] to further improve the sensitivity of low molecular mass EGF detection by Western blot beyond that demonstrated by Karey and Sirbasku. There is an increase in sensitivity from a lower limit of 30 ng loaded protein with nontreated membranes to a lower limit of 0.1–0.3 ng with treated ones. Interestingly, even after formaldehyde treatment, EGF transferred to PVDF is still susceptible to displacement by detergent, in this case 0.1% (w/v) Tween 20. A likely explanation is that under these conditions, the formaldehyde fixed the EGF to the larger gelatin molecules that coated the membrane rather than to the membrane itself, and detergent treatment may have removed the gelatin-EGF complex.

10.3 DETECTION

10.3.1 DIRECT PROTEIN STAIN ON THE BLOT

As an indication of transfer quality, the transferred proteins can be visualized with stains such as amido black,[26] Coomassie blue[1] or the reversible stain Ponceau S.[27] Ponceau S has the advantage of being water soluble and can therefore be washed away before the blot is probed. Coomassie blue and amido black stain protein irreversibly. The complete removal of Ponceau S before blocking can be accelerated by washing for 1–2 min with 0.14 M NaCl, 2.7 mM KCl, 9.6 mM NaKPi, 0.1% Tween 20, pH 7.3 (Bendickson, unpublished). Blots stained with amido black or Coomassie blue can be destained with 90% methanol and 2% acetic acid.[1,26] Recently, a reversible staining method using a reactive fabric dye (Reactive brown 10) was described.[28] The authors report a detection limit approximately similar to colloidal gold (1 ng). The method is very rapid (5–10 s staining followed by ~30 s destain in water) and the stain can be completely removed by incubating under alkaline conditions (e.g., 0.1N NaOH for ~10 min).

In some cases, such as two-dimensional gel analysis and microsequencing, it is advantageous to produce images of specific immunostaining and total protein staining from a single blot. Several recent procedures accomplish this. In one method, the blot is first subjected to enzyme-linked immunodetection employing an enhanced chemiluminescence system, then stained for total protein with a bovine serum albumin–gold conjugate.[29] Gold staining of the blot was similar to silver staining of gels in terms of sensitivity, but was incompatible with ovalbumin blocking before immunodetection and incompatible with gelatin coating of the membrane before electrotransfer.[29] A disadvantage of this method is that the antibodies attached to the membrane-bound antigens need to be removed before gold staining.[30] Protein stains that do not interfere with subsequent immunodetection obviate the need for antibody removal. Examples are India ink[31,32] and the fluorogenic dye 2-methoxy-2,4-diphenyl-3(2H)-furanone.[33]

10.3.2 Antigen Detection

Western blotting relies on the specificity between the membrane-bound antigen and an antigen-specific probe. The probe can be linked directly to the signal (i.e., ^{125}I-labeled immunoglobulin (Ig)G), can be conjugated to an enzyme that catalyzes the color change of a chromogenic substrate, or can act as an intermediate in a sandwich arrangement to which a third "signaling" molecule can bind (e.g., horseradish peroxidase–conjugated protein A). In each case, the presence of the antigen is inferred from the signal recorded on the blot itself, a piece of film, or a phosphor storage plate.

Specificity and sensitivity are paramount to optimizing antigen detection. Specificity is a function of the primary detection molecules. Do they recognize only the epitopes of the protein being studied or do they confound the data by recognizing other similar epitopes (cross-reactivity) or binding nonspecifically to a number of unrelated sites? Specificity can be improved by "blocking" the membrane with a blocking agent, such as nonfat milk or Tween 20. Blocking is intended to prevent nonspecific interactions between the detection antibodies and the membrane. Specificity can also be improved by immunoprecipitating the protein of interest before resolving it by PAGE.[34] In this case, specificity is increased by first allowing the antibody to react with native epitopes in solution where cross-reacting epitopes may be inaccessible. However, combining IP with Western blotting is potentially disadvantageous because the antibodies used to precipitate the antigen during the IP remain in the sample during gel electrophoresis and are eventually electrotransferred with the antigen. Separated into their heavy and light chain components by electrophoresis, these antibodies are likely to be detected during the Western blot and may obscure the signal corresponding to the protein of interest. Despite this disadvantage, the combination of IP with one antibody followed by Western blotting has become an extremely popular means of determining whether two proteins form a complex.

Detection of the antibody used for IP by the antibody used for Western blotting becomes a problem when the antigen detected by the second antibody is in the molecular mass range of either the light or heavy immunoglobulin chains. A good example of the problem, coupled with a strategy for solving it, was demonstrated in study of lipoprotein lipase, a protein of about 57 kDa.[34] The chicken and rabbit IgG heavy chains (66 kDa, 50 kDa) of the immunoglobulins used for immunoprecipitation produced signals on the subsequent Western blot that interfered with detection of the lipoprotein lipase antigen. The strategy for solving the problem consisted of two parts and can be generally applied to situations in which the target protein is similar in size to IgG heavy chain. First, instead of using whole immunoglobulin as the primary antibody during the IP, they used Fab fragments created by digesting the primary antibody with papain and removing the Fc fragment. Fab fragments contain the antigen-binding domains of the original antibody but are much smaller (~25 kDa) than the full-length heavy chain. Next, the secondary antibody used in the IP was cross-linked to killed *Staphylococcus aureus*, thus preventing its migration into the gel during electrophoresis. As a result, lipoprotein lipase could be studied by Western blot without interference from IgG heavy chains.

Sensitivity, the second parameter for optimizing antigen detection, depends on the visualization threshold of the system used and the avidity of the antibody-antigen complex.[35] Most optimized Western blot assays can detect nanogram amounts of protein. Sensitivity can be improved by increasing the proportion of the target protein in the sample. For example, the protein of interest can be immunoprecipitated before it is resolved on a gel, as discussed previously. Other means of selectively concentrating the protein of interest to increase sensitivity of the assay include affinity capture using beads with covalently bound molecules such as ligand or a lectin that binds the carbohydrate portion of a glycoprotein. The protein, bound to the beads, is selected for by centrifugation (or magnetic attraction if the beads are magnetic). In another approach, "immunofiltration" lowered the detection threshold from 50 ng to 10 ng.[36] The antibody and wash solutions were filtered through the blot instead of simply being incubated with the blot on a shaker table. Sensitivity can also be improved by amplifying the signal, as discussed in the next section.

10.3.3 DETECTING ANTIBODIES ON THE BLOT

After they are bound to antigens, antibodies are typically visualized by the interaction of a secondary binding protein that is either another antibody or a bacterial protein (protein A or protein G) that specifically interacts with the Fc region of the antibody. Thus the primary antibody is "sandwiched" between the antigen bound to the blot and the secondary protein. This second protein is labeled so that it can be detected visually. Labels include ^{125}I for visualization by autoradiography and enzymes for color visualization via chromogenic substrates and for fluorographic visualization via chemiluminescent substrates. Signal generation from enzyme-linked systems involves the use of an enzyme that catalyzes the conversion of a substrate to an insoluble product that is either colored or chemiluminescent. For example, with the alkaline phosphatase:BCIP-NBT system, the signal is recorded directly on the blot as a stain in the form of a colored precipitate. However, the colored precipitates from this system create a purple stain that is often not very intense and is more difficult to capture in a photographic record of the stained blots compared with procedures that use radioisotopes or chemiluminescent products. The most frequently used enzymes are horseradish peroxidase (HRP) or alkaline phosphatase.

The advantage of most of these sandwich techniques is that they amplify the signal. For example, each primary antibody molecule usually has more than one epitope that the secondary antibodies recognize and bind, which increases the secondary antibody/primary antibody ratio. Similarly, in the streptavidin-biotin system, which exploits the extremely high affinity of streptavidin, a bacterial protein, for biotin, biotinylation of the secondary antibody results in multiple streptavidin molecules attached to a single secondary antibody. This further increases the ratio of primary antibody to signaling molecule (streptavidin). Enzymes linked to the end of the chain (i.e., to streptavidin) provide another point of amplification because they catalyze the conversion of many molecules of substrate to insoluble chromogenic or chemi-luminescent products. Subscribing to the "more is better" philosophy, Fukuda et al.[37] applied a polymer immunocomplex method to Western blotting to even further increase the amplification factor. With the secondary antibody conjugated to com-mercially available HRP polymers, they were able to decrease the exposure time from 2 h (for biotin-streptavidin/HRP detection) to 1 min and achieve similar signal intensity while reducing the nonspecific signal. One must be aware that the sensitivity of the assay varies with the method of signal generation and depends on the extent of amplification and the sensitivity of the detection method for the signal.

It is important to note that certain conditions can affect the activity of the linked enzyme. For example, NaN_3, a common additive used for preserving buffers from microbial contamination, inhibits the activity of horseradish peroxidase.[38] Conditions that result in loss of enzyme activity or other problems with the assay are easily detected if a positive control is included. The positive control is a sample that contains a known amount of the protein of interest and that will give a positive signal if all of the components of the system are functional. A common positive control is the antigen preparation used to produce the primary antibody. If the primary antibody is from a commercial source, then the company will often provide a sample containing the antigen to use for a positive control. Frequently, if the protein has been cloned, expression vectors are available and a positive control sample can be created by transiently transfecting COS cells or other eukaryotic cells with the expression vector.

10.3.4 Alternatives to Antibodies for Blotting

Other types of detection molecules with high specificity and affinity are being devel-oped that may complement or even replace polyclonal and monoclonal antibodies in Western blots. Oligonucleotide and peptide aptamers are two of these new classes. Oligonucleotide (DNA and RNA) aptamers, sometimes referred to as oligonucleotide ligands, are selected using systematic evolution of ligands by exponential enrichment from a large random pool of oligonucleotides.[39,40] Aptamers exhibit high affinities and high specificities for their protein targets: their dissociation constants are usually in the picomolar to nanomolar range. As more become available and blotting methods for aptamers are optimized, they are likely to be increasingly used in Western blot applications. Importantly, aptamers can be targeted with greater specificity than antibodies. For example, an RNA aptamer might bind vascular endothelial growth factor 165[41] specifically, whereas a monoclonal antibody to vascular endothelial growth factor might recognize both isoforms.[42] The tunability of aptamers and their application in Western blots were also illustrated by an aptamer selected to differentiate between native and denatured Erk2.[43]

Engineered protein scaffolds represent yet another class of detection molecule.[44] The sequences of strategically selected regions of small proteins, such as [10]Fn3, the 94-amino acid tenth fibronectin type III domain[45,46] or the 174-amino acid bilin-binding protein, a member of the lipocalin family characterized by an eight stranded β-barrel motif,[47] are randomized to produce a large library of molecules that are selected or "panned" against an antigen of interest using phage display[48,49] or mRNA display[50] technologies. These protocols have so far produced a variant of [10]Fn3 that binds tumor necrosis factor-alpha with K_D of 20 pM[46] and a bilin-binding protein variant that binds fluorescein with a K_D of 35 nM.[47]

10.3.5 QUANTIFICATION

Although Western blotting is often used in a qualitative way to determine the presence or absence of antigen in a particular sample, quantification of antigens is also possible. A linear relationship between the amount of protein and its corresponding signal is required when attempting to quantify an unknown. For quantifications to be valid, the unknown values must fall within this linear range. Linearity can be established by including a range of protein quantities on the same blot as the unknown. Care must be taken to use an excess of probe to preclude the possibility of exhausting the probe before fully saturating the target molecule. The quantification method depends on the type of signal generating molecule(s) bound to the primary antibody, as illustrated in the following examples.

10.3.5.1 ^{125}I-Protein A

Griswold et al.[51] quantified blotted protein by first visualizing the signal generating molecule, ^{125}I-protein A, on X-ray film to locate the bands of interest. Corresponding regions of the blot were cut out and counted in a gamma counter. A linear relationship was demonstrated over a range of 2.5–30 ng protein. Alternatively, a less destructive way to quantify the signal is to expose the blot to either film or a phosphor storage plate to determine the amount of radioactivity. Fang et al.[52] used this technique to detect a specific protein in serum and tissue extracts. Linearity was demonstrated between 0.5 and 15 ng of protein.

10.3.5.2 HRP-2° IgG

Uhl and Newton[53] visualized the proteins by staining the blot with 4-chloro-1-naphthol and H_2O_2. Peroxidase activity in the excised bands was measured spectrophotometrically by exposing them to the peroxidase chromogenic substrate o-phenylenediamine in the presence of H_2O_2. The reaction rate was linearly related to amount of protein over a range of 10–1000 ng.

Huang and Amero[54] used HRP-2° IgG in conjunction with enhanced chemilu-minescence (ECL) to visualize two RNA-binding proteins found in nuclear extracts. The signal, recorded on X-ray film, was measured and plotted against total protein. In this case, the data from one of the proteins fit an asymptotic curve, and the other was linear. This led the authors to conclude that "...the relationship between the amount of total protein and the antigen signal is specific and must be determined empirically." A simple explanation of an asymptotic curve is that the amount of probe is limiting relative to the amount of antigen on the blot. This example highlights the necessity for having a standard curve with reference samples that encompass the range of values derived from the unknown samples. We have found that detection of bovine serum albumin, using an ECL signal generated by a secondary antibody linked to horseradish peroxidase (Amersham Biosciences/GE Healthcare Life Sciences, Piscataway, NJ), is linear between 25 and 375 ng protein, being the entire range tested (Bendickson and Nilsen-Hamilton, unpublished).

10.3.5.3 Fluorescent Tags

Another nonradioactive means of visualizing and quantifying proteins on Western blots is through the use of fluorescent compounds. Diamandis et al.[55] used a streptavidin-based macromolecular complex labeled with a fluorescent europium chelate as the signal generating molecule. Quantification was done by scanning with a time-resolved fluorometer. The authors demonstrated a "near-linear" relationship between fluores-cence and protein and also showed that this method is as sensitive as the BCIP-NBT method. A more direct method, used by Fradelizi et al.,[56] conjugated the fluorescent molecule, Cy5, directly to the secondary antibody. Using this method, the authors showed a linear relationship between the signal detected by a Storm Phosphorimager and protein over a range of 6.25–100 ng and that this method is comparable to the alkaline phosphatase:BCIP-NBT method.

10.3.6 MOLECULAR MASS STANDARDS

It is often desirable to characterize the sizes of proteins on a Western blot by estimating their molecular masses. This can be accomplished by comparing the relative electro-phoretic mobility of the protein to the mobility of a set of protein standards of known molecular mass that are included on the same gel. Because the Western blot technique is designed to be specific for the protein(s) of interest, the standards are typically visualized by other means. The appropriate means depends on the method used to visualize the protein of interest. For example, if a chromogenic substrate is used to directly stain the blot, a direct staining method such as India ink, Ponceau S, Coomassie blue, or gold staining would be called for to visualize the standards. On the other hand, the protein standards can be marked with radioactive or luminescent compounds if the signal from the protein of interest is recorded on film or by phosphor storage media.

10.3.6.1 Prestained Markers

Prestained molecular mass standards are available from several companies. Fluores-cently labeled standards can be prepared in the laboratory with dansyl chloride,[57] dabsyl chloride,[58] or fluorescamine.[59] If the blot is to be cut into several pieces after transfer, these standards are also useful as lane indicators.[60] Multicolored prestained standards, such as the "Kaleidoscope" standards (Bio-Rad Laboratories, Hercules, CA), provide a means of estimating apparent molecular mass with the added benefit of unambiguous identification of each standard.

10.3.6.2 Luminescent Markers

Our laboratory uses an adhesive backed luminescent paper (Diversified Biotech, Newton Centre, MA) as a means of locating standards on Western blot films, as well as recording other pertinent information. After the blot is stained with Ponceau S, the locations of the protein standards are marked directly on the blot with a graphite pencil. Before exposure to film, a piece of luminescent paper is cut out and applied to the blot near the standards. A mark is then made on the paper that precisely indicates the location of the pencil mark recorded earlier and hence the location of the standard. This light on dark image will be recorded in reverse on the subsequent film exposure. We have also tried luminescent fabric paint (Ghostly Glo #SC 350; Duncan, Fresno, CA) commonly available at craft stores. Similarly, ZnS in suspension has been suggested as an alternative marking substance.[61] Although the fabric paint and ZnS methods work well for autoradiography, they are less convenient to use with visualization by chemiluminescence because their required drying time (4–6 h with paint) are not compatible exposure of the blot to film during the period of peak chemiluminescent emission.

10.3.6.3 Visualizing Standards with Antibodies

Molecular mass standards can also be visualized by antisera developed against them.[62] In this case, the standards can be seen using the same protocol as for detecting the protein of interest except that the anti-standard antibodies are included with the primary antibody for the protein of interest. A disadvantage of this technique, however, is that each antiserum may contribute to nonspecific background and might cross-react with one or more proteins in the samples. An alternative approach of producing a series of protein A fusion proteins in *Escherichia coli* enables the standards to be detected simultaneously with the protein of interest with no additional steps, provided an antibody that binds protein A is used in the detection protocol.[63]

10.4 WESTERN BLOT PROTOCOL

The following procedure describes the electrotransfer of protein from a SDS-PAGE gel in the step-by-step procedure that is used in our laboratory. Based on the procedure of Burnette,[1] it works well for proteins of less than 80 kDa and results in uniform transfer of protein to the membrane. We use polyclonal primary antibodies with this method to detect proteins of interest in mammalian and bacterial cell lysates, conditioned medium from cultured cells, tissue extracts, and tissue fluids. Although we have found the conditions listed to be optimal for our needs, modifications may need to be made where Western blotting is used for other applications.

There are many methods for preparing and running the SDS-PAGE gels from which proteins are to be blotted. The most popular SDS-PAGE system is that described by Laemmli.[75] We prefer a system based on that of Ornstein[76] and Davis[16] that provides better resolution than the Laemmli gel system.[77] However, these gels take somewhat longer to run than the Laemmli system (about 4.5 h vs. 3 h for a 10-cm gel).

The Western blot procedure requires transfer membrane, primary antibody, enzyme-conjugated secondary molecule, and a means to record the signal. The necessary equipment is a transfer apparatus and power supply. The transfer apparatus generally used in our laboratory is the Hoefer TE50 electrotransfer tank or a semi-dry unit (Panther Model HEP-1) from Owl Scientific (Woburn, MA). The transfer cassette requires 15–30 min to set up. Transfer times with this procedure are 2–12 h. Note that transfer time depends on the type of transfer device used, the size of the protein(s) to be transferred, and the thickness of the gel. For example, transfers using the very thin gels from the Phastgel system can be completed in about 20 min. Semidry transfer devices can result in transfer times on the order of 30 min. In our experience, the semidry system provides an adequate transfer, although we have occasionally observed evidence of nonuniform transfer across the gel when using a semidry apparatus. After transfer is complete, subsequent antibody incubations and washes will take about 4 h. If quantification is required of the results, a densitometer, fluorescence reader, or similar device should also be available. A digital image of the stained Western blot can also be quantified using the Image J program available from the National Institutes of Health (http://rsb.info.nih.gov/ij/).

10.4.1 PROCEDURE

1. Cut one sheet of nitrocellulose membrane and two sheets of absorbent filter paper (Whatman 3MM) to the size of the gel.
2. Soak the gel in transfer buffer for 30 min. Wet the membrane, filter papers, and support pads in the same buffer.
3. To assemble the transfer sandwich, place the gel on a glass plate and then place a wetted filter paper (cut to size) on top. Use three filter papers if you are doing semidry transfer. Invert the glass plate and set it on one of the support pads. Using a spatula, gently pry the gel/filter paper from the glass. Place the membrane on top of the gel. Remove any air bubbles by gently rolling a cylinder such as a pencil or glass rod over the membrane. Place the second piece of filter paper on the stack and remove air bubbles. Again, use three filter papers if you are doing semidry transfer. Air bubbles left between the gel and membrane will prevent transfer in the region of the bubble.
4. Place the complete sandwich in the transfer tank with the membrane closest to the positive electrode (red) and the gel to the negative (black) electrode.
5. Electrotransfer tank. Fill the tank with chilled buffer and, if necessary, set up the buffer cooling system. For the Hoefer system at least 4 l of buffer are needed. The buffer can be reused at least three times. However, it may be necessary to add fresh buffer before reuse to counter the evaporation that occurs during transfer.

Semidry transfer: The buffer is held by the filter papers that are discarded after transfer.

6. Electrotransfer tank. Transfer at 4°C with a constant 90 volts (0.6–0.8 Amps) for 4 h or 30 volts for 12 h while stirring the buffer with a stir bar. Between these two extremes, other volt-hour combinations can be used as long as the volt × hour value equals 360.

   Semidry transfer: Transfer at constant 400 mA at room temperature (~23°C) for 1–2 h. The transfer time should be optimized for your protein of interest. During transfer, the voltage may vary but should be at 10–20 V.

7. After transfer, separate the transfer stack, place the gel in Coomassie blue stain solution and place the membrane (blotted side up) in a separate container for staining. Add 50 ml of Ponceau S staining solution on top of the membrane and incubate for 1–2 min with gentle agitation. The Ponceau S can be reused many times.

8. Wash the membrane briefly with water to remove excess background stain.

9. Mark the position of the molecular mass markers on the membrane with a pencil.

10. Photograph the stained membrane.

11. Continue washing the membrane until all the stain is gone. If stain persists where protein is abundant, wash the membrane in NaKPS + Tween 20 for 1–2 min and then rinse with water for 1–2 min with several changes.

12. Incubate the membrane in 200 ml blocking buffer at 37°C with gentle shaking for 30 min or overnight at 4°C. After blocking, wash as follows with shaking:

    Wash procedure:
    (a) 100 ml NaKPS + Tween 20, one wash for 15 min
    (b) 100 ml NaKPS, two washes for 15 min each time

13. Incubate the membrane with primary antibody incubation solution for 1.5 h at room temperature with shaking.

14. Wash as in step 12.

15. Incubate with the HRP conjugated protein (either protein A or secondary antibody) solution for 30 min at room temperature with shaking.

16. Wash as in step 12.

17. Blot the membrane dry then incubate with ECL substrate for 1 min. Blot excess substrate then place membrane onto filter paper and wrap with saran wrap. Expose to film.

10.4.2 SOLUTIONS

1. Transfer buffer (4 l)

   12.11 g Trizma (Tris) (final = 25 mM)
   57.65 gm glycine (final = 190 mM)
   800 ml methanol (final = 20%)
   Bring to volume with distilled water.

Note: Check to make sure the pH is about 8.3. Do not adjust the pH. If it is not close to pH 8.3, the solution has been made incorrectly and must be remade. Degas this solution before using it.

2. Coomassie stain solution (1 l)

   0.47 g Coomassie brilliant blue R (Sigma-Aldrich, St. Louis, MO)
   234 ml isopropanol
   94 ml glacial acetic acid
   Bring to 1 l with distilled water.

3. Destain solution (1 l)

   94 ml isopropanol
   94 ml glacial acetic acid
   Bring to 1 l with distilled water.

4. Ponceau S staining solution

   2 g Ponceau S (final = 0.2%) obtained from SERVA (FeinBiochemica, Heidelberg, NY; catalog #33427)
   30 g TCA (final = 3%)
   Bring to 1 l with distilled water; this solution can be reused at least 10 times.

5. Blocking buffer

   25 g nonfat dry milk (final = 5% w/v)
   0.1 g NaN_3 (final = 0.02%)
   500 ml TD solution
   Stir the solution overnight at room temperature to completely dissolve the milk.

6. Primary antibody incubation solution

   0.2 ml antiserum (for 1/200 dilution. The optimum dilution needs to be established for each antiserum. For whole antiserum, we use dilutions between 1/100 and 1/500. Commercially available affinity purified IgG antibodies (polyclonal and monoclonal) are typically used between 0.1 and 1µg/ml final IgG concentration.
   40 ml blocking buffer

   Note: This is enough antibody solution for one gel (~13 × 10 cm by 2 mm thick) and can be reused at least four times.

7. HRP-conjugated protein, secondary probe

   40 µl HRP*secondary antibody diluted 1:100 with NaKPS (for 1/100,000 dilution)*
   Or
   4 µl 1mg/ml HRP*protein A (for 1/10,000 dilution)*
   40 ml NaKPS

*The optimum dilution needs to be determined for each application. The optimal value can vary greatly, but is usually found somewhere between 5000- and 100,000-fold dilutions.

8. 4X TD (4 l)

    128 g NaCl (final = 0.14 *M*)
    6.08 g KCl (final = 5 mM)
    1.60 g Na_2HPO_4 (final = 0.4 mM)
    48 g Trizma (final = 25 mM)

    Bring to about 3.5 l with distilled water. Then add approximately 32 ml of concentrated HCl to adjust the pH to 7.4–7.5 (at room temperature) using a pH meter. Bring to 4 l volume. Pour into a clean 4 l bottle and store at 4°C. To make TD, dilute this stock 1:3 with distilled water.

9. 10X NaKPS (2 l)

    160 g NaCl (final = 1.37 M)
    4 g KCl (final = 27 mM)
    4 g KH_2PO_4 (final = 15 mM)
    142 g Na_2HPO_4 (final = 81 mM)

    Dissolve in about 1500 ml of distilled water. Adjust pH to 7.3 with 10 N NaOH then bring to volume. For long-term storage, autoclave in aliquots of 300–400 ml.

10. NaKPS (1 l)

    100 ml 10X NaKPS
    Bring to volume with distilled water.

11. NaKPS + Tween 20 (1 l)

    100 ml 10X NaKPS
    10 ml 10% Tween 20 (final = 0.1%)

    Bring to volume with distilled water.

12. ECL substrate

    Mix equal parts of detection solution #1 and #2 as per manufacturer's instructions (ECL Western blotting reagents, Amersham Bioscience/GE Healthcare Life Sciences).
    Once made, the substrate can be reused up to 48 h if kept at 4°C.

10.5 ARTIFACTS AND TROUBLESHOOTING

One must be aware of potential artifacts when conducting and interpreting any assay, and Western blotting is no exception. Using protein A in concert with an enhanced chemiluminescent detection system, we periodically see spurious, amorphous signals that can usually be disregarded because they are not consistently associated with specific protein bands. These signals can, on occasion, obscure areas of authentic signal. We speculate that these signals may be due to static discharge, the presence of air bubbles between the gel and membrane during transfer, or manufacturing defects in the membrane itself.

Another significant problem, protein band distortion, can occur if the gel is overloaded. For example, we have found that when dealing with samples containing growth medium supplemented with bovine serum, the presence of serum albumin at a final concentration greater than 1–2% distorts the band pattern of other proteins of similar molecular mass, making identification of these proteins difficult. If possible, it is best to avoid loading samples that contain greater than 2% serum albumin.

Nonspecific or lower affinity interactions of the primary or secondary antibodies with other proteins or the membrane can be the cause of background signal. The blocking agent might also be a nonspecific target for one of the antibodies. For example, increased background signals of blots blocked with bovine milk when using protein A as the secondary molecule is attributed to the weak affinity of protein A for IgG2 found in bovine milk.[30,64]

When detection molecules are conjugated directly to enzymes, such as alkaline phosphatase or horseradish peroxidase, one needs to be concerned with the presence of endogenous activity on the blot. For example, peroxidase in membrane preparations from rabbit gastric mucosa is able to survive SDS-PAGE separation and produce a signal when the blot is incubated with a chemiluminescent substrate.[65] This activity can be quenched by treating the blot with 3% H_2O_2 before processing. Endogenous biotinylated proteins can contribute to nonspecific signal when the streptavidin-biotin detection system is used. While looking for nuclear factor kappa B (NF-κB) proteins in cell extracts, Vaitaitis et al.[66] observed four additional bands even when the blot was incubated with streptavidin-HRP alone and developed with a chemiluminescent substrate. To solve the problem, the endogenous biotin was first blocked with 0.25 g/ml streptavidin, followed by a wash with 50 ng/ml of d-biotin (to block the bound streptavidin). Biotinylated secondary antibody was then used with streptavidin-HRP to detect only NF-κB.

10.6 ANTIBODY ARRAYS

The antibody array format has the potential for analysis of many analytes in parallel given the availability of appropriate antigen capture reagents.[67,68] By contrast, the immunoassays discussed so far in this chapter are designed to assay for one analyte at a time. If information about several analytes is desired, these assays must be run sequentially. This not only takes time but requires that additional sample be available. Antibody arrays being developed today are typically constructed using robotic spotting equipment and include a set of carefully selected antibodies chosen to provide information about the protein complement of the cell. Nitrocellulose membranes and chemically treated glass microscope slides are the substrata most often used to immobilize the antibodies.

Detection of bound antigen is accomplished by labeling the antigen directly or with a sandwich-type system in which the captured antigen is recognized by a second antibody. The second antibody can itself be fluorescently labeled and measured or can serve as the recognition molecule for a second layer of signaling molecules. The many variations of this sandwich type of detection system are reviewed by Nielsen and Geierstanger.[67]

Antibody arrays are showing promise as research and diagnostic tools but are still in the developmental stage. Knight et al. developed a 16-plex microarray immunoassay designed to measure cytokines involved in the human inflammatory response.[69] Compared with a standard ELISA, the microarray was just as sensitive (detection limit of 4–12 pg/ml) while less expensive. On a per analyte basis, the ELISA costs $4.50, whereas the microarray costs $0.30. Aptamers that can be linked to their protein targets by photochemistry (photoaptamers) have been used as an alternative to antibodies in a 17-element array designed to measure clinically relevant proteins, such as basic fibroblast growth factor, vascular endothelial growth factor, and von Willebrand factor.[70] In this format, very stringent wash conditions can be used to reduce background signal because the photoaptamer is covalently linked to its target by irradiation with light at 308 nm before washing.

In addition to the multianalyte arrays discussed, antibody arrays can be produced (and are commercially available) that include hundreds of antibodies. These arrays are useful for studying protein:protein interactions where information about putative binding partners to a single protein of interest is desired. The array is incubated with an extract containing a complex mixture of proteins that include the protein of interest, washed thoroughly, then probed with an antibody against the protein of interest. Signal on the blot associated with the array address of an antibody that does not recognize the protein of interest indicates a potential binding partner.

Although antibody arrays hold great potential for proteomics research, antibody cross-reactivity is a problem that must be overcome if large scale protein profiling is to become a reality. Some have concluded that, given the relatively small number of available antibody pairs that exhibit both high affinity and high specificity, the number of analytes that can be measured per array is limited to ~50.[67,71] The development of alternative antigen capture reagents such as aptamers will undoubtedly increase this number.

10.7 CLINICAL APPLICATIONS

A complete review of the clinical applications of procedures, such as Western blotting, immunoblotting, and antibody arrays, is beyond the scope of this chapter. The following, however, will serve as a brief introduction. Immunoassays, including Western blotting, are part of the toolkit used by clinicians to diagnose infectious disease. Blood banks also use these assays to screen donors, as well as the blood supply, for the presence of antibodies against infectious agents such as HIV, hepatitis C virus, and human T-cell lymphotrophic virus.

In practice, sera are screened with an enzyme immunoassay (EIA, ELISA), and positive results are confirmed by Western blot. Most diagnostic Western blots are essentially antibody capture assays in which the patient's serum is tested for antibodies against pathogens. Antigenic proteins from these pathogens are electrophoretically separated and blotted onto nitrocellulose or PVDF, sliced into narrow strips, and sold as a kit. Specific antibodies in sera from individuals suspected of being infected will bind to the antigens and are detected by applying an appropriate signal-generating molecule. The results from these blots are interpreted using criteria specifically developed for the set of antigens included on the blot. In some cases, recombinant antigens are applied to the blot to improve the specificity of the assay. Because the majority of recombinant proteins are produced by bacteria and are therefore not glycosylated, they can react differently with patients' antibodies that were developed against the native glycosylated proteins. Thus, as for immunoblots that use native protein preparations, immunoblots and Western blots based on recombinant proteins or peptides as the antigens for capturing antibodies from serum sources should be carefully tested for specificity, sensitivity, and reliability.

10.8 IMMUNOPRECIPITATION AND RELATED APPLICATIONS

A feature of many applications of antibodies is that they or the target antigen are immobilized on a solid surface. For example, antibody arrays are immobilized on nitrocellulose membranes and ELISAs use antibodies that are immobilized on plastic surfaces. Immobilization on solid supports, such as synthetic beads or fixed bacteria, allows the antibody and captured antigen to be immunoprecipitated. For IP, the solid support is generally a type of synthetic bead made of a material, such as agarose, Sepharose, or a polymer with a magnetic core. Antibodies can be fixed to these substrata by chemical means. Alternatively, the beads can be decorated with covalently linked protein A or protein G, which are proteins that have high affinities for the Fc region of immunoglobulins. An older alternative to synthetic beads are fixed *S. aureus*, the source of protein A. Bacterial adsorbants are now infrequently used, presumably because of the lower reliability of the bacterial preparations as adsorbants and the greater frequency of nonspecific adsorption of proteins to the bacterial surfaces compared with the synthetic supports. Centrifugation or magnets are used to separate the antibody-linked supports and associated proteins from the nonbound proteins and other components of the supernatant. After several washes, the proteins that remain associated with the beads can be denatured by heating in an SDS-containing sample buffer and resolved by electrophoresis to identify the proteins captured by the support-linked antibody.

If the immunoprecipitated proteins are labeled, such as with a radioisotope or a fluorescent tag, they can be detected by this means. The immunoprecipitated proteins can also be visualized by Western blotting. Before separation by gel electrophoresis, the antigen can be released with 100–200 mM glycine (pH 2.5–3) for 5 min, leaving the capturing antibody in the precipitate. By contrast, protocols for coimmunoprecipitation, which use an SDS sample gel buffer to dissolve the proteins in the immunoprecipitated pellet, also remove the capturing antibody with the antigen from the precipitate. In this latter case, depending on the specificity of the secondary antibody used for the Western blot protocol, the capturing antibody may be detected on the blot. A discussion of how to deal with interference by the staining of immunoprecipitating antibody can be found in the section Antigen Detection.

IP combined with Western blotting has many applications. We will describe two: (1) co-IP and the related pull-down assays and (2) chromatin immunoprecipitation assays (ChIPs). In both assays, IP is used to establish evidence for the interaction between the protein that is captured by the precipitating antibody and another suspected partner molecule. For co-IP assays the suspected partner molecule is almost always a protein. Thus the immunoprecipitate is analyzed by Western blot with an antibody that recognizes the suspected partner protein. In the pull-down assay, a binding element other than an antibody is used to capture the target protein. For example, instead of a linked antibody, the beads might be derivatized with glutathione to capture a glutathione S-transferase (GST) fusion protein or with streptavidin to capture a biotinylated protein.[72] As for IP, a Western blot is often used to identify potential partner proteins in the pull-down assay. However, other means including mass spectrometry can also be used for identification of potential partners.[73]

ChIP assays provide a means of identifying transcription factors and other proteins associated with particular regions of the chromatin that are generally sequences in gene promoters.[74] In this assay, cells that express the target protein are lightly fixed by treatment with formaldehyde. This procedure cross-links the target protein to its partner macromolecules and to the DNA if this is one of the molecules with which it is interacting at the time of fixation. The DNA is then extracted from the cell and sheared into fragments of 400–1200 bp. The sheared DNA (with cross-linked proteins) is immunoprecipitated with an antibody to the target protein or with a control antibody (such as nonimmune IgG or serum) in a parallel immunoprecipitation. The immunoprecipitate is analyzed by PCR with primers that straddle the region of the chromosome with which the target protein is suspected to interact. If PCR results in the amplification of the expected fragment from the antibody immunoprecipitate (and not from the control immunoprecipitate), the target protein is interpreted to have been binding to that region of the chromosome when the cells were fixed. The ChIP assay is used to analyze the effects of a variety of treatments of cells (for example, with hormones or growth factors) on the interaction of proteins such as transcription factors and cofactors with specific genes in the intact cell.

10.9 CONCLUSION

With its use starting in the early 1960s, IP was the first application of antibodies used to selectively identify a protein in a cell or tissue lysate. The procedure rapidly gained popularity in the mid-1970s. By the 1980s, flow cytometry, ELISAs, and Western blot assays had been introduced. Because IP is more technically difficult and more cumbersome than these other assays, the use of IP decreased in frequency during the 1980s and 1990s. The beginning of the 21st century saw a rejuvenation of the use of IP with more analyses in which it was coupled with Western blots in co-IP and with PCR in the ChIP assay.

First described in 1981, the Western blot gained its name as a play on words. The means of detecting DNA fragments on a gel after blotting to nitrocellulose established in the mid 1970s was named after its inventor, Ed Southern. Soon afterwards, the protocol for transferring RNA from a gel and similarly detecting it by hybridization after blotting to nitrocellulose was therefore called a Northern blot. Following the trend, the parallel protocol but blotting proteins for detection by antibodies on a nitrocellulose membrane was called a Western blot. The main advantage of the Western blot over other immunologic detection techniques is the ability to visualize the protein of interest after it has been separated from other proteins and potential antigens. As well as separating the protein from potential contaminating antigens, the gel separation provides an independent means, other than the ability to interact with the antibody, for the investigator to identify the specified protein by its molecular mass. The protein-antibody interaction on the Western blot can be used to detect a specific protein in a mixture such as from a tissue extract, or it can be used to detect a particular group of antibodies among a mixture such as in serum. Because it includes a powerful protein separation procedure, relies on high-affinity antibody-antigen interactions and the detection systems involve amplification of the signal, the Western blot is a highly sensitive and very specific assay. In addition, with the appropriate controls and standards, the Western blot assay is linear with the amount of protein added to the system. For these reasons, the Western blot assay is widely used in research and clinical laboratories.

REFERENCES

1. Burnette, W.N., "Western blotting": electrophoretic transfer of proteins from sodium dodecyl sulfate–polyacrylamide gels to unmodified nitrocellulose and radiographic detection with antibody and radioiodinated protein A, *Anal. Biochem.*, 112, 195, 1981.
2. Cassab, G.I., Localization of cell wall proteins using tissue-print western blot techniques, *Methods Enzymol.*, 218, 682, 1993.
3. Towbin, H., Staehelin, T., and Gordon, J., Electrophoretic transfer of proteins from polyacrylamide gels to nitrocellulose sheets: procedure and some applications, *Proc. Natl. Acad. Sci. U. S. A.*, 76, 4350, 1979.
4. Nagy, B., Costello, R., and Csako, G., Downward blotting of proteins in a model based on apolipoprotein(a) phenotyping, *Anal. Biochem.*, 231, 40, 1995.
5. Zeng, L., Tate, R., and Smith, L.D., Capillary transfer as an efficient method of transferring proteins from SDS-PAGE gels to membranes, *Biotechniques*, 26, 426, 1999.
6. Kurien, B.T. and Scofield, R.H., Heat-mediated, ultra-rapid electrophoretic transfer of high and low molecular weight proteins to nitrocellulose membranes, *J. Immunol. Methods*, 266, 127, 2002.
7. Kyhse-Andersen, J., Electroblotting of multiple gels: a simple apparatus without buffer tank for rapid transfer of proteins from polyacrylamide to nitrocellulose, *J. Biochem. Biophys. Methods*, 10, 203, 1984.
8. Tovey, E.R. and Baldo, B.A., Comparison of semi-dry and conventional tank-buffer electrotransfer of proteins from polyacrylamide gels to nitrocellulose membranes, *Electrophoresis*, 8, 384, 1987.
9. Southern, E.M., Detection of specific sequences among DNA fragments separated by gel electrophoresis, *J. Mol. Biol.*, 98, 503, 1975.
10. Lichtenstein, A.V., Moiseev, V.L., and Zaboikin, M.M., A procedure for DNA and RNA transfer to membrane filters avoiding weight-induced gel flattening, *Anal. Biochem.*, 191, 187, 1990.
11. Braun, W. and Abraham, R., Modified diffusion blotting for rapid and efficient protein transfer with phastsystem, *Electrophoresis*, 10, 249, 1989.
12. Kurien, B.T. and Scofield, R.H., Multiple immunoblots after non-electrophoretic bidirectional transfer of a single SDS-PAGE gel with multiple antigens, *J. Immunol. Methods*, 205, 91, 1997.

13. Olsen, I. and Wiker, H.G., Diffusion blotting for rapid production of multiple identical imprints from sodium dodecyl sulfate polyacrylamide gel electrophoresis on a solid support, *J. Immunol. Methods*, 220, 77, 1998.

14. Ranganathan, V. and De, P. K., Western blot of proteins from Coomassie-stained polyacrylamide gels, *Anal. Biochem.*, 234, 102, 1996.

15. Gruber, C. and Stan-Lotter, H., Western blot of stained proteins from dried polyacrylamide gels, *Anal. Biochem.*, 253, 125, 1997.

16. Davis, B.J., Disc electrophoresis—II method and application to human serum proteins, *Ann. N. Y. Acad. Sci.*, 404, 1964.

17. Tovey, E.R. and Baldo, B.A., Protein binding to nitrocellulose, nylon and PVDF membranes in immunoassays and electroblotting, *J. Biochem. Biophys. Methods*, 19, 169, 1989.

18. Jacobson, G. and Karsnas, P., Important parameters in semi-dry electrophoretic transfer, *Electrophoresis*, 11, 46, 1990.

19. Fairbanks, G., Steck, T.L., and Wallach, D.F., Electrophoretic analysis of the major polypeptides of the human erythrocyte membrane, *Biochemistry*, 10, 2606, 1971.

20. Bolt, M.W. and Mahoney, P.A., High-efficiency blotting of proteins of diverse sizes following sodium dodecyl sulfate-polyacrylamide gel electrophoresis, *Anal. Biochem.*, 247, 185, 1997.

21. Chen, L.M., Liang, Y., Tai, J.H., and Chern, Y., Comparison of nitrocellulose and PVDF membranes in GTP-overlay assay and western blot analysis, *Biotechniques*, 16, 600, 1994.

22. Yom, H.C. and Bremel, R.D., Xerographic paper as a transfer medium for western blots: quantification of bovine alpha s1-casein by western blot, *Anal. Biochem.*, 200, 249, 1992.

23. Karey, K.P. and Sirbasku, D.A., Glutaraldehyde fixation increases retention of low molecular weight proteins (growth factors) transferred to nylon membranes for western blot analysis, *Anal. Biochem.*, 178, 255, 1989.

24. Too, C.K., Murphy, P.R., and Croll, R.P., Western blotting of formaldehyde-fixed neuropeptides as small as 400 daltons on gelatin-coated nitrocellulose paper, *Anal. Biochem.*, 219, 341, 1994.

25. Nishi, N., Inui, M., Miyanaka, H., Oya, H., and Wada, F., Western blot analysis of epidermal growth factor using gelatin-coated polyvinylidene difluoride membranes, *Anal. Biochem.*, 227, 401, 1995.

26. Schaffner, W. and Weissmann, C., A rapid, sensitive, and specific method for the determination of protein in dilute solution, *Anal. Biochem.*, 56, 502 1973.

27. Salinovich, O. and Montelaro, R. C., Reversible staining and peptide mapping of proteins transferred to nitrocellulose after separation by sodium dodecylsulfate-polyacrylamide gel electrophoresis, *Anal. Biochem.*, 156, 341, 1986.

28. Yonan, C.R., Duong, P.T., and Chang, F.N., High-efficiency staining of proteins on different blot membranes, *Anal. Biochem.*, 338, 159, 2005.

29. Exner, T. and Nurnberg, B., Immuno- and gold staining of a single Western blot, *Anal. Biochem.*, 260, 108, 1998.

30. Kaufmann, S.H., Ewing, C.M., and Shaper, J.H., The erasable Western blot, *Anal. Biochem.*, 161, 89, 1987.

31. Eynard, L. and Lauriere, M., The combination of Indian ink staining with immunochemiluminescence detection allows precise identification of antigens on blots: application to the study of glycosylated barley storage proteins, *Electrophoresis*, 19, 1394, 1998.

32. Glenney, J., Antibody probing of Western blots which have been stained with India ink, *Anal. Biochem.*, 156, 315, 1986.

33. Alba, F.J. and Daban, J.R., Rapid fluorescent monitoring of total protein patterns on sodium dodecyl sulfate-polyacrylamide gels and western blots before immunodetection and sequencing, *Electrophoresis*, 19, 2407, 1998.

34. Doolittle, M.H., Ben-Zeev, O., and Briquet-Laugier, V., Enhanced detection of lipoprotein lipase by combining immunoprecipitation with Western blot analysis, *J. Lipid Res.*, 39, 934, 1998.

35. Harlow, E. and Lane, D. *Antibodies: A Laboratory Manual,* Cold Spring Harbor Laboratory, Cold Spring Harbor, NY, 1988.

36. Clark, C.R., Kresl, J.J., Hines, K.K., and Anderson, B.E., An immunofiltration apparatus for accelerating the visualization of antigen on membrane supports, *Anal. Biochem.*, 228, 232, 1995.

37. Fukuda, T., Tani, Y., Kobayashi, T., Hirayama, Y., and Hino, O., A new western blotting method using polymer immunocomplexes: Detection of tsc1 and tsc2 expression in various cultured cell lines, *Anal. Biochem.*, 285, 274, 2000.

38. Harris, R.Z., Liddell, P.A., Smith, K.M., and Ortiz de Montellano, P.R., Catalytic properties of horseradish peroxidase reconstituted with the 8-(hydroxymethyl)- and 8-formylheme derivatives, *Biochemistry*, 32, 3658, 1993.

39. Tuerk, C. and Gold, L., Systematic evolution of ligands by exponential enrichment: RNA ligands to bacteriophage t4 DNA polymerase, *Science*, 249, 505, 1990.

40. Ellington, A.D. and Szostak, J.W., *In vitro* selection of RNA molecules that bind specific ligands, *Nature*, 346, 818, 1990.

41. Green, L.S. et al., Nuclease-resistant nucleic acid ligands to vascular permeability factor/vascular endothelial growth factor, *Chem. Biol.*, 2, 683, 1995.

42. Drolet, D.W., Moon-McDermott, L., and Romig, T.S., An enzyme-linked oligonucleotide assay, *Nat. Biotechnol.*, 14, 1021, 1996.

43. Bianchini, M. et al., Specific oligobodies against erk-2 that recognize both the native and the denatured state of the protein, *J. Immunol. Methods*, 252, 191, 2001.

44. Nygren, P.A. and Skerra, A., Binding proteins from alternative scaffolds, *J. Immunol. Methods*, 290, 3, 2004.

45. Koide, A., Bailey, C.W., Huang, X., and Koide, S., The fibronectin type iii domain as a scaffold for novel binding proteins, *J. Mol. Biol.*, 284, 1141, 1998.

46. Xu, L. et al., Directed evolution of high-affinity antibody mimics using mRNA display, *Chem. Biol.*, 9, 933, 2002.

47. Beste, G., Schmidt, F.S., Stibora, T., and Skerra, A., Small antibody-like proteins with prescribed ligand specificities derived from the lipocalin fold, *Proc. Natl. Acad. Sci. U. S. A.*, 96, 1898, 1999.

48. Winter, G., Griffiths, A.D., Hawkins, R.E., and Hoogenboom, H.R., Making antibodies by phage display technology, *Annu. Rev. Immunol.*, 12, 433, 1994.

49. Smith, K.A. et al., Demystified…recombinant antibodies, *J. Clin. Pathol.*, 57, 912, 2004.

50. Wilson, D.S., Keefe, A.D., and Szostak, J.W., The use of mRNA display to select high-affinity protein-binding peptides, *Proc. Natl. Acad. Sci. U. S. A.*, 98, 3750, 2001.

51. Griswold, D.E., Hillegass, L., Antell, L., Shatzman, A., and Hanna, N., Quantitative western blot assay for measurement of the murine acute phase reactant, serum amyloid p component, *J. Immunol. Methods*, 91, 163, 1986.

52. Fang, Y. et al., Signaling between the placenta and the uterus involving the mitogen-regulated protein/proliferins, *Endocrinology*, 140, 5239, 1999.

53. Uhl, J., and Newton, R.C., Quantitation of related proteins by western blot analysis, *J. Immunol. Methods*, 110, 79, 1988.

54. Huang, D. and Amero, S.A., Measurement of antigen by enhanced chemiluminescent western blot, *Biotechniques*, 22, 454, 1997.

55. Diamandis, E.P., Christopoulos, T.K., and Bean, C.C., Quantitative western blot analysis and spot immunodetection using time-resolved fluorometry, *J. Immunol. Methods*, 147, 251, 1992.

56. Fradelizi, J., Friederich, E., Beckerle, M.C., and Golsteyn, R.M., Quantitative measurement of proteins by western blotting with cy5^{tm}-coupled secondary antibodies, *Biotechniques*, 26, 484, 1999.

57. Lubit, B.W., Dansylated proteins as internal standards in two-dimensional electrophoresis and western blot analysis, *Electrophoresis*, 5, 358, 1984.

58. Tzeng, M.C., A sensitive, rapid method for monitoring sodium dodecyl sulfate-polyacrylamide gel electrophoresis by chromophoric labeling, *Anal. Biochem.*, 128, 412, 1983.

59. Strottmann, J.M., Robinson, J.B., Jr., and Stellwagen, E., Advantages of preelectrophoretic conjugation of polypeptides with fluorescent dyes, *Anal. Biochem.*, 132, 334, 1983.

60. Tsang, V.C., Hancock, K., and Simons, A.R., Calibration of prestained protein molecular weight standards for use in the "Western" or enzyme-linked immunoelectrotransfer blot techniques, *Anal. Biochem.*, 143, 304, 1984.

61. Seto, D., Rohrabacher, C., Seto, J., and Hood, L., Phosphorescent zinc sulfide is a nonradioactive alternative for marking autoradiograms, *Anal. Biochem.*, 189, 51, 1990.

62. Carlone, G.M., Plikaytis, B.B., and Arko, R.J., Immune serum to protein molecular weight standards for calibrating Western blots, *Anal. Biochem.*, 155, 89, 1986.

63. Lindbladh, C., Mosbach, K., and Bulow, L., Standard calibration proteins for Western blotting obtained by genetically prepared protein a conjugates, *Anal. Biochem.*, 197, 187, 1991.

64. Lindmark, R., Thoren-Tolling, K., and Sjoquist, J., Binding of immunoglobulins to protein A and immunoglobulin levels in mammalian sera, *J. Immunol. Methods*, 62, 1, 1983.

65. Navarre, J., Bradford, A., Calhoun, B.C., and Goldenring, J.R., Quenching of endogenous peroxidase in western blot, *Biotechniques*, 21, 990, 1996.

66. Vaitaitis, G.M., Sanderson, R.J., Kimble, E.J., Elkins, N.D., and Flores, S.C., Modification of enzyme-conjugated streptavidin-biotin Western blot technique to avoid detection of endogenous biotin-containing proteins, *Biotechniques*, 26, 854, 1999.

67. Nielsen, U.B. and Geierstanger, B.H., Multiplexed sandwich assays in microarray format, *J. Immunol. Methods*, 290, 107, 2004.

68. Pavlickova, P., Schneider, E.M., and Hug, H., Advances in recombinant antibody microarrays, *Clin. Chim. Acta*, 343, 17, 2004.

69. Knight, P.R. et al., Development of a sensitive microarray immunoassay and comparison with standard enzyme-linked immunoassay for cytokine analysis, *Shock*, 21, 26, 2004.

70. Bock, C. et al., Photoaptamer arrays applied to multiplexed proteomic analysis, *Proteomics*, 4, 609, 2004.

71. Schweitzer, B. et al., Multiplexed protein profiling on microarrays by rolling-circle amplification, *Nat. Biotechnol.*, 20, 359, 2002.

72. Harris, M., Use of GST-fusion and related constructs for the identification of interacting proteins, *Methods Mol. Biol.*, 88, 87, 1998.

73. Loomis, J.S., Courtney, R.J., and Wills, J.W., Binding partners for the ul11 tegument protein of herpes simplex virus type 1, *J. Virol.*, 77, 11417, 2003.

74. Wells, J. and Farnham, P.J., Characterizing transcription factor binding sites using formaldehyde crosslinking and immunoprecipitation, *Methods*, 26, 48, 2002.

75. Laemmli, U.K., Cleavage of structural proteins during the assembly of the head of bacteriophage t4, *Nature*, 227, 680, 1970.
76. Ornstein, L., Disc electrophoresis—I background and theory, *Ann. N. Y. Acad. Sci.*, 321, 1964.
77. Nilsen-Hamilton, M. and Hamilton, R., Detection of proteins induced by growth regulators, *Methods Enzymol.*, 147, 427, 1987.

11 Immunohistochemical Methods

José A. Ramos-Vara and Julie Ackerman Saettele

CONTENTS

273

11.1 INTRODUCTION

Immunohistochemistry (IHC) is the detection of antigens in tissue sections by means of specific antibodies. IHC has the unique advantage over other techniques for demonstrating proteins, enabling the correlation of antigens with their location within a tissue or cell. This is very important for the study of cell function in normal and pathologic tissues. IHC was introduced by Coons in the late 1940s[1] and, since then, has been applied extensively to multiple areas of biology, including cell function, characterization of neoplastic processes, and detection of infectious diseases. As its name indicates, IHC bridges three major disciplines: morphology, histochemistry, and immunology. Although this chapter will focus on the detection of antigens in tissue sections, we will briefly mention IHC methods on cytologic preparations (otherwise named immunocytochemistry).

11.1.1 OVERVIEW OF METHODOLOGY

In the past, IHC involved fluorescent labels and frozen sections. Technologic advances have permitted the use of formalin-fixed, paraffin-embedded tissues for IHC, allowing correlation of morphology with antigen detection and the use of archived material. In IHC, the binding of the antibody to an antigen on tissue sections is typically detected with a colored histochemical reaction. Fixation can hinder detection of some antigens, and numerous "antigen retrieval" methods have been developed to overcome this problem. The sensitivity of the reaction is critical for the detection of some antigens present in very small amounts. In the last decade, highly sensitive IHC methods have been developed. Various combinations of enzyme-substrate-chromogen producing different colors have permitted the detection of multiple antigens in the same section.

11.1.2 A PRACTICAL APPROACH TO THE STANDARDIZATION OF A NEW IMMUNOHISTOCHEMICAL TEST

Standardization of a new IHC test is a challenging process in which many factors affect the outcome. Most commercial antibodies have been tested in human tissues and rarely in other species. Information from the manufacturer of the antibody is essential to determine the suitability of the antibody for a particular species as well as the IHC detection of a particular antigen by Western blot or in frozen sections. However, many times this information is unavailable, and the researcher needs to develop a standard protocol to test new antibodies. The following protocol assumes that tissues have been fixed in formalin and embedded in paraffin.

11.1.2.1 Method

1. Select the tissue that is most likely to be used as a positive control. Ideally, it should have areas known to lack the antigen of interest.
2. Processing (e.g., fixation, embedding) of the control tissue should be done in the same manner as for test tissues. Control tissue should be from the same species as test tissues. An additional control tissue from a species known to react with that antibody, if available, is also recommended.
3. Prepare dilutions of the primary antibody (usually four twofold dilutions will suffice).
4. To determine optimal antigen retrieval (AR), prepare three sets of slides; one will not have any type of AR; another will have enzymatic AR (e.g., proteinase K); the third will have heat-induced epitope retrieval (HIER) with citrate buffer at pH 6.0.
5. Run the IHC test following a standard procedure. Incubation of the primary antibody (time, temperature) will depend on the available information. If there is no information, incubate 60 min at room temperature.
6. After the IHC test is done, examine the slides to determine staining quality.

11.1.2.2 Notes

If specific staining is achieved, the concentration of the primary antibody, incubation times, incubation temperature, and AR procedures may need to be optimized to obtain the best signal-to-noise ratio. If there is no staining, the concentration of the primary antibody, incubation and temperature time (including overnight or longer incubation at 4°C), and AR procedures (use of other enzymes or buffers at different pH) may need to be modified. Keep in mind that the detection of the antigen may not be possible because of fixation, animal species variability, and lack or low amount of the epitope recognized by the antibody.

11.1.3 Antibodies in Immunohistochemistry

Antibodies are made by immunizing animals (e.g., mice, rabbit, goat, horse) with purified antigen.

Polyclonal antibodies are produced in multiple animal species, particularly rabbit, horse, goat, and chicken. Polyclonal antibodies have higher affinity and wider reactivity but lower specificity when compared with monoclonal antibodies. The ability of polyclonal antibodies to detect multiple epitopes in fixed tissues contributes to the increased sensitivity of polyclonal antibodies compared with monoclonal antibodies. However, a polyclonal preparation may be very heterogeneous because of the presence of antibodies to strong and weakly immunogenic epitopes in the same preparation. The "immunologic promiscuity" of polyclonal antibodies can be an advantage (e.g., more possibilities of detecting an antigen due to multiplicity of epitope recognition). However, the likelihood of cross-reactivity with similar epitopes in other proteins and, thus, false positives may be problematic.

Monoclonal antibodies are produced mostly in mice by technology developed by Köhler and Milstein.[2] The advantage of monoclonal antibodies is their higher specificity when compared with polyclonal antibodies. This specificity reduces (although does not eliminate) the possibility of cross-reactivity with other antigens. One reason for cross-reactivity is that monoclonal antibodies are directed against epitopes consisting of a small number of amino acids, which can be part of several types of proteins and peptides. In some instances, it is difficult to determine whether immunoreactivity is due to shared epitopes (cross-reactivity) or epitopes resulting from protein crosslinking during fixation with aldehydes. Under certain circumstances, background staining of monoclonal antibodies due to nonspecific immunoglobulins is reduced (ascites fluid) or nonexistent (cell culture supernatant).

Commercially available *rabbit monoclonal antibodies* have been made possible by the development of antibody libraries. The advantage of rabbit monoclonals over mouse monoclonals is their higher affinity, suitability for use on mouse tissues without special procedures, increased specificity in some cases, and avoidance of AR methods.[3]

11.1.3.1 Presentation of Commercial Antibodies

It is becoming more common, particularly with companies specialized in the production of antisera, to include the following information in their catalogs: format of the antibody (e.g., purified, whole serum, supernatant, ascites, immunoglobulin isotype), host in which the antibody was produced (e.g., mouse, rabbit), protein concentration, immunogen used (including epitope and molecular weight, if known), species reactivity (e.g., human, mouse; others not known), cellular localization (e.g., cytoplasmic, membrane, nuclear), positive tissue control recommended, applications (e.g., immunoprecipitation, Western blotting, enzyme-linked immunosorbent assay, immunohistology with formalin/paraffin and frozen tissues), and pertinent literature. However, when working with a new antibody, specific information about a particular species or tissue might not be available, and contact with the manufacturer of the antibody for additional information is advisable before attempting to use it. Other data not commonly available from manufacturers of antisera are the affinity constant of antibodies and the possibility of cross-reactivity with other antigens (e.g., serotypes of viruses, other related viruses or bacteria, cell antigens) that may be a result of fixation, antigen retrieval, or shared epitopes among different antigens (Figure 11.1).[3]

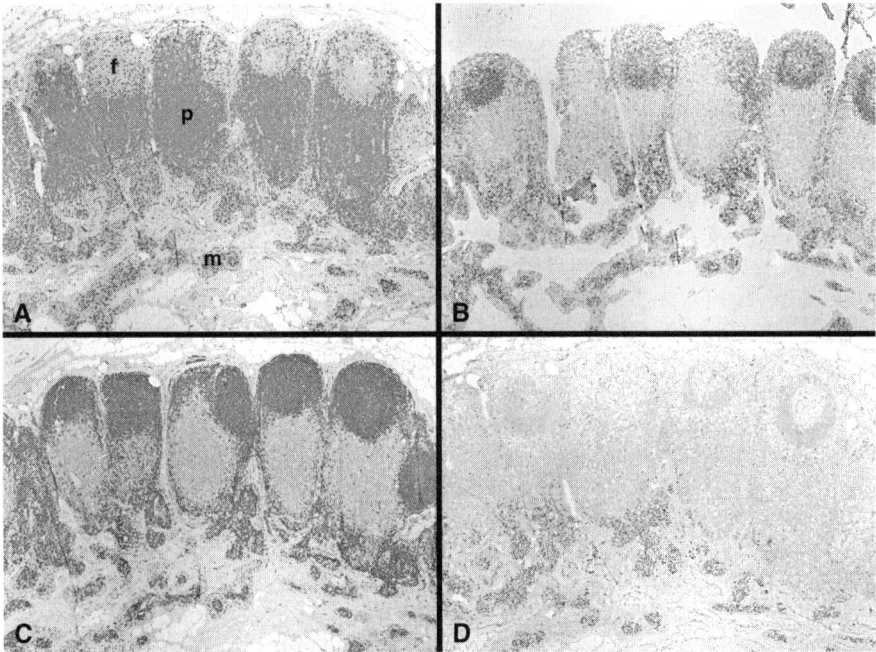

FIGURE 11.1 (See color insert following page 338.) Knowing the location of the antigen of interest is paramount to determining the quality of the immunostaining. In this lymph node, four antigens to CD3 (A), CD79a (B), CD20 (C), and MUM1 (D) were compared. The follicular areas (f) are mostly positive for CD79a and CD20 with a few positive cells for CD3 and fewer for MUM1. The paracortex (p) is mostly populated by CD3-positive cells with few positive cells for the other markers. The medulla (m) contains mostly CD79a, CD20, and MUM1-positive cells with fewer T cells. Immunoperoxidase-DAB, hematoxylin counterstain.

11.2 TISSUE PREPARATION

In great measure, tissue preparation will determine the success of the IHC reaction. Tissue preparation will depend in some ways on the type of antigen to be detected (e.g., some antigens will be destroyed during fixation in formaldehyde and tissues have to be frozen or fixed in a different fixative). One rule to follow, regardless of the antigen to be detected, is to collect and process samples before tissues undergo autolysis.

If tissues are not frozen immediately, they must be fixed. Fixation is necessary to: (1) preserve cellular components, including soluble and structural proteins; (2) prevent autolysis and displacement of cell constituents, including antigens and enzymes; (3) stabilize cellular materials against deleterious effects of subsequent procedures; and (4) facilitate conventional staining and immunostaining. Two types of fixatives are used in histopathology: cross-linking (noncoagulating) fixatives and coagulating fixatives.[3]

Formaldehyde is the gold standard fixative for routine histology and IHC. It preserves mainly peptides and the general structure of cellular organelles. It also interacts with nucleic acids but has little effect on carbohydrates. It is a good preservative of lipids if the fixative contains calcium. The basic mechanism of fixation with formaldehyde is the formation of addition products between the formalin and uncharged reactive amino groups ($-NH$ or $-NH_2$), finally forming methylene bridges (cross-links) with modification of the tertiary and quaternary structure of proteins. Formalin fixation is a progressive, time- and temperature-dependent process. Over-fixation can produce false negative results in IHC because of excessive cross-links. However, underfixation can also produce unexpected results. There is no optimal standard fixation time for every antigen.

Many formalin substitutes are coagulating fixatives that precipitate proteins by breaking hydrogen bonds in the absence of protein crosslinking. Typically, the non–cross-linking fixative is ethanol. Other fixatives used in IHC are glyoxal (dialdehyde), a mixture of glyoxal and alcohol, 4% paraformaldehyde, and zinc formalin. The conformational changes of proteins produced by these types of fixatives may not be as profound as with formalin. A trial-and-error approach starting with the recommendations noted on the manufacturer's antibody specification sheet should be used. In the authors' experience, 4% paraformaldehyde or zinc formalin fixative, followed by paraffin embedding, is appropriate for initial investigation when testing in a research environment with primary antibodies that have only been characterized in frozen tissue. By employing this method, the ease of use of archival tissue may be obtained. Because different fixatives affect immunoreactivity in different ways, paraffin blocks from tissues fixed in fixatives other than formaldehyde should be labeled appropriately before being archived.

11.2.1 GLASS SLIDES

Quality IHC results require a strong bond between the tissue section and the glass slide to prevent tissue adhesion problems that may cause pooling of reagents and staining artifacts. Commercially available glass slides with a special coating intended for IHC offer standardized coated slide quality. Here, we include two methods to coat regular slides for IHC.

11.2.1.1 Poly-ʟ-Lysine–Coated Slides

1. Prepare a 0.1% aqueous w/v solution of poly-ʟ-lysine (Sigma P-8920).
2. Dip the slides in the solution.
3. Allow the slides to dry at room temperature.
4. Store slides at room temperature.

11.2.1.2 3-Aminopropyltriethoxy Silane–Coated Slides

1. Prepare a solution of 2% (v/v) silane (Sigma A-3648) in acetone.
2. Immerse slides in acetone for 1–5 min.
3. Immerse slides in 2% silane solution for 1–5 min.
4. Rinse in two consecutive baths of acetone for 5 min.
5. Dry and store at room temperature.

11.2.2 Smears/Cytospins

1. Prepare smears/cytospins onto silane/poly-ʟ-lysine-coated slides.
2. Air dry at room temperature for 30 min.
3. Fixed in precooled, fresh acetone at –20°C for 20 min.
4. Air dry at room temperature for at least 15 min.
5. Stain smears/cytospins.

11.2.2.1 Notes

1. Alternatively, 100% ethanol may be used in place of fresh acetone; the choice of fixative depends on the target antigen being stained in the tissue.
2. Unstained slides may be wrapped in aluminum foil and stored at –20°C. When ready to stain: unwrap slides and allow them to come to room temperature, postfix for at least 30 sec using the same fixative as previously used before storage, air dry sections, rehydrate in buffer for 5 min, and stain slides.
3. Stability of unstained slides is antigen dependent; some antigens are labile and will not stain appropriately unless stained immediately.

11.2.3 Frozen Sections

1. Cut four 7-μm-thick unfixed frozen sections.
2. Place sections on silane/poly-ʟ-lysine–coated slides.
3. Air dry sections for 30 min.
4. Fix in precooled, fresh acetone at 4°C for 20 min.
5. Air dry sections at room temperature for at least 15 min.
6. Stain sections.

11.2.3.1 Notes

1. Alternatively, 100% ethanol may be used instead of fresh acetone; the choice of fixative depends on the target antigen being stained in the tissue.
2. Unstained slides may be wrapped in aluminum foil and stored at –20°C. When ready to stain: unwrap slides and allow them to come to room

temperature, postfix for at least 30 sec with the same fixative as used before storage, air dry sections, rehydrate in buffer for 5 min, and stain slides.

3. Stability of unstained slides depends on the antigen; some antigens are labile and will not stain appropriately unless stained immediately.

11.2.4 PARAFFIN WAX SECTIONS

1. Fix tissues in 10% formaldehyde (other fixatives might be more appropriate) for 12–48 h. The actual length of fixation will depend on tissue thickness, temperature, and antigen examined.
2. Paraffin embedding.
3. Cut 3- to 5-μm thick sections.
4. Place sections on silane/poly-L-lysine-coated slides, using an adhesive free water bath.
5. Dry the sections on edge.
6. Place sections in a 60°C oven for 30–60 min.
7. Stain the sections after deparaffinizing.

11.2.4.1 Notes

1. Prolonged drying (overnight at room temperature) may be necessary to obtain optimal tissue section adhesion for tissues rich in fat (brain, mammary gland).
2. Drying temperatures higher than 60°C may have deleterious effects on detection of some antigens or produce an increase in the background staining.[4]
3. Prolonged storage of paraffin sections may reduce immunoreactivity for some antigens regardless of storage conditions.[5-7] In the authors' experience, estrogen receptor, progesterone receptor, cyclin-associated protein (e.g., Ki-67, PCNA, cyclin D1, or bcl-6), and thyroid transcription factor-1 antigens are labile, thus are not stabile for prolonged storage after sections are cut from the paraffin block.
4. When inconsistent staining happens with archived paraffin sections, repeat the test with freshly cut sections.

11.2.5 CYTOBLOCK

This procedure is used for fragments of tissue that are too small for normal processing and embedding in paraffin. The reagents required to form what is called a Cytoblock (Thermo Electron Corporation 7401150) are supplied with a commercial kit. You will also need a Cytospin. Avoid using formol calcium or solutions containing phosphate ions because these inhibit polymerization of the gel.

1. Fix tissues in formalin.
2. Transfer tissues to a centrifuge tube.
3. Centrifuge at 2000 rpm for 5 min ($\sim$1260g).
4. Discard most of the supernatant.
5. Add four drops of the blue Cytoblock fluid (reagent 2) to two drops of sediment (specimen) and mix. Each block should have two drops of specimen or less.

6. Assemble Cytoblock cassette into Cytoclip with specimen chamber.
7. Add three drops of reagent 1 into the center of the well in the board insert.
8. Place the assembled Cytoclip into the Cytospin.
9. Place the mixed cell suspension in a Cytofunnel and centrifuge at 1500 rpm for 5 min.
10. Remove cassette containing Cytoblock from assembly.
11. Add one more drop of colorless Cytoblock fluid (reagent 1) to the gel block.
12. Enclose block in the cassette and place in fixative.
13. Embed in paraffin.
14. Follow standard procedure for paraffin-embedded tissues.

11.2.5.1 Note

The reagents in the kit are proprietary and the chemical names for them are not public.

11.3 ENZYME LABELS

Although a variety of enzymes have been used in IHC, the most widely used are horseradish peroxidase (HRP) and alkaline phosphatase (AP).

11.3.1 HORSERADISH PEROXIDASE

The substrate for HRP is hydrogen peroxide. Chromogens are oxidized by peroxide reaction products, resulting in a precipitate at the site of bound antibody. Of several chromogens, 3,3′ diaminobenzidine tetrachloride (DAB) is the most commonly used. This chromogen results in a brown precipitate that is insoluble in organic solvents. When endogenous peroxidase activity is very high or melanin pigment is prominent, DAB is not the optimal chromogen, and other chromogens or AP should be used. Another chromogen for peroxidase, 3-amino-9-ethylcarbazole (AEC), produces a red precipitate. However, this chromogen is soluble in organic solvents, and slides must be mounted in water-soluble medium. Other chromogens produce a similar red color and are resistant to organic solvents. Another chromogen, 4-chloro-1-naphthol, precipitates as a blue product that is soluble in organic solvents. There are different procedures to intensify the color of the DAB end-product.

11.3.2 ALKALINE PHOSPHATASE

AP is isolated from calf intestine. The rationale for the AP methods is based on the hydrolysis of phosphates that contain substituted naphthol groupings to produce an insoluble naphthol derivative, which then couples to a suitable diazonium salt to produce a colored, insoluble azo dye at the site of enzyme activity. The chromogens most commonly used are 5-bromo-4-chloro-3-indolylphosphate/nitro blue tetrazoliumchloride (blue, permanent media), Fast red (red, aqueous mounting media), and new fuchsin (fuchsia, aqueous mounting media). AP is recommended for immunocytochemistry of cytologic specimens or tissues with high content of endogenous peroxidase. AP is used for double IHC in combination with HRP.

11.4 CAUSES OF BACKGROUND AND NONSPECIFIC REACTIVITY

Background is one of the most common problems in IHC and can seriously affect the interpretation of the immunologic reaction. The most common causes of background are listed below.

11.4.1 ENDOGENOUS ENZYME ACTIVITY

11.4.1.1 Endogenous Peroxidase Activity

Enzyme activity naturally present in red blood cells (pseudoperoxidase), granulocytes (myeloperoxidase), and neurons can react with DAB to produce a brown product indistinguishable from specific immunostaining. Pretreatment of tissue sections with a diluted solution (0.03–3%) of H_2O_2 in water with 0.1% sodium azide will weaken or abolish pseudoperoxidase activity of red blood cells and peroxidase activity in myeloid cells in frozen sections. In routine paraffin-embedded tissue sections and those with abundant hemorrhage or with acid hematin, a stronger solution of H_2O_2 is needed (3% H_2O_2 in water with 0.1% sodium azide) to remove this endogenous activity or a longer incubation in less concentrated solutions. The use of H_2O_2-methanol is not recommended for routine use, particularly when cell surface antigens are to be detected.

Methanol should be avoided due to its deleterious effects on antigenicity and immunoreactivity. However, should methanol-hydrogen peroxide use be necessitated, limit the incubation time to no more than 15 min.

11.4.1.2 Endogenous Alkaline Phosphatase

There are two isoenzymes of AP in mammalian tissues that can produce background staining when using AP methods: intestinal and nonintestinal forms. The nonintestinal form is easily inhibited by 1 mM levamisole (L-tetramisole), whereas the isoenzyme from calf intestine, used as a reporter molecule in immunoalkaline methods, is unaffected. The intestinal isoform can be blocked with 1% acetic acid, but this can damage some antigens.

11.4.2 OTHER ENDOGENOUS ACTIVITIES

11.4.2.1 Endogenous Avidin–Biotin Activities

Endogenous biotin is widely dispersed in mammalian tissues, particularly in liver, lung, spleen, adipose tissue, mammary gland, kidney, and brain. Background from endogenous biotin is greatly diminished after formalin fixation, but can be pronounced in frozen sections. The high ionic attraction of basic egg white avidin for oppositely charged cellular molecules, such as nucleic acids, phospholipids, and the glycosaminoglycans in the cytoplasm of mast cells, may result in nonspecific binding. Egg white avidin has an isoelectric point (pI) of 10.0 and has a basic positive charge at the almost-neutral pH used in immunostaining. Harsh heat-based AR methods expose endogenous biotin in formalin-fixed tissues.

Binding of avidin used in detection systems to endogenous biotin can produce strong background and needs to be inhibited. This nonimmune binding can be prevented by preparing streptavidin biotin complex (ABC) or labeled biotin solutions at pH 9.4 instead of 7.6. Also, a 5% solution of nonfat dry milk can prevent nonspecific binding to nuclei, cytoplasm, and cell membranes by avidin-fluorochrome conjugates. Substituting avidin from egg white with streptavidin (from *Streptomyces avidinii*), which has a pI of 5.5–6.5, significantly reduces the nonspecific binding in IHC methods. Streptavidin is the most common type of avidin preparation in commercially available detection kits. There are commercial reagents containing 0.1% streptavidin and 0.01% biotin that block this endogenous activity. Highly sensitive detection methods such as those based on tyramide-streptavidin need a previous blocking step to avoid excessive background. Alternatively, commercial detection systems are now available that are polymer based, thus negating the use of avidin-based secondary detection antibodies and avidin-biotin blocking reagents.

11.4.2.2 Endogenous Immunoglobulin Activity

This cause of background results when the primary antibody used in the IHC method binds with immunoglobulin on the surface of the tissue being stained. Nonspecific blocking agents, such as 5% bovine serum albumin, may be used before the primary antibody incubation step to block this nonspecific binding of the primary antibody to unrelated antigens in the tissue.

11.4.3 PRIMARY ANTIBODIES AS CAUSE OF NONSPECIFIC REACTIVITY OR BACKGROUND

11.4.3.1 Antibody Titration

An inadequate dilution (titer) of the primary antibody can be a cause of background staining. The ideal titer should be that which produces a strong signal with minimal or no background. To accomplish the latter, every antibody should be standardized using checkerboard titrations or at least multiple dilutions of the primary antibody. Multiple variables affect the optimal primary antibody titer, including the primary antibody incubation time and temperature, AR parameters (method, time, temperature, pH, chemical composition), buffer (phosphate-buffered saline [PBS] vs. tris-buffered normal saline [TBS], pH), and detection method (sensitivity, specificity, time, and temperature) used (Figure 11.2).

11.4.3.2 Antigen "Specific Cross-Reactivity"

This term refers to the binding of antibodies specific for a particular antigen to different cells or tissues than the one intended. This is the result of common epitopes shared among different cells or tissues. Examples include the strong labeling of smooth muscle cells by CD79a antibodies as well as the intended B-lymphocytes. CD15 stains Reed-Sternberg cells in Hodgkin's lymphoma; however, this antibody also cross-reacts with macrophages. A particular type of specific cross-reactivity

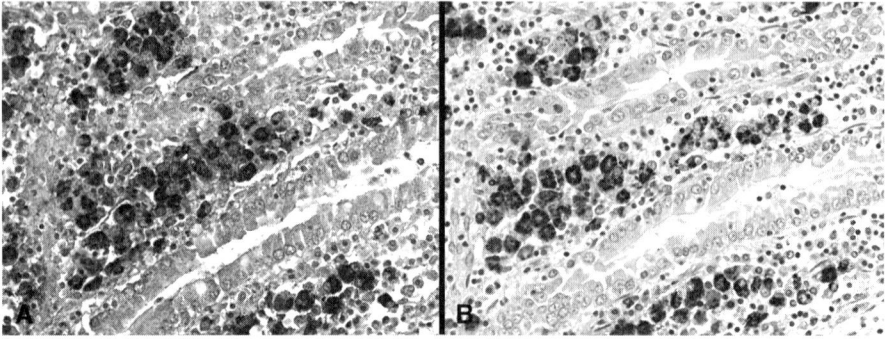

FIGURE 11.2 (See color insert following page 338.) From the intestine of a cow with paratuberculosis produced by *Mycobacterium avium* subsp. *paratuberculosis*. Optimal dilution of the primary antibody is critical. Insufficiently diluted primary antibody (A) results in nonspecific staining in intestinal epithelium and lamina propria. The optimal dilution (B) detects only antigen in infected macrophages (m) without staining epithelium (e). Immunoperoxidase-DAB, hematoxylin counterstain.

is "species cross-reactivity" in which the same antigen (epitope) is present in multiple species (e.g., cytokeratins in dogs, cats, horses; vimentin in multiple species; coronavirus detected in ferret, dogs, cats, and pigs by a monoclonal antibody produced against porcine coronavirus). When using secondary antibodies that might recognize tissue antigens (e.g., secondary rabbit-anti rat on mouse tissues) the secondary antibody needs to be adsorbed with immunoglobulins from the same species as the tissue examined (e.g., mouse immunoglobulin [Ig]G or mouse serum proteins).

11.4.3.3 Antibody "Nonspecific Cross-Reactivity"

This cause of background results when antibodies (primary or secondary) bind by nonimmune mechanisms to unrelated proteins or as a result of changes in the conformation of proteins from fixation or antigen retrieval.

11.4.4 BACKGROUND FROM HYDROPHOBIC AND IONIC INTERACTIONS

11.4.4.1 Hydrophobic Interactions of Proteins

Hydrophobic forces are key for successful antigen–antibody (Ag–Ab) binding, but can also produce unacceptable background. Tissue proteins are rendered more hydrophobic by fixation with aldehyde-containing fixatives as a result of cross-linking of reactive epsilon- and alpha-amino acids, both within and between adjacent tissue proteins. The increased hydrophobicity of proteins during fixation increases the background staining in immunohistochemical procedures and therefore, prolonged fixation in formalin or other aldehyde-based fixatives should be avoided. This background staining due to overfixation may be remedied by postfixation with Bouin's, Zenker's, or B5 fixatives. Immunoglobulins are also very hydrophobic proteins, particularly antibodies of the IgG_1 and IgG_3 subclasses. Aggregation and polymerization leading to increase in hydrophobicity is another problem observed during storage of immunoglobulins. Protein–protein interactions between conjugates and polar groups in tissue sections also produce background.

There are several methods to reduce hydrophobic binding of immunoglobulins and tissue proteins, including diluent buffers with a pH different from the isoelectric point of the antibody (particularly for monoclonal antibodies); diluents with low ionic strength (low salt concentration); addition of nonionic detergents (e.g., Tween 20, Triton X) or ethylene glycol to the diluent; or raising the pH of the diluent used for polyclonal antibodies. But probably the most common method to reduce background because of hydrophobic interactions is the use of blocking proteins before incubation with the primary antibody. Classically, this has been done with immunoglobulins of the same species to the secondary link or labeled antibody; however, bovine serum albumin, fish gelatin, fetal calf serum, nonfat dry milk, and, more recently, casein can be used.[3] Casein appears to be more effective than normal serum to block hydrophobic background staining.

11.4.4.2 Ionic Interactions

Ionic interactions are one of the main forces that control Ag–Ab interactions, but they also contribute to nonspecific background. The pI of the majority of antibodies ranges from 5.8 to 8.5. At the pH commonly used in diluent buffers, antibodies can have either net negative or positive surface charges and ionic interactions between immunoglobulins and tissue proteins can be expected if the latter possess opposite net surface charges. It is important to allow diluent buffer to reach room temperature before use to help prevent opposite net charge production. Nonimmune binding of immunoglobulins to tissues or cells with negative charge (e.g., endothelium, collagen) can be effectively blocked by using diluent buffers with high ionic strength. Solving nonspecific background staining in IHC becomes more complicated when we realize that many times this nonspecific staining is the result of a combination of ionic and hydrophobic interactions and, as previously mentioned, remedies for one type of interaction may aggravate the other.

11.5 ANTIGEN RETRIEVAL TECHNIQUES

Fixation with cross-linking antigens modifies the tertiary and quaternary structure of many antigens making them undetectable by specific antibodies. AR methods are intended to retrieve the loss of antigenicity, theoretically returning proteins to their prefixation conformation. Approximately 85% of antigens fixed in formalin require some type of AR to optimize the immunoreaction.[8] The need for AR depends not only on the antigen examined, but also the antibody used. Polyclonal antibodies are more likely to detect antigens than monoclonal antibodies in the absence of AR due to their sensitivity.[8] Two main groups of AR methods are used: enzyme digestion and heat-induced epitope retrieval.

11.5.1 DETERGENTS AND CHAOTROPIC SUBSTANCES

Detergents form micelles in aqueous solutions and are also surfactants because they decrease the surface tension of water. Detergents solubilize membrane proteins by mimicking the lipid-bilayer environment, forming mixed micelles consisting of lipids and detergents and detergent micelles containing proteins (usually one protein molecule per micelle). The more common detergents used in IHC are the nonionic type, which are better suited for breaking lipid-lipid and lipid-protein interactions than protein-protein interactions; therefore, they are considered nondenaturant. Nonionic detergents are TritonR-X100, Tween 20, saponin, BRIJR, and Nonidet P40; these are usually added to rinsing buffers (e.g., 0.05% v/v for Tween 20).

Chaotropic substances, including guanidine hydrochloride, sodium thiocyanate, and cesium, are thought to work by protein denaturation and/or partial opening/reversal of formalin bonds by dissociation of protein complexes. AR with detergents or chaotropic substances is in general less effective than enzymatic AR or AR by heat (HIER) and usually is used in combination with either enzymes or heat.

11.5.2 ENZYMATIC ANTIGEN RETRIEVAL

Protease-induced epitope retrieval (PIER) was the most commonly used AR method before the advent of heat-based AR methods. Many enzymes have been used for this purpose with the most common being trypsin, proteinase K, pronase, ficin, and pepsin.[3] The mechanism of action of PIER is most likely digestion of protein cross-linkages introduced during formalin fixation, but this cleavage is nonspecific and some antigens may be negatively affected by this treatment. The effect of PIER depends on the concentration and type of enzyme, incubation parameters (time, temperature, and pH) as well as on the duration of fixation. The enzyme digestion time is inversely related to the fixation time. It is our and others' preference to optimize a few enzymes rather than to use a broad range of enzymes. We use a commercially available ready-to-use solution of proteinase K that has good activity at room temperature and, therefore, can be used with automatic stainers. The disadvantages of PIER are the rather low number of antigens for which it is the optimal AR method, possible alteration of tissue morphology, and possible destruction of epitopes. Below are three enzymatic digestion methods.

11.5.2.1 Trypsin (see Reagents Preparation)

1. Immerse slides in 0.1% solution of trypsin at pH 7.8 and 37°C for 10–30 min.
2. Rinse the slides well in running tap water.
3. Proceed with the immunohistochemical method.

The AR results will depend on the activity of the trypsin solution among different companies and batches.

11.5.2.2 Protease

1. Add drops of a 0.05–0.5% (w/v) warm solution of protease XXIV to the slides and incubate at 37°C for 10–25 min.
2. Wash well in running tap water.
3. Proceed with the immunohistochemical method.

11.5.2.3 Pepsin

1. Immerse slides in a 0.4% (w/v) solution of pepsin.
2. Incubate at 37°C for 10–25 min.
3. Wash well in running tap water.
4. Proceed with immunohistochemical method.

11.5.3 HEAT-INDUCED EPITOPE RETRIEVAL

This group of AR methods has revolutionized the immunohistochemical detection of antigens fixed in cross-linking fixatives (e.g., formaldehyde). HIER is based on a concept developed by Fraenklen-Conrat and collaborators[9] who documented that the chemical reactions between proteins and formalin may be reversed, at least in part, by high-temperature heating or strong alkaline hydrolysis.[3] The mechanism involved in HIER is unknown, but its final effect is to revert conformational changes produced during fixation. Heating may unmask epitopes by hydrolysis of methylene cross-links, but it also acts by other less known mechanisms because it enhances immunostaining of tissue fixed in ethanol, which does not produce crosslinks. Other hypotheses proposed are extraction of diffusible blocking proteins, precipitation of proteins, and rehydration of the tissue section allowing better penetration of antibody, and heat mobilization of trace paraffin. Tissue-bound calcium ions may be important in masking some antigens during fixation. Calcium chelating agents (e.g., EDTA) are sometimes more effective than citrate buffer in antigen retrieval. In essence, applying heat to formalin-fixed, paraffin-embedded tissue sections by different means (microwave, pressure cooker, steamer, water bath) will revert the deleterious effects of formaldehyde fixation. Heating at high temperature (100°C) for a short duration (10 min) gives better results than those achieved with a comparatively low temperature for a longer time. However, we obtain satisfactory results using a steamer (90–95°C) with a 20-min incubation for the majority of antigens needing HIER. A universal AR solution is not available. Thus several HIER solutions made of different buffers (e.g., citrate, Tris) and pH (3–10) have been used. The pH of the solution is important. Some antigens will be retrieved with low pH solutions, others with only high pH solutions and a third group will be retrieved with solutions with a wide pH range.[10] For most antigens, HIER with 10 mM sodium citrate buffer (pH 6.0) will give satisfactory results and good cell morphology when compared with buffers with higher pH or solutions containing EDTA.

The degree of fixation can dramatically modify the response of antigens to antigen retrieval. Unfixed proteins are denatured at temperatures of 70–90°C, whereas such proteins do not exhibit denaturation at the same temperatures when they have been fixed in formaldehyde. Caution must be taken when testing new antibodies using different AR protocols. The possibility of unexpected immunostaining should always be considered when using HIER, particularly with buffers at low pH. When possible, it is recommended to do a comparative study of immunoreactivity using fresh frozen and routinely processed paraffin tissue sections.

11.5.3.1 HIER in a Microwave Oven

1. Using a microwave with revolving plate, timer and choice of watt settings, heat 500 ml of buffer in a plastic jar that will hold the slides and a plastic beaker with 200 ml of water for 2 min at 750 W. At the end of this time, remove the beaker of water.
2. Immerse the dewaxed slides completely in the warm buffer. Cover the container loosely with its lid.
3. Microwave at 750 W for 5 min (the solution needs to boil).
4. Check the level of the solution and make up to the original volume with the warm water.
5. Repeat steps 3 and 4 for the required length of time (usually 15–20 min).
6. Remove the container from the microwave oven and place it in cold tap water for 15 min.
7. Rinse the slides in distilled water and continue with the immunostaining method.

11.5.3.1.1 Note

Care should be taken when selecting this method for AR because of intraoven temperature variability, which may lead to inconsistent staining results/patterns.

11.5.3.2 HIER in a Pressure Cooker

11.5.3.2.1 Reagents and Equipment

Decloaker (Biocare Medical, DC2002)
Tissue Tek containers or plastic Coplin jars
HIER buffer
Distilled water

11.5.3.2.2 Method

1. Plug in the unit and place the pan into the body.
2. Fill the pan with 500 ml of deionized water and turn on the red toggle switch.
3. Place slides to be HIERed into Tissue Tek containers filled with 250 ml of retrieval buffer. Alternatively, plastic Coplin jars may be used filled with 50 ml of retrieval buffer.
4. Place containers with slides into center of pan.
5. Place the heat shield in the center of the pan.

6. Place the monitor Steam Strip on top of the staining dish, put the lid on and secure.
7. Put the weight on the Vent Nozzle.
8. Push the Display Set and check each of the displayed parameters.
9. Set the SP1 function (heating time) between 30 sec and 5 min, depending on the antigen.
10. Push the Display Set button to SP1 and push start.
11. When the timer goes off, push the Start/Stop button.
12. When the temperature reaches 90°C, the timer will sound off again.
13. Push the Start/Stop button to end the program. The pressure should read 0.
14. Open the lid and let the slides cool for several minutes.
15. Remove slide container and slowly rinse the slides in running tap water.
16. Transfer slides to rinse buffer.

11.5.3.2.3 Notes

1. Do not let the slides dry out at any time after being dewaxed and rehydrated.
2. Be careful handling the slides inside the decloaker because of the hot water.

11.5.3.3   HIER in a Steamer

1. Fill bottom of steam container with water.
2. Place Tissue Tek containers filled with 250 ml of retrieval buffer into steamer basket. Alternatively, plastic Coplin jars may be used filled with 50 ml of retrieval buffer.
3. Turn steamer on and preheat the AR buffer in the Tissue Tek containers/Coplin jars. Place thermometer through steam holes in the lid of the steamer suspended in the AR buffer. Bring the buffer up to 95°C.
4. When the temperature reads 95°C, quickly place the slides in the buffer (avoid touching it with bare hands) and bring up the temperature of the buffer to 95°C.
5. Steam and count 20 min or the required time for AR only after the temperature reaches 95°C.
6. Remove the beaker with the slides from the steamer and let it cool at room temperature for 20 min.
7. Wash in tap water for 10 min.
8. Put the slides in rinse buffer and continue with immunohistochemical staining.

11.6   IMMUNOENZYME TECHNIQUES

The number of immunohistochemical techniques available is large, and modifications to the classic methods are very common to optimize the immune reaction and detection of antigens. The final aim of any immunohistochemical method is to detect the maximum amount of antigen with the least possible background (maximum signal-to-noise ratio). In the past, direct techniques (the primary antibody is labeled with an enzyme or fluorochrome) were the only ones available, which provided a quick detection of antigens but low sensitivity. Indirect methods include those in which the primary antibody is not labeled. These methods consist of two or more layers of reagents, with the last layer usually containing the label. These methods are more laborious, but also more sensitive (reaching several hundred times the sensitivity of direct methods) and the ones currently used in IHC. As a rule, with some exceptions, the more complex the technique the more sensitive it is. This section is focused on formalin-fixed, paraffin-embedded sections, but similar procedures have been developed for tissues fixed in other fixatives, frozen sections, or cytospin preparations. The choice of method will depend on the amount of antigen present, the level of sensitivity required, and the technical capabilities of the laboratory. There are numerous detection kits; we will describe the following techniques using some of these commercial kits applied in an automatic stainer. The number of steps that can be done in the automatic stainer will depend on the model.

11.6.1 Direct Methods

11.6.1.1 Method

1. Dewax sections in two changes of xylene or substitute.
2. Hydrate sections using 100% ethanol (two changes, 5 min each) and 95% ethanol (two changes, 5 min each). Rinse slides in water to adequately remove alcohol.
3. If HIER or AR with enzymes at 37°C is needed, do it at this point.
4. Place the slides in buffer for 5 min and transfer to the autostainer.
5. Block endogenous peroxidase.
6. Rinse in buffer.
7. AR with protease K at RT (if needed), 5 min.
8. Incubate sections with nonspecific binding blocking solution for 10–20 min.
9. WITHOUT rinsing, blow (blot if done manually) the fluid off the slide.
10. Incubate with primary labeled antibody for 30 min.
11. Rinse with rinsing buffer three times.
12. Incubate sections in the DAB solution for 5–10 min.
13. Rinse sections in distilled water.
14. Counterstain with Mayer's hematoxylin.
15. Rinse sections in distilled water. Blue sections using dilute ammonium hydroxide solution.
16. Dehydrate using 95% ethanol (two changes, 3–5 min), 100% ethanol (two changes, 3–5 min), and xylene or substitute (two changes, 3–5 min).
17. Mount in synthetic mounting medium.

11.6.1.2 Results

Sites of antigen/antibody reactivity: Brown
Nuclei: Blue

11.6.2 Indirect Methods

11.6.2.1 Indirect Immunoenzyme Method

This is the first of two two-step methods.

11.6.2.1.1 Method

1. Dewax sections in two changes of xylene or substitute.
2. Hydrate sections using 100% ethanol (two changes, 2 min each) and 95% ethanol (two changes, 2 min each). Rinse slides in water to adequately remove alcohol.
3. If HIER or AR with enzymes at 37°C is needed, do it at this point.
4. Place the slides in buffer for 5 min and transfer to the autostainer.
5. Block endogenous peroxidase.
6. Rinse in buffer.
7. AR with proteinase K at RT (if needed), 5 min.
8. Incubate sections with nonspecific binding blocking solution for 10–20 min.
9. WITHOUT rinsing, blow (blot if done manually) the fluid off the slide.
10. Incubate with primary unlabeled antibody for 30 min.
11. Rinse with rinsing buffer three times.
12. Incubate with labeled secondary antibody for 30 min.
13. Rinse with rinsing buffer three times.
14. Incubate sections in the DAB solution for 5–10 min.
15. Rinse sections in distilled water.
16. Counterstain with Mayer's hematoxylin.
17. Rinse sections in distilled water. Blue sections using diluted ammonium hydroxide solution.
18. Dehydrate using 95% ethanol (two changes, 3–5 min), 100% ethanol (two changes, 3–5 min), and xylene or substitute (two changes, 3–5 min).
19. Mount in synthetic mounting medium.

11.6.2.1.2 Results

Sites of antigen/antibody reactivity: Brown
Nuclei: Blue (mixture of brown and blue if the antigen is nuclear)

11.6.2.1.3 Note

Although inhibition of endogenous peroxidase is usually done before incubation with the primary antibody, some antigens (especially CDs) might be sensitive to this treatment, particularly if the hydrogen peroxide solution is concentrated. In such cases, this treatment can be done after incubation of the primary antibody and always before adding reagents containing peroxidase.

11.6.2.2 Polymer-Based Immunoenzyme Method

This method is much more sensitive than the indirect method. It is based in the capability of binding many molecules of label (e.g., peroxidase) to an inert backbone of polymer (e.g., dextran) to which molecules of immunoglobulin (e.g., goat anti-rabbit immunoglobulins) recognizing the primary antibody (in this case, rabbit immunoglobulins) are also bound. Another advantage of this method is the lack of avidin or biotin molecules involved in the reaction and therefore the lack of endogenous avidin-biotin background. Detection kits based on the polymer technology are usually more expensive than avidin-biotin methods.

11.6.2.2.1 Method

1. Dewax sections in two changes of xylene or substitute.
2. Hydrate sections using 100% ethanol (two changes, 5 min each) and 95% ethanol (two changes, 5 min each).
3. If HIER or AR with enzymes at 37°C is needed, do it at this point.
4. Place the slides in buffer for 5 min.
5. Block endogenous peroxidase.
6. Rinse in buffer.
7. AR with proteinase K at RT (if needed), 5 min.
8. Incubate sections with nonspecific binding blocking solution for 10–20 min.
9. WITHOUT rinsing, blow (blot if done manually) the fluid off the slide.
10. Incubate with primary unlabeled antibody for 30 min.
11. Rinse with rinsing buffer three times.
12. Incubate with polymer-immunoglobulin-enzyme complex for 30 min.
13. Rinse with rinsing buffer three times.
14. Incubate sections in the DAB solution for 5–10 min.
15. Rinse sections in distilled water.
16. Counterstain with Mayer's hematoxylin (usually done outside the autostainer).
17. Rinse sections in distilled water. Blue sections using dilute ammonium hydroxide solution.
18. Dehydrate using 95% ethanol (two changes, 3–5 min), 100% ethanol (two changes, 3–5 minutes), and xylene or substitute (two changes, 3–5 min).
19. Mount in synthetic mounting medium.

11.6.2.2.2 Results

Sites of antigen/antibody reactivity: Brown
Nuclei: Blue (mixture of brown and blue if the antigen is nuclear)

11.6.3 MULTIPLE-STEP METHODS

Multiple-step methods are more laborious than indirect methods but usually are more sensitive. They are based on the high affinity of avidin (glycoprotein found in the egg white) or streptavidin (glycoprotein from *Streptomyces avidinii*) for biotin (glycoprotein present in egg yolk). Avidin contains four subunits that form a tertiary structure with four biotin binding hydrophobic sites. Avidin has oligosaccharide residues that have some affinity for tissue components resulting in nonspecific binding. Streptavidin produces less background because of its lack of oligosaccharide residues and its neutral isoelectric point. Biotin is usually conjugated to the secondary antibody or to enzymes (up to 150 molecules of biotin per molecule of immunoglobulin). The label in avidin-biotin methods is usually attached or complexed to molecules of avidin (tertiary reagent). Avidin-biotin methods are highly sensitive and are currently the most widely used IHC methods (Figure 11.3).

11.6.3.1 Labeled Streptavidin/Biotin Method

1. Dewax sections in two changes of xylene or substitute.
2. Hydrate sections using 100% ethanol (two changes, 5 min each) and 95% ethanol (two changes, 5 min each).
3. If HIER or AR with enzymes at 37°C is needed, do it at this point.
4. Place the slides in buffer for 5 min.

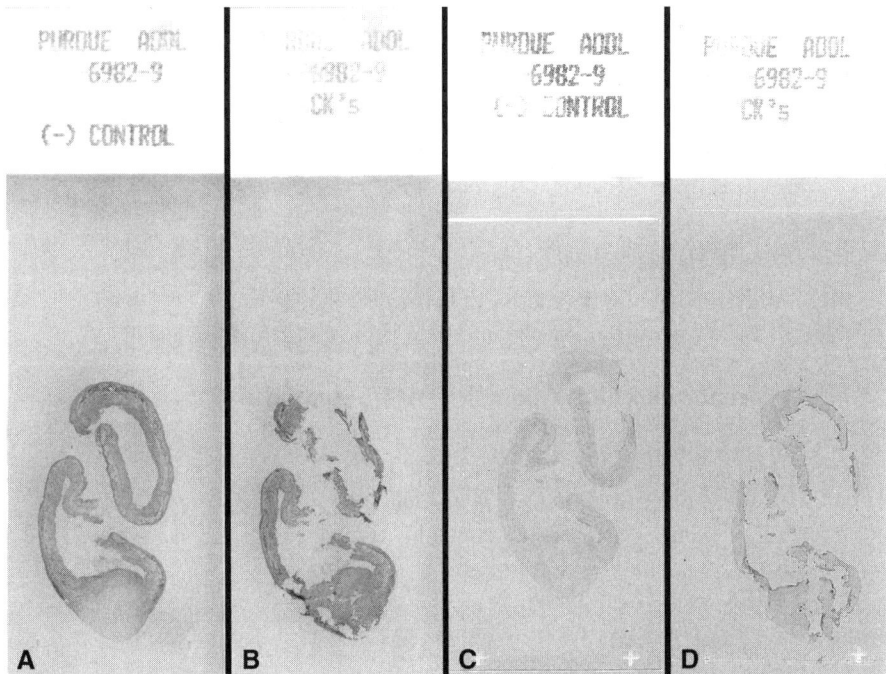

FIGURE 11.3 (See color insert following page 338.) Multiple-step methods are more sensitive than direct methods but may cause unexpected reactions. The urinary bladder from a goat was stained with monoclonal antibody to cytokeratins. A and B were stained with LSAB+ link antibody that recognizes rabbit, mouse, and goat primary antibodies. The identical diffuse staining in the section with the negative reagent control section (A) and the anti-cytokeratin antibody (B) is caused by nonspecific binding of the link antibody to endogenous (tissue) goat immunoglobulins. To avoid this problem a detection system that does not recognize goat immunoglobulins (ENVISION+) was used in images C and D. The negative reagent control (C) does not show staining, and the positive reagent tissue (D) has distinct staining in the expected areas of the tissue. Immunoperoxidase-DAB, hematoxylin counterstain.

5. Block endogenous peroxidase.
6. Rinse in buffer.
7. AR with proteinase K at RT (if needed), 5 min.
8. Incubate sections with nonspecific binding blocking solution for 10–20 min.
9. WITHOUT rinsing, blow (blot if done manually) the fluid off the slide.
10. Incubate with primary unlabeled antibody for 30 min.
11. Rinse with rinsing buffer three times
12. Incubate with biotinylated secondary antibody for 30 min.
13. Rinse with rinsing buffer three times.
14. Incubate with tertiary reagent (peroxidase labeled avidin/streptavidin) for 30 min.
15. Rinse with rinsing buffer three times.
16. Incubate sections in the DAB solution for 5–10 min.
17. Rinse sections in distilled water.
18. Counterstain with Mayer's hematoxylin (usually done outside the autostainer).
19. Rinse sections in distilled water. Blue sections using dilute ammonium hydroxide solution.
20. Dehydrate using 95% ethanol (two changes, 3–5 min), 100% ethanol (two changes, 3–5 min), and xylene or substitute (two changes, 3–5 min).
21. Mount in synthetic mounting medium.

11.6.3.1.1   Results

Sites of antigen/antibody reactivity: Brown
Nuclei: Blue (mixture of brown and blue if the antigen is nuclear)

11.6.3.2   Streptavidin Biotin Complex (ABC) Method

11.6.3.2.1   Method

1. Dewax sections in two changes of xylene or substitute.
2. Hydrate sections using 100% ethanol (two changes, 5 min each) and 95% ethanol (two changes, 5 min each).
3. If HIER or AR with enzymes at 37°C is needed, do it at this point.
4. Place the slides in buffer for 5 min.
5. Block endogenous peroxidase.
6. Rinse in buffer.
7. AR with proteinase K at RT (if needed), 5 min.
8. Incubate sections with nonspecific binding blocking solution for 10–20 min.
9. WITHOUT rinsing, blow (blot if done manually) the fluid off the slide.
10. Incubate with primary unlabeled antibody for 30 min.
11. Rinse with rinsing buffer three times
12. Incubate with biotinylated secondary antibody for 30 min.
13. Rinse with rinsing buffer three times.
14. Incubate with tertiary reagent (preformed avidin-streptavidin-peroxidase complex) for 30 min.

15. Rinse with rinsing buffer three times.
16. Incubate sections in the DAB solution for 5–10 min.
17. Rinse sections in distilled water.
18. Counterstain with Mayer's hematoxylin (usually done outside the autostainer)
19. Rinse sections in distilled water. Blue sections with dilute ammonium hydroxide solution.
20. Dehydrate using 95% ethanol (two changes, 3–5 min), 100% ethanol (two changes, 3–5 min), and xylene or substitute (two changes, 3–5 min).
21. Mount in synthetic mounting medium.

11.6.3.2.2   Results

Sites of antigen/antibody reactivity: Brown
Nuclei: Blue (mixture of brown and blue if the antigen is nuclear)

11.6.3.2.3   Note

Avidin and peroxidase solutions are preincubated *in vitro* to form a complex before adding to the slides.

11.6.3.3   Tyramide-Based Methods (Avidin-Biotin or Fluorescein-Based)

Tyramide-based methods amplify the immune reaction 100 to 1000-fold when compared with a conventional ABC method. Also the dilution of the primary antibody can be increased several hundred-fold. These methods are based on the deposition of molecules of labeled (biotin, fluorescein) tyramide followed by a secondary reaction with peroxidase conjugated to streptavidin or peroxidase-conjugated antifluorescein. Originally, this method was based on an avidin-biotin binding. The high sensitivity of this method also increases the chances of background from endogenous avidin-biotin. A nonavidin-biotin, fluorescein-based method has recently being developed to avoid this problem but its sensitivity is reduced. Background is still possible with the fluorescein-based method. An improved tyramine-based method with reduction of the background has been recently published.[11]

11.6.3.3.1   Method

1. Dewax sections in two changes of xylene or substitute.
2. Hydrate sections using 100% ethanol (two changes, 5 min each) and 95% ethanol (two changes, 5 min each). Rinse slides in water to adequately remove alcohol.
3. If HIER or AR with enzymes at 37°C is needed, do it at this point.
4. Place the slides in buffer for 5 min.
5. Block endogenous peroxidase.
6. Rinse in buffer.
7. AR with proteinase K at RT (if needed), 5 min.
8. Incubate sections with nonspecific binding blocking solution for 10–20 min.

9. WITHOUT rinsing, blow (blot if done manually) the fluid off the slide.
10. Incubate with primary unlabeled antibody for 30 min.
11. Rinse with rinsing buffer three times
12. Incubate with F(ab')$_2$ biotinylated secondary antibody (avidin-biotin method) or anti-mouse IgG-peroxidase (fluorescein method), 15 min.
13. Rinse with rinsing buffer three times.
14. Incubate with primary peroxidase-streptavidin-biotin complex (avidin-biotin method) or fluorescyl-tyramide amplification reagent (fluorescein method), 15 minutes.
15. Rinse with rinsing buffer three times.
16. Incubate with biotinyl-tyramide amplification reagent, (avidin-biotin method), 15 min or anti-fluorescein-peroxidase (fluorescein method), 15 min.
17. Rinse with rinsing buffer three times.
18. Incubate with peroxidase-streptavidin complex (avidin-biotin method) or proceed to step 20 (fluorescein method), 15 min.
19. Rinse with rinsing buffer three times.
20. Incubate with DAB substrate-chromogen, 5 min.
21. Rinse with distilled water.
22. Counterstain with Mayer's hematoxylin.
23. Rinse sections in distilled water. Blue sections with dilute ammonium hydroxide solution.
24. Dehydrate with 95% ethanol (two changes, 3–5 min), 100% ethanol (two changes, 3–5 min), and xylene or substitute (two changes, 3–5 min).
25. Mount in synthetic mounting medium.

11.6.3.3.2   Results

Sites of antigen/antibody reactivity: Brown
Nuclei: Blue (mixture of brown and blue if the antigen is nuclear)

11.6.3.4   Immunohistochemical Detection of Multiple Antigens

In the past, detection of multiple antigens in the same tissue section has been challenging. There are detection kits that allow the detection of at least two antigens using two different enzymes as labels (e.g., HRP and AP). These detection kits are highly effective, but also very expensive. The key issue in detection of multiple antigens is their location (different or same tissues, cells, or cellular compartments). A careful selection of the chromogen for each antigen is also necessary to achieve the best distinction between antigens. Examples of enzyme substrate combinations in double immunostaining are included in the catalog from Vector Laboratories (www.vectorlabs.com). The chance of good visualization of both antigens is reduced if they are anatomically close to each other (e.g., both antigens are within the nucleus of the same cell type). Double immunodetection is also complicated by the variety of AR methods used for different antigens. In other words, an AR necessary for one

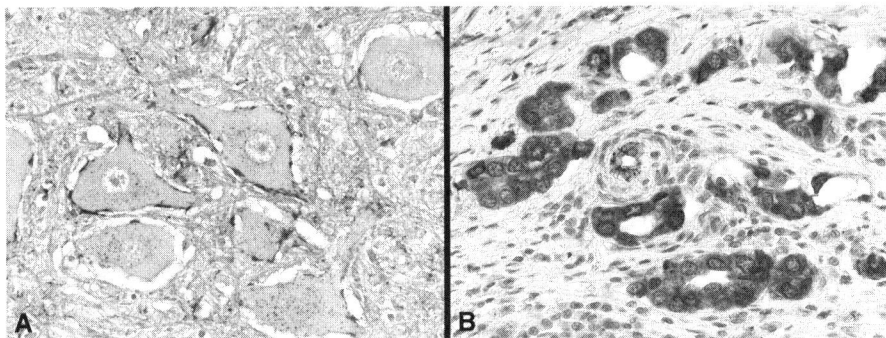

FIGURE 11.4 (See color insert following page 338.) (A) Immunoalkaline phosphatase staining for prion protein in the brainstem of a sheep. The staining is fuchsia against a blue counterstain. (B) Double immunostaining for COX-2 (fuchsia) and CD31 (brown). Transitional cell carcinoma cells expressing COX-2 are detected with alkaline phosphatase, and endothelial cells expressing CD31 (arrowhead) are detected with peroxidase.

antigen might have deleterious effects for the second antigen to be detected. For a comprehensive review of multiple immunostaining read van der Loos's monograph[12] on multiple immunoenzymatic staining. Here we include two double immunoenzymatic protocols; one for primary antibodies from different species and another for primary antibodies from the same species (Figure 11.4).[13]

11.6.3.4.1 Double Immunoenzymatic Staining with Primary Antibodies from Different Species[12]

This method uses the indirect/indirect concept with two primary antibodies from different species.

11.6.3.4.1.1 Method

1. Dewax sections in two changes of xylene or substitute.
2. Hydrate sections with 100% ethanol (two changes, 5 min each) and 95% ethanol (two changes, 5 min each). Rinse slides in water to adequately remove alcohol.
3. If HIER or AR with enzymes at 37°C is needed, do it at this point.
4. Place the slides in buffer for 5 min.
5. Block endogenous peroxidase.
6. Rinse in buffer.
7. AR with proteinase K at RT (if needed), 5 min.
8. Incubate sections with nonspecific binding blocking solution for 10–20 min.
9. WITHOUT rinsing, blow (blot if done manually) the fluid off the slide.
10. Incubate with both primary unlabeled antibodies for 30 min.
11. Rinse with rinsing buffer three times
12. Incubate with a mixture of secondary reagent (HRP-labeled goat anti-mouse 1 and AP-labeled goat anti-rabbit 2), 30 min.

13. Rinse with rinsing buffer three times.
14. First enzyme activity (AP chromogen) detection, 5 min.
15. Rinse with rinsing buffer three times.
16. Second enzyme activity (HRP chromogen) detection, 5 min.
17. Rinse with distilled water.
18. Incubate with DAB substrate-chromogen, 5 min
19. Counterstain with Mayer's hematoxylin (usually done outside the autostainer).
20. Rinse sections in distilled water. Blue sections with dilute ammonium hydroxide solution.
21. Mount in appropriate mounting medium.

11.6.3.4.1.2   Results

For DAB-HRP and Fast Red-AP, the colors of the reaction should be brown and red, respectively.

11.6.3.4.1.3   Notes

1. The concentrations of the primary antibodies should be at least double concentrated to those used in separate IHC methods.
2. It is critical to choose the right color combination, which will depend greatly on the amount of antigen and its location in the tissue section.
3. The counterstain should not mask the color of the immune reaction.

11.6.3.4.2   Sequential Double Immunoenzymatic Staining with Primary Antibodies from Same or Different Species[12]

The advantage of this method is that the primary antibodies can be from the same species. However, this method is much more laborious and needs an elution or blocking step between the primary and second immune reactions. There are commercially available kits based on polymer-based technology (e.g., ENVISION, Dako USA, Carpinteria, CA).

11.6.3.4.2.1   Method

1. Dewax sections in two changes of xylene or substitute.
2. Hydrate sections using 100% ethanol (two changes, 5 min each) and 95% ethanol (two changes, 5 min each). Rinse slides in water to adequately remove alcohol.
3. If HIER or AR with enzymes at 37°C is needed, do it at this point.
4. Place the slides in buffer for 5 min and transfer to the autostainer.
5. Block endogenous peroxidase.
6. Rinse in buffer.
7. AR with proteinase K at RT (if needed), 5 min.
8. Incubate sections with nonspecific binding blocking solution for 10–20 min.
9. WITHOUT rinsing, blow (blot if done manually) the fluid off the slide.
10. Incubate with primary unlabeled antibody for 30 min.
11. Rinse with rinsing buffer three times
12. Incubate with ENVISION/peroxidase reagent, 30 min.
13. Rinse with rinsing buffer three times.
14. Develop peroxidase activity with DAB-chromogen reagent, 5 min.

15. Elution step with DAKO double staining block or alternatively, 5 min. boiling in citrate pH 6.0.
16. Rinse with rinsing buffer three times.
17. Incubate sections with nonspecific binding blocking solution for 10 min.
18. WITHOUT rinsing, blow (blot if done manually) the fluid of the slide.
19. Mouse or rabbit antibody 2, 30 min.
20. Rinse with rinsing buffer three times.
21. Incubate with ENVISION/AP reagent, 30 min.
22. Rinse with rinsing buffer three times.
23. Develop AP activity with Fast Red, 5–30 min.
24. Rinse with distilled water.
25. Light counterstain with Mayer's hematoxylin.
26. Mount in aqueous mounting medium.

11.6.3.4.2.2   Results

For DAB-peroxidase and Fast Red-AP systems, the colors of the reaction should be brown and red, respectively.

11.6.3.4.2.3   Note

It is recommended to develop HRP activity with DAB first. This chromogen has the potential for sheltering the first set of reagents, preventing cross-reactions between the first and second staining sequences.

11.7 IMMUNOFLUORESCENCE TECHNIQUES

Although the use of immunofluorescence techniques has diminished in the last two decades with the advent of IHC, this technique is still a valuable method for the *in situ* detection of antigens and particularly when multiple antigens need to be detected simultaneously. Numerous laboratories routinely use immunofluorescence to characterize autoimmune disorders. Immunofluorescence is also widely used in the detection of neuropeptides in the nervous system and in the tridimensional distribution of antigens with confocal microscopy.

Immunofluorescence is based in the use of antibodies labeled with fluorochromes, substances that require excitation with light of a specific wavelength to make them emit visible light. The advantage of immunofluorescence techniques over IHC is that the colored reaction is observed against a dark, nonfluorescent background. The main disadvantage of immunofluorescence methods is the need of using special microscopes to observe the immune reaction and the rather rapid fading of the fluorescence, particularly with fluorescein. Newer fluorochromes (e.g., Oregon Green, Molecular Probes, Inc., Eugene, OR) and commercial, proprietary mounting media are more resistant to fading.

11.7.1 Fluorochromes

11.7.1.1 Fluorescein

It is widely used in the isothiocyanate form (TRITC). It emits a bright apple-green fluorescence (520 nm) when excited by light with a wavelength of 495 nm.

11.7.1.2 Rhodamine

Rhodamine derivatives (e.g., Texas Red) fade less rapidly than TRITC, fluoresce red at an excitation maximum of 555–596 nm. It is widely used with TRITC for colocalization of two antigens.

11.7.1.3 Phycoerythrin

It is a fluorochrome from seaweed and a rhodamine derivative, fluorescing at the same excitation range as fluorescein. It emits a weak orange light and is used in combination with TRITC to detect two antigens in different structures because there is no need to change the filter to observe both colors.

11.7.1.4 7-Amino-4-Methyl-Coumarin-3-Acetic Acid

7-Amino-4-methyl-coumarin-3-acetic acid produces blue fluorescence and is used for multiple labeling.

11.7.2 INDIRECT IMMUNOFLUORESCENCE TECHNIQUE

11.7.2.1 Method

1. Hydrate paraffin sections or frozen sections in PBS, pH 7.2, 5 min.
2. Wipe off the excess buffer and incubate with primary antibody, 30 min.
3. Rinse with rinsing buffer three times.
4. Incubate with secondary antibody conjugated to a fluorochrome, 30 min.
5. Rinse with rinsing buffer three times.
6. Mount in suitable mountain medium according to manufacturer recommendations.
7. Examine immediately or keep slides in a dark and cool place.

11.7.2.2 Results

Sites of antigen-antibody reaction fluoresce with a color related to the fluorochromes used.

APPENDIX

A.1 REAGENTS FOR IMMUNOHISTOCHEMICAL STAINING

A.1.1 ENDOGENOUS PEROXIDASE BLOCKING SOLUTION

A.1.1.1 Paraffin Sections

A.1.1.1.1 Method
1. Dewax and hydrate the slides.
2. Immerse sections in 3% hydrogen peroxide (10 ml of 30% aqueous (v/v) hydrogen peroxide per 100 ml of distilled water) for 5–15 min.
3. Rinse in water, then buffer, and proceed with immunostaining.

A.1.1.1.2   Notes

1. Buffer or methanol can be used instead of water to dilute the stock solution of hydrogen peroxide.
2. Pretreatment of the sections for AR can be done before or after the endogenous peroxidase blocking step.
3. For some antigens, particularly in the cytoplasmic membrane, this treatment is deleterious and should be done after the incubation with the primary antibody.
4. Store stock solution of 3% hydrogen peroxide at 4°C and in the dark. Stable for 1 week.
5. There are commercially available solutions that simultaneously block both endogenous peroxidase and AP activities.

A.1.1.2 For Sections with Abundant Red Blood Cells

A.1.1.2.1   Reagents

Periodic acid (Sigma P 7875)
Potassium borohydride (Sigma P 4129)

A.1.1.2.2   Method

1. Immerse dewaxed slides in 1.8% hydrogen peroxide (6 ml of 30% aqueous hydrogen peroxide in 100 ml of water) for 10–20 min.
2. Rinse in tap water.
3. Immerse in 2.5% aqueous periodic acid solution for 5 min.
4. Rinse in tap water.
5. Immerse in 0.02% aqueous potassium (or sodium) borohydride solution for 2 min.
6. Rinse in tap water and proceed with immunostaining.

A.1.1.2.3   Notes

1. Potassium borohydride solution must be made fresh.
2. This treatment can be deleterious for some antigens.

A.1.1.3 Frozen Sections and Cytologic Preparations

A.1.1.3.1   Method

Immerse sections in 0.03% hydrogen peroxide diluted in water containing 0.1% sodium azide for 5 min.

A.1.2 ENDOGENOUS ALKALINE PHOSPHATASE BLOCKING SOLUTION

A.1.2.1 Reagent

Levamisole (tetramisole hydrochloride) (Sigma, L 9756), 1 mM (12 mg/50 ml buffer)

A.1.2.2 Method

This solution can be added before or mixed with the AP developing solution.

A.1.2.3 Note

Commercial blocking solutions are available, and some block endogenous
peroxidase and AP activities simultaneously.

A.1.3 NONSPECIFIC BINDING BLOCKING SOLUTION

A.1.3.1 Reagents

Normal serum from same species as secondary antibody diluted 5–20% in Tris buffer.

A.1.3.2 Method

1. Rinse the slides in buffer.
2. Blot or blow slides and incubate sections with nonspecific binding blocking solution for 10–20 min.
3. WITHOUT rinsing, blot or blow slides.
4. Incubate with primary unlabeled antibody.

A.1.3.3 Notes

Alternatively, a solution of bovine serum albumin can be used instead of
normal serum but it is more expensive. There are also commercially available blocking solutions.

A.1.4 ENDOGENOUS AVIDIN/BIOTIN BLOCKING SOLUTION[14]

A.1.4.1 Reagents

Avidin (Sigma, A 9275)/streptavidin (Sigma, S 4762), 1 mg/ml of buffer
Biotin (Sigma, B 4501), 0.1 mg/ml of buffer

A.1.4.2 Method

1. Incubate dewaxed and hydrated sections in avidin for 20 min.
2. Wash in buffer, 5 min.
3. Incubate with biotin for 20 min.
4. Rinse in buffer, 5 min.
5. Continue with immunostaining procedure.

A.1.4.3 Notes

1. There are commercially available blocking kits that can be used with
   automatic stainers. These solutions tend to be expensive. If endogenous
   avidin-biotin is common in the tissues tested, consider using non-avidin-
   biotin detection systems.

2. Commercially available biotin can be replaced with egg white (one egg white per 100 ml of water and 0.1% of sodium azide). Store at 4°C.
3. Commercially available avidin can be replaced with skimmed milk or 5% powdered, fat free, dried milk in PBS with 0.05% Tween 20 and 0.1% sodium azide. Store at –20°C.

A.1.5 ANTIGEN RETRIEVAL BUFFERS

A.1.5.1 Citrate Buffer, pH 6.0

A.1.5.1.1 Reagents

100 mM citric acid monohydrate (Sigma C0706), 2.1 g
2 N NaOH solution

A.1.5.1.2 Method

1. Dissolve the citric acid in 1 l of distilled water.
2. Add the sodium hydroxide solution until the pH reaches 6.0 (about 13 ml).
3. This buffer can be kept at room temperature for 5 days.

A.1.5.2 EDTA Buffer, pH 9.0, 10X

A.1.5.2.1 Reagents

Tris(hydroxymethyl)methylamine, 12 g
Ethylene-diamine tetra-acetic acid, disodium salt, dehydrate (Sigma E4884), 1 g
1 M hydrochloric acid, 500 ml
Distilled water, 500 ml

A.1.5.2.2 Method

1. Dissolve the Tris and EDTA salts in the water.
2. Add the hydrochloric acid.
3. Check the pH and adjust if necessary.

A.1.5.2.3 Working Solution
Store the 10× solution at 4°C. Working dilution is prepared dilution one part of stock solution in nine parts of distilled water (v/v).

A.1.5.3 High pH Tris Buffer (pH 10.0)

A.1.5.3.1 Reagents

Tris base (Calbiochem 648310), 1.21 g
Distilled water, 1000 ml
1 N sodium hydroxide
Tween 20

A.1.5.3.2   Method

1. Mix Tris with water to dissolve.
2. Adjust pH to 10 using 1 N NaOH.
3. Add 0.5 ml of Tween 20 and mix well.

A.1.6   TRYPSIN SOLUTION FOR ANTIGEN RETRIEVAL

A.1.6.1   Reagents

Trypsin (crude porcine) (ICN, 150213), approximately 435 USP units/mg
Calcium chloride (Sigma C 4901), 0.1% in water or 0.005 M Tris/HCl buffer, pH 7.6–8.0
0.1 M sodium hydroxide

A.1.6.2   Preparation

Dissolve trypsin in the calcium chloride to make a 0.1% solution. Trypsin will not dissolve completely until pH nears 7.8 (adjusted with sodium hydroxide).

A.1.6.3   Note

Trypsin is commercially available in preweighed tablets.

A.1.7   PROTEASE SOLUTION FOR ANTIGEN RETRIEVAL

A.1.7.1   Reagents

Proteinase XXIV, bacterial, 7–14 U/mg (Sigma, P 8038)
PBS, pH 7.2–7.4, prewarmed at 37°C

A.1.7.2   Method

Make a 0.05–0.5% solution of protease XXIV in PBS (pH does not need to be adjusted).

A.1.8   PEPSIN SOLUTION FOR ANTIGEN RETRIEVAL

A.1.8.1   Reagents

Pepsin, from porcine stomach, 600–1800 U/mg (Sigma, P 7125)
Hydrochloric acid, 10 mM at 37°C

A.1.8.2   Method

Make a 0.4% solution of pepsin in HCl.

A.1.9 RINSE BUFFER TBS 500 mM, pH 7.6

A.1.9.1 Reagents

Trizma base (Sigma, T 1503)
Trizma HCl (Sigma, T 3253)
Sodium chloride (Sigma, S 3014)
Hydrochloric acid 12 N
Tween 20

A.1.9.2 Method

1. Dissolve 13.9 g of Trizma base, 60.6 g Trizma HCl, and 87.66 g of NaCl in 500 ml of distilled water with a magnetic stir plate.
2. Add 12 N HCl until the pH is 7.6.
3. Add 5 ml of Tween 20.
4. Adjust the final volume of buffer to 1 l with distilled water.

A.1.9.3 Notes

1. There are commercially available packages of TBS that only need addition of water.
2. Do not use sodium azide.
3. The final solution is 0.5 M Tris, 1.5 M NaCl, with 0.5% of Tween 20.
4. To make a working solution of Tris buffer (50 mM), mix 900 ml of distilled water and 100 ml of 0.5 M Tris buffer normal saline.
5. Storage of TBS. Average shelf life at room temperature is 4 days and at 4°C is 7 days.

A.1.10 PBS, 20 mM, pH 7.0

A.1.10.1 Reagents

Sodium chloride (Sigma, S 3014), 8.7 g
Sodium phosphate dibasic (anhydrous) (Sigma, S0876)
Sodium phosphate monobasic (anhydrous) (Fisher, S369)
Tween 20 (Sigma, P 9416)
Sodium hydroxide, 50% w/w solution (Baker, 3727)

A.1.10.2 Method

1. To 500 ml of distilled water, add 5.38 g of sodium phosphate monobasic, 8.66 g of sodium phosphate dibasic, and 8.77 g of NaCl. Dissolve using a magnetic stir plate.
2. Adjust the solution to pH 7.0 with NaOH.

3. Once pH 7.0 is reached, add 0.5 ml of Tween 20 and mix.
4. Make up to 1 l with distilled water.

A.1.10.3 Note

PBS solutions should be stored at 2–8°C to inhibit bacterial growth.

A.1.11 ANTIBODY DILUTION BUFFER

There are commercially available solutions. However, you can prepare a solution with PBS or TBS containing 0.1% bovine serum albumin (Sigma, A 4503). Do not add sodium azide as preservative to enzyme-labeled reagents as it inhibits enzyme activity. Use protexidase (ICN, 980631) as a diluent.

A.1.12 PREPARATION OF SUBSTRATES AND CHROMOGENS

A.1.12.1 Horseradish Peroxidase

A.1.12.1.1 Peroxidase-DAB Solution
There are DAB tablets commercially available as well as stable solutions. The method indicated here is for homemade DAB.

A.1.12.1.1.1 Reagents

  3,3′-diaminobenzidine tetrahydrochloride (DAB), 50 mg (Sigma, D5637)
  TBS, 100 ml
  Hydrogen peroxide (30%), 30 µl (final concentration 0.01–0.03%)

A.1.12.1.1.2 Method

1. Add the DAB to the buffer and mix. This solution should be clear or light straw color.
2. Add hydrogen peroxide just before use.
3. Immerse the slides (previously in buffer) in DAB solution, 5 min or longer.
4. Rinse the slides in distilled water.
5. Counterstain slides lightly with Mayer's hematoxylin and mount in synthetic medium.

A.1.12.1.1.3 Notes

1. The incubation time will depend on the amount of antigen. Longer incubation times tend to increase background.
2. Disposal of DAB. DAB is considered toxic or potentially carcinogenic so proper disposal is necessary. Each institution has its own regulations so contact your waste hazard office. In general, to inactivate DAB, add several drops of household bleach or sodium hypochlorite. The solution will turn black (from oxidation of DAB) and can be washed down the sink with plenty of water. A Food and Drug Administration–approved alternative method consists of preparing a solution

containing 15 ml of concentrated sulfuric acid added slowly to 85 ml of water. Add 4 g of potassium permanganate. Add this solution to the used DAB and leave overnight. Neutralize with sodium hydroxide, then discard.

A.1.12.1.2 Peroxidase-DAB Enhanced Solution

The use of metals after the standard DAB procedure or during development of the DAB incubation can improve the intensity of the reaction or change the color of the reaction. We include here three different methods. There are also commercially available enhancing solutions of the DAB precipitate.

A.1.12.1.2.1   Copper Sulfate (Modified from Hanker et al.)[15]

A.1.12.1.2.1.1   Reagents

Copper sulfate ($CuSO_4$) (Sigma, C1297)
0.85% sodium chloride

A.1.12.1.2.1.2   Method

1. Dissolve 0.25 g of copper sulfate in 50 ml of 0.85% sodium chloride (0.5% solution).
2. After doing the regular DAB procedure, immerse slides for 2–10 min. in the copper sulfate solution. Check under the microscope.
3. Rinse, counterstain, and mount in synthetic medium.

A.1.12.1.2.1.3   Notes

The color of the DAB reaction will darken with this method.
Use 50 mM Tris-HCl buffer, pH 7.6, instead of sodium chloride to prepare the copper solution.

A.1.12.1.2.2   Cobalt (Modified from Hsu and Soban)[11]

A.1.12.1.2.2.1   Reagents

Cobalt chloride (Sigma, C8661), 0.5% aqueous
DAB (Sigma, D 5637), 50mg/100 ml of buffer
Hydrogen peroxide, 30%

A.1.12.1.2.2.2   Method

1. Add 1 ml of cobalt chloride solution to 100 ml of DAB solution with continuous stirring.
2. Incubate slides for 5 min.
3. Mix with 10 μl hydrogen peroxide and incubate for an additional 1–5 min, checking under the microscope. The reaction product will be dark blue or bluish black.
4. Rinse in tap water, counterstain, and mount as usual.

A.1.12.1.2.2.3   Note

This method is done during the development of the DAB reaction. Because of the color of the reaction, use nuclear red or methyl green as nuclear counterstain.

The concentration of cobalt can be increased up to 2 ml of 1% of cobalt.

A.1.12.1.2.3   Peroxidase-AEC¹²

A.1.12.1.2.3.1   Reagents

AEC (Sigma, A5754).
N,N-Dimethyl formamide (Aldrich, 31,993-7)
AEC stock solution: 1% AEC in formamide (store at room temperature in the dark)
0.05 M sodium acetate/acetic acid buffer, pH 5.2
30% hydrogen peroxide
Working solution: 2.5 ml of stock AEC solution mixed with 47.5 ml of acetate buffer. Filter the turbid solution, and, just before use, add 20 µl of hydrogen peroxide.

A.1.12.1.2.3.2   Method

1. Wash tissue sections in water and incubate in acetate buffer for 3 min.
2. Incubate with AEC working solution for 10–20 min. Check under the microscope at intervals after rinsing in buffer and reapply if more time is needed.
3. Rinse in acetate buffer and then in water.
4. Counterstain with Mayer's hematoxylin and mount in aqueous mountant.

A.1.12.1.2.3.3   Note

This method gives a red precipitate that is alcohol soluble (do not use synthetic or organic mountants).

A.1.12.2   Alkaline Phosphatase Naphthol AS-MX Phosphate/Fast Blue BB (Modified from Burstone)¹⁶

A.1.12.2.1   Reagents

Naphthol AS-MX phosphate, sodium salt (Sigma, N5000), 20 mg
N,N-Dimethyl formamide (Aldrich, 31,993-7), 0.4 ml
0.1 M Tris/HCl buffer, pH 8.2, 19.6 ml
Levamisole hydrochloride (Sigma, L 9756)
Fast Blue BB salt (Sigma, F 3378)

A.1.12.2.2   Method

1. Dissolve the naphthol AS-MX phosphate in formamide using a glass container.
2. Add the Tris buffer quickly, stirring.

3. Add levamisole to give a 1 mM solution (approximately 1.2 mg per 5 ml of solution).
4. Just before use, add Fast Blue BB, 1 mg/ml. Mix well and filter on to the preparations.
5. Incubate at room temperature for 5–15 min. At 37°C, the incubation time is shorter.
6. Wash in running tap water.
7. Counterstain in carmalum and mount in aqueous mountant.

A.1.12.2.3 Notes

1. To give a red end product, use Fast Red TR salt (Sigma, F8764) instead of Fast Blue BB, and Mayer's hematoxylin as a counterstain.
2. The stock solution of naphthol AS-MX mixed with dimethylformamide and Tris buffer can be stored at 4°C for several weeks.
3. Levamisole is used to inhibit endogenous AP activity except that of the intestinal isoenzyme.
4. There are commercially available endogenous AP inhibitors. If used, levamisole does not need to be added to the substrate.

A.1.12.3 Counterstains

Use of an optimal counterstain is critical in IHC. As a rule, the color of the immunologic reaction must be clearly distinguished from the color of the counterstain. Three main counterstains are used in IHC: Mayer's hematoxylin (blue), Nuclear Fast Red (red), and methyl green (green). There are commercially available counterstains, some especially adapted to immunohistochemical procedures. The methods given here can be used in manual staining. A blue counterstain (hematoxylin) is compatible with DAB (brown), AEC (red), and Fast Red TR (red). Methyl green is compatible with DAB and incompatible with Fast Red TR and AEC. Nuclear Fast Red is compatible with DAB. For a more complete list of compatibility between counterstains and chromogens, consult the Vector Laboratories catalog or at www.vectorlabs.com.

A.1.12.3.1 Mayer's Hematoxylin

A.1.12.3.1.1 Reagents

Aluminum potassium sulphate (alum) (Sigma, A7210), 50 g
Hematoxylin (Sigma, H3136), 1 g
Sodium iodate (Sigma, S4007), 0.2 g
Glacial acetic acid (Fluka, 45726), 20 ml
Distilled water, 1 l

A.1.12.3.1.2 Stock Solution

1. Dissolve alum in distilled water.
2. Add hematoxylin and dissolve completely.
3. Add sodium iodate and acetic acid.
4. Bring to a boil and cool. Filter.

A.1.12.3.1.3   Staining Method

1. Bring sections to distilled water after immunohistochemical staining.
2. Stain in stock solution for 2–5 min. at room temperature.
3. Rinse in distilled water.
4. Dehydrate quickly through 95% and two changes of 100% alcohol.
5. Clear in xylene.
6. Mount in regular mounting medium.

A.1.12.3.1.4   Results

Nuclei stain blue.

A.1.12.3.2   Nuclear Fast Red

A.1.12.3.2.1   Reagents

Nuclear Fast Red (Fluka, 60700), 0.1 g
Ammonium sulfate (Sigma, A2939), 5 g
Distilled water, 100 ml

A.1.12.3.2.2   Stock Solution

1. Dissolve ammonium sulfate in distilled water.
2. Add Nuclear Fast Red and slowly heat and bring to boil for 5 min.
3. Cool and filter.
4. Add a grain of thymol as a preservative.

A.1.12.3.2.3   Staining Method

1. Bring sections to distilled water after immunohistochemical staining.
2. Stain in stock solution for 5 min at room temperature.
3. Rinse in distilled water.
4. Dehydrate through 95% and two changes of 100% alcohol.
5. Clear in xylene.
6. Mount in regular mounting medium.

A.1.12.3.2.4   Results
Nuclei stain red.

A.1.12.3.3   Methyl Green

A.1.12.3.3.1   Reagents

A.1.12.3.3.1.1   0.1 M Sodium Acetate Buffer, pH 4.2

1. Sodium acetate, trihydrate (MW 136.1) (Sigma, S7670), 1.36 g
2. Distilled water, 100 ml
3. Mix to dissolve and adjust pH to 4.2 with concentrated glacial acetic acid

A.1.12.3.3.1.2   Methyl Green Solution (0.5%)

1. Methyl green (ethyl violet free) (Aldrich, 19,808-0), 0.5 g
2. 0.1 M sodium acetate buffer, pH 4.2, 100 ml
3. Mix to dissolve

A.1.12.3.3.2   Method

1. Bring sections to distilled water after immunohistochemical staining.
2. Stain in methyl green solution for 5 min at room temperature.
3. Rinse in distilled water (sections will look blue).
4. Dehydrate quickly through 95% (10 dips, sections will turn green), and two changes of 100% alcohol.
5. Clear in xylene.
6. Mount in synthetic mounting medium.

A.1.12.3.3.3   Results

Nuclei stain green.

REFERENCES

1. Coons, A.H. et al., Immunological properties of an antibody containing a fluorescent group. *Proc. Soc. Exp. Biol. Med.*, 47, 200, 1941.
2. Köhler, G. and Milstein, C., Derivation of specific antibody-producing tissue culture and tumor lines by cell fusion. *Eur. J. Immunol.*, 6, 511, 1976.
3. Ramos-Vara, J.A., Technical aspects of immunohistochemistry. *Vet. Pathol.*, 42, 409, 2005.
4. Henwood, A.F., Effect of slide drying at 80°C on immunohistochemistry. *J. Histotechnol.*, 28, 45, 2005.
5. Atkins, D. et al., Immunohistochemical detection of EGFR in paraffin-embedded tumor tissues: variation in staining intensity due to choice of fixative and storage time of tissue sections. *J. Histochem. Cytochem.*, 52, 893, 2004.
6. Fergenbaum, J.H. et al., Loss of antigenicity in stored sections of breast cancer tissue microarrays. *Cancer Epidemiol. Biomarkers Prev.*, 13, 667, 2004.
7. Mirlacher, M. et al., Influence of slide aging on results of translational research studies using immunohistochemistry. *Mod. Pathol.*, 17, 1414, 2004.
8. Ramos-Vara, J.A. and Beissenherz, M.E., Optimization of immunohistochemical methods using two different antigen retrieval methods on formalin-fixed, paraffin-embedded tissues: experience with 63 markers. *J. Vet. Diagn. Invest.*, 12, 307, 2000.
9. Fraenkel-Conrat, H. et al., Reaction of formaldehyde with proteins. VI. Cross-linking of amino groups with phenol, imidazole, or indole groups. *J. Biol. Chem.*, 174, 827, 1948.
10. Shi, S.R. et al., Antigen retrieval immunohistochemistry under the influence of pH using monoclonal antibodies. *J. Histochem. Cytochem.*, 43, 193, 1995.
11. Hasui, K. and Murata, F., A new simplified catalyzed signal amplification system for minimizing non-specific staining in tissues with supersensitive immunohistochemistry. *Arch. Histol. Cytol.*, 68, 1, 2005.

12. van der Loos, C.M., *Immunoenzyme Multiple Staining Methods*, Bios Scientific Publishers, Oxford, 1999.
13. Malik, N.J. and Daymon, M.E., Improved double immunoenzyme labeling using alkaline phosphatase and horseradish peroxidase. *J. Clin. Pathol.*, 35, 1092, 1982.
14. Wood, G.S. and Warnke, R., Suppression of endogenous avidin-binding activity in tissues and its relevance to biotin-avidin detection systems. *J. Histochem. Cytochem.*, 29, 1196, 1981.
15. Hanker, J.S. et al., Facilitated light microscopic cytochemical diagnosis of acute myelogenous leukemia. *Cancer Res.*, 39, 1635, 1979.
16. Burstone, M.S., Histochemical demonstration of phosphatases in frozen sections with naphthol AS-phosphates. *J. Histochem. Cytochem.*, 9, 146, 1961.

12 Immunoelectron Microscopy

Sara E. Miller and David N. Howell

CONTENTS

12.6 INTRODUCTION

This chapter provides an overview of methods for applying immunolabeling to the realm of electron microscopy (EM). A wide range of fixation, embedment, sectioning, and staining techniques are available for ultrastructural immunostaining to which the limited space available allows only a brief introduction. For more detailed

315

information, the reader is encouraged to consult several excellent texts devoted exclusively to immunoelectron microscopy (IEM).[1-6]

Several technical hurdles must be surmounted to achieve successful immunolabeling at the ultrastructural level. Some of these are shared with light microscopic immunostaining methods; others are peculiar to EM. As with light microscopic immunolabeling, primary antibodies used for IEM (or secondary/tertiary reagents directed against them) must be tagged in a fashion that allows them to be visualized. For the electron microscope, this means labeling them with metals (or substances containing metals) capable of partially or totally blocking the electron beam. Spherical particles of colloidal gold are the most common choice, but other options, such as the iron-binding protein ferritin and iron-containing synthetic polymers, are also available. Horseradish peroxidase (HRP), used in light microscopic immunohistochemistry, can also be employed for ultrastructural immunolabeling. Though the enzyme and the products of the reactions it catalyzes are electron-lucent, the reaction products can be stained with heavy metals, allowing ultrastructural detection.

A second problem for electron microscopic immunolabeling is that tissue fixation, necessary for preservation of ultrastructure, has a progressive deleterious effect on the integrity of many antigenic epitopes. This is also a problem for light microscopic immunolabeling, but is a particularly complex issue for EM. Insufficient fixation of samples for EM often leads to loss of ultrastructural detail undetectable at the light microscopic level. However, excessive fixation can easily abrogate the reactivity of antibodies with ultrathin sections, where epitope density may be sparse and immunolabeling reactions are often subtle.

Conventional embedment of specimens in epoxy resins poses a problem for immunolabeling that is unique to EM. Unlike paraffin, which is removed from light microscopic tissue sections with organic solvents and subsequently replaced with an aqueous buffer, the polymerized resin typically remains with an EM section for the duration of its life, severely limiting access to antigenic determinants by antibody reagents. Methods exist for etching the resin to allow access to tissue antigens, but these are generally quite harsh and may themselves cause damage to reactive epitopes. Hydrophilic resins are also available as substitutes for epoxy and, in some cases, allow successful immunostaining without recourse to section-etching techniques.

At the light microscopic level, damage of epitopes by fixation and the necessity of tissue embedment can both be circumvented by preparation of frozen sections of unfixed tissue (albeit with some loss of histologic detail). Frozen-sectioning methods are also a cornerstone of IEM, though they require considerable skill and specialized equipment. Unlike frozen sections for light microscopy, ultracryosections cannot be produced in the absence of some degree of prior tissue fixation without unacceptable loss of ultrastructural preservation. Fortunately, the fixatives used in this process are often gentler than those used for routine EM.

Another means of circumventing the damage and inaccessibility of antigenic epitopes caused by fixation and embedment is to perform the immunolabeling before these steps. Unless special permeabilization techniques are employed, this approach is limited to applications where the antigen in question is on the surface of the thing being labeled. Care must be taken during embedment and sectioning to produce sections that display the labeled surfaces of the specimen (e.g., by sectioning perpendicular to the labeled face of a flat piece of tissue or by making carbon replicas).

A final problem unique to IEM is the necessarily small size of ultrastructural tissue blocks and sections. Tissue specimens for conventional epoxy embedment have an upper limit of approximately 1 mm in any dimension, and frozen tissue blocks for ultracryotomy are generally smaller still. This is not usually a problem for homogeneous samples (e.g., embedded pellets produced from uniform cell suspensions or tissues such as muscle with a relatively uniform structure), but can cause considerable difficulty for complex, nonuniform samples (e.g., kidney or liver tissue), in which the structure of interest (for example, glomeruli or portal tracts) may represent only a small fraction of the total tissue volume. The conventional approach to this problem for routine EM is to produce multiple tissue blocks and use light microscopic examination of semithin "survey" sections to select blocks containing the feature of interest. This method can be used for IEM, but is labor intensive.

12.7 ANTIBODIES

12.7.1 Primary Antibodies

Selecting a primary antibody for IEM has much in common with the selection process for antibodies used in light microscopic immunolabeling. Because this topic is dealt with at length elsewhere in Chapter 11, it will not be covered in detail here.

When more than one antibody against a ligand of interest is available, several factors influence the choice, including the relative merits of various antibody types (monoclonal vs. polyclonal), preparations (serum, ascites, purified immunoglobulin fraction, affinity purified reagent), and cost. Published literature and manufacturer's information can sometimes be used to predict the performance of an antibody in ultrastructural immunolabeling, but many of the antibodies available have been validated only for biochemical applications (e.g., Western blotting, immunoprecipitation) and are untested for IEM.

In general, the most practical route is often to obtain one or more antibodies and test them on tissues known or strongly suspected to contain the ligand in the hope of finding a reagent that works. Ideally, a test tissue expressing the ligand at high density should be selected, even if it is not the tissue ultimately targeted for study. Pretesting of antibodies for possible use in IEM with a visible-light immunolabeling method, such as immunoperoxidase or immunofluorescent staining, is strongly recommended, because these techniques are in general less labor-intensive and more sensitive. Though it is clearly impossible to duplicate all of the conditions of the anticipated ultrastructural labeling experiment in light microscopic testing, an attempt should be made to standardize as many parameters as possible (e.g., choice, concentration, and duration of fixation). An antibody that gives weak, subtle, or variable staining in such a test system will usually be a poor choice for IEM. Antibodies that react robustly with antigenic epitopes expressed in a very focal manner provide one exception to this rule. The reactivity of such antibodies may be difficult to distinguish from nonspecific background at the light microscopic level, but easily identified as specific at the ultrastructural level, particularly if it is associated reproducibly with a defined substructural feature.

12.7.2 TROUBLESHOOTING AND STORAGE

Failure of an antibody to provide expected staining results can stem from a number of causes. In some cases, the antigenic epitope may have been masked by embedment resin or lost/destroyed during specimen/section preparation. The antibody may also have been produced against a form of the antigen different from that represented in the specimen as a result of fixation, reduction or other chemical alterations. Such problems can sometimes be remedied by changes in the fixation/embedment procedure or use of antigen-retrieval techniques. For antibodies that give the desired result initially but stop working over time, a common culprit is improper storage. Antibodies are best stored at relatively high concentrations rather than the final working dilution. If dilution is necessary, inclusion of a carrier protein in the diluent, such as bovine serum albumin or fetal bovine serum, is recommended. For prolonged refrigerator storage, 0.02% sodium azide can be added to prevent bacterial growth. Antibodies can also be stored frozen, but repeated freeze–thaw cycles are not advisable, because they can damage immunoglobulin molecules and induce formation of precipitates. Glycerol (20–30%) can be added as a cryoprotectant. Antibodies can be stored frozen at –20°C for several months, but storage for more than a year is best achieved at –70°C.

12.7.3 SECONDARY REAGENTS

IEM can be performed using a primary antibody directly conjugated with an electron-dense label, such as colloidal gold (direct immunolabeling). For a number of reasons, however, it is usually advantageous to use an indirect staining method in which the primary antibody does not bear a directly detectable tag. These techniques include detection of unlabeled primary antibodies with metal-labeled anti-immunoglobulin reagents or bacterial products with immunoglobulin-binding properties (staphylococcal protein A, streptococcal protein G). Primary antibodies (and other analyte-specific reagents, such as nucleic acid probes) derivatized with small molecules, including dinitrophenol,[7,8] fluorescein,[9] digoxigenin,[10,11] and biotin,[12] can be detected using antibodies against the relevant tag. Biotinylated primary antibodies can also be detected using avidin tagged with a heavy metal. Secondary antibodies and avidin can react with more sites on the primary antibody than protein A or G; this may lead to augmented signal when the former methods are employed, but complicates quantitation of antibody binding (see the following section). The utility of protein A is also limited to some extent by its inability to bind certain subclasses of immunoglobulin; this drawback can be circumvented by interposing a "bridging" antibody that binds to the primary and is subsequently bound by labeled protein A. Other sandwich techniques with more than two layers, including the avidin–biotin–peroxidase complex[13] and peroxidase–antiperoxidase[14] methods, can be used to enhance the sensitivity of ultrastructural immunolabeling.

Indirect staining methods offer several potent advantages. Chief among these is that each reagent layer theoretically allows an amplification of signal. This is a particular boon if the antigenic determinant in question is sparsely distributed or the primary antibody is of suboptimal avidity. Because high-quality antibodies conjugated with colloidal gold are often expensive or difficult to produce, use of a single conjugated secondary reagent with several different primary antibodies also offers considerable economy. (In many cases, a directly conjugated primary antibody may simply be unavailable.) Validation of an unconjugated or biotinylated primary antibody's reactivity with a tissue of interest and establishment of an approximate working titer can often be accomplished with relative ease at the light microscopic level, then applied at the more labor-intensive ultrastructural level with a simple switch of secondary reagents. Colloidal gold conjugates can be visualized at the light microscopic level by silver enhancement techniques, in principle allowing light microscopic testing of both primary and secondary reagents as a prelude to ultrastructural labeling. The same thing can be accomplished using secondary antibodies with combined fluorescent/gold tags (FluoroNanogold, Nanoprobes, Yaphank, NY).

On the negative side, indirect staining techniques are of necessity more complex and time-consuming than direct ones. They are difficult (but not impossible) to employ in double-labeling experiments with two primary antibodies of the same species origin and isotype. Finally, spatial resolution is decreased somewhat with each reagent layer, a drawback if precise localization of antigenic epitopes is desired.

12.7.4 TITRATION OF REAGENTS

Establishing optimal working titers for antibody reagents for IEM is of paramount importance. The primary aim is to identify the lowest concentration of a reagent that, for a given set of conditions (e.g., incubation time, temperature) saturates the available binding sites. Achieving this goal is often more of an art than a science, but a few general principles are of value.

Suggested working titers for use of a given reagent in IEM may be available from the published literature, manufacturer's specifications, or discussions with colleagues. These provide a useful starting point, though results may vary. If recommendations for IEM are unavailable, published working titers for visible-light immunostaining methods may be a reasonable first approximation, particularly if the tissues to be labeled are prepared in similar ways (e.g., light-microscopic frozen section, ultrathin cryosection). Suggested titers for Western blotting (a preferred method used by antibody producers to validate the reactivity and specificity of their wares) may bear minimal relation to the optimal titer for ultrastructural immunostaining; Western blotting, particularly with radioligand or chemiluminescent detection systems, is an extraordinarily sensitive technique and can often be accomplished at antibody concentrations that yield little or no signal in tissue staining, particularly in the electron microscope.

To establish a working titer, a reasonable approach is to stain several replicate tissue sections with twofold serial dilutions of the reagent in question in a range bracketing the "best guess." If no information is available from the literature, peers, or manufacturers, very rough starting points used in our laboratories are undiluted for supernatants of cultured hybridoma cell lines, 1:100 for hybridoma ascites or antiserum and 0.2 µg/ml for purified immunoglobulin preparations. In an ideal experiment, intensity of staining will rise with increasing reagent concentration until the point of ligand saturation, beyond which staining intensity will be relatively constant (absent increased nonspecific background staining). After such a titer has been established, there is little to be gained by increasing reagent concentration in an attempt to enhance staining of tissues with sparse antigen expression; in general, this will only potentiate nonspecific, artifactual staining.

When working out an indirect staining procedure for a new, untested primary antibody, it is advisable to establish ideal working concentrations for the secondary (and tertiary) reagents using another primary antibody with known and robust binding characteristics. Most laboratories that perform IEM with regularity maintain stocks of secondary and tertiary reagents characterized in this manner. Trying to establish ideal working concentrations for several untested reagents simultaneously using a "checkerboard" titration scheme is a recipe for trouble.

12.7.5 Nonspecific Background Staining

In addition to ligand-specific interactions, one or more of the reagents used in ultrastructural immunolabeling often binds to cell or tissue components in a nonspecific manner (e.g., via electrostatic or hydrophobic interactions, covalent binding with residual reactive aldehyde groups from fixative). This leads to "background" staining that must be distinguished rigorously from antigen-specific staining. Employment of negative controls (see the following section) is one important method for recognizing the presence of background staining. The presence of labeling in unexpected locations (e.g., areas of embedment medium without specimen, incongruous locations within the specimen) is another helpful clue.

Total eradication of background staining is virtually impossible, but it can often be minimized by careful attention to experimental detail. (A small amount of background staining actually serves as a reassurance that the secondary antibody has been applied successfully to the sections.) Titration of each labeling reagent to the lowest satisfactory concentration is often helpful. If the primary antibody is in the form of an impure preparation such as whole serum or ascites fluid, adsorption with cells or tissues lacking the antigen in question may remove contaminating antibodies directed against other specificities. All reagents except colloidal-gold conjugates (which can be pelleted by modest centrifugal force) should be microcentrifuged before use to remove precipitates. Gold conjugates should be stored in concentrated form and diluted just before use into freshly microcentrifuged buffer containing a protein carrier.

Preincubation of sections with a dilute solution of a nonreactive protein (e.g., 2–5% bovine serum albumin, 5–10% fetal bovine serum, 1% ovalbumin, or 0.1–1% cold-water fish skin gelatin) and subsequent inclusion of the protein as a carrier in the reagent diluent can reduce background substantially. Some authors refer to this maneuver as high-molecular-weight blocking. Nonimmune serum from the same species as the secondary antibody (and thus unlikely to react with it) is another good choice of carrier protein. Inclusion of a small amount of nonionic detergent (e.g., 0.05% Tween-20) in the diluent can also help reduce background staining.

The presence of free reactive aldehydes that can bind covalently to protein reagents is a particular problem for tissues fixed in glutaraldehyde, which has two aldehyde groups. This feature makes it an excellent cross-linker and fixative, but leaves a free aldehyde whenever only one of the two aldehydes forms a covalent bond with the specimen. These groups can be covalently modified in a manner that renders them nonreactive by treatment of tissue sections with 0.1% $NaBH_4$ or 50 mM glycine, lysine, or ammonium chloride. The first of these compounds reduces the offending groups; the others block them via formation of Schiff bases. These manipulations, which can also be used to abrogate the autofluorescence of free aldehyde groups in immunofluorescent staining, are sometimes referred to as low-molecular-weight blocking.

12.7.6 CONTROLS

Controls for IEM are similar in principle to those for other forms of immunostaining. Every experiment should have a negative control, in which a parallel sample (a serial section or sections on a separate grid) is stained with a series of reagents in which the primary antibody is replaced with something predicted to have no antigen-specific reactivity with the tissue, but otherwise resembling the primary as closely as possible. If the primary antibody is in the form of whole serum, control serum from an unimmunized animal of the same species (ideally preimmune serum from the animal used to produce the antiserum) is the usual choice. For purified immuno-globulin primary antibodies, control immunoglobulins matched as closely as possible for species of origin, isotype, and concentration should be used. If a directly con-jugated primary antibody is being used for staining, the negative control should be conjugated in a similar manner.

Occasionally, investigators employ a "buffer control," in which the primary antibody is replaced by diluent devoid of immunoglobulin. This option should be avoided, as it fails to control for non–antigen-specific (e.g., electrostatic) interactions of the primary antibody with tissue elements. Specificity controls for the primary antibody, in which the staining procedure (complete with primary) is performed on tissue devoid of the cognate antigen, though useful in their own right, should also not be mistaken or substituted for negative controls.

If the immunogen used to produce the primary antibody is available in pure and soluble form, an alternative and highly desirable form of negative control can be performed by preincubating the primary antibody with a saturating concentration of its ligand, effectively blocking its ability to bind to tissue epitopes but having no effect on non–antigen-specific interactions. This works particularly well with anti-peptide antibodies, some of which are supplied with "blocking peptide" by their manufacturers for this specific purpose. The concentration of peptide and length of incubation necessary to achieve thorough blocking are not well standardized and differ from antibody to antibody. As a rule of thumb, we typically use a 10-fold molar excess of peptide and incubate for 30 min at 4°C.

If the tissue or cell of interest is not known to express the antigen being examined at high density, and if an alternative sample that expresses high levels of the antigen is available, it may be worthwhile to use the second tissue as a "positive control," staining it in parallel with the same sequence of reagents used for the experimental sample. Failure of the positive control to produce staining in an appropriate pattern suggests a transient or intrinsic flaw in one or more steps of the staining procedure.

This control is considered mandatory for clinical diagnostic immunostaining and is often of significant utility in ultrastructural immunolabeling procedures, particularly those with subtle results.

12.8 LABELS

12.8.1 COLLOIDAL GOLD

The label employed in the vast majority of modern IEM is colloidal gold.[15–20] Under appropriate conditions, gold salts form stable, spherical aggregates that can be coated with a variety of proteins, including immunoglobulins and immunoglobulin-binding ligands. Binding of reagent-conjugated gold particles is most frequently analyzed by conventional transmission EM. They can be detected by a number of other ultrastructural imaging modalities, however, including both conventional secondary electron imaging[21–23] and backscatter imaging[24] in the scanning electron microscope, energy-dispersive X-ray analysis,[25] and atomic force microscopy.[26,27]

Colloidal gold particles can be produced in a wide range of sizes, from <1 nm to approximately 250 nm, though particles in the range of 5–20 nm are employed most frequently for ultrastructural immunolabeling. Particles of different sizes conjugated with different reagents can be distinguished in the electron microscope, allowing simultaneous staining for more than one ligand in a manner analogous immunofluorescence with multiple fluorochromes. Particle size can be controlled to some extent by the conditions under which the colloid is formed.[19,20] Reduction of gold chloride (hydrogen tetrachloroaurate, or $HAuCl_4$) with yellow or white phosphorus produces small particles (~5 nm); sodium ascorbate produces particles of about 12 nm, and sodium citrate can be used to produce particles of 16–150 nm.[19] Purification of gold particles within a narrower size range can be accomplished by differential ultracentrifugation, either by banding in a glycerol or sucrose gradient[17] or selective pelleting at carefully chosen speeds and durations of centrifugation.

Though the reagents required for production of colloidal gold particles are fairly inexpensive, and the reactions not prohibitively difficult to perform, some experience with chemical synthesis is needed. Producing particles of a uniform size introduces an additional complexity; this is a particular problem if multiple-label experiments are desired. After coating with antibodies or other reagents has been accomplished, extensive testing is necessary to assure that the finished product performs properly. For these reasons, a majority of laboratories utilize standardized commercial gold preparations.

Colloidal gold particles can be coated with a wide variety of antibodies and antibody-binding proteins, which are adsorbed noncovalently to the particle surface. (Conjugation with other ligand-binding substances, including toxins, hormones, polysaccharides, lectins, and other proteins, is also possible). The coating not only mediates specific binding of the particles to ligands, but also prevents their coagulation. All but the smallest particles will typically have multiple immunoglobulin molecules bound to their surfaces, in a quantity roughly proportional to the particle surface area. Commercial colloidal gold particles are usually supplied already coated. However, stable suspensions of uncoated colloidal gold particles are available from commercial suppliers (e.g., Structure Probe, Inc., West Chester, PA) for laboratories wishing to produce custom conjugates.

Gold particles smaller than 5 nm in diameter can be made and have the advantage of enhanced specimen penetration. Conventional colloidal gold particles measuring 1–3 nm in diameter, often termed "ultrasmall" gold, are produced and coated in a manner similar to their larger counterparts.[28] A second family of particles consists of a core of gold atoms covalently coupled to a shell of organic molecules. These include "Nanogold" (a 1.4-nm particle containing approximately 67 gold atoms)[29,30] and "undecagold" (an even smaller particle containing 11 gold atoms),[30,31] both supplied by Nanoprobes. Unlike conventional colloidal gold particles, reagents are coupled to Nanogold and undecagold via covalent bonding to organic linkers on the particle surface, with potential advantages of stability and uniform orientation of the antibody with its antigen-binding regions accessible. The specimen penetration of small gold particles can be further enhanced by substituting antibody cleavage products, such as Fab and $F(ab)'_2$ fragments[29,30] or even smaller single-chain variable fragments (scFv),[32] for whole immunoglobulin molecules.

Gold particles in the sub–5-nm range are difficult to resolve in the electron microscope, but can be rendered visible after binding to their targets by deposition of layers of silver around the gold particle nucleus.[33,34] This technique, referred to as "silver enhancement" or "autometallography," was initially developed for light microscopic immunostaining as a modification of a photographic method termed "physical enhancement."[35] In addition to rendering small gold particles visible, silver enhancement can be used to increase the size of larger gold particles. Interposition of a silver enhancement step between applications of two reagents conjugated to gold particles of the same size (e.g., 5 nm) allows distinction of the two because of the augmented size of the first[36] and has the added benefit of rendering the first set of antibody reagents sterically inaccessible, facilitating double labeling with two primary antibodies raised in the same species. Silver-enhanced gold particles can also be detected in the scanning electron microscope by both secondary electron[37] and backscatter[38] imaging. The processing time for silver enhancement can be decreased by microwaving.[39]

Commercially available colloidal gold conjugates come in a wide array of reagent coatings and sizes. Most are supplied in a buffered saline solution, often supplemented with a protein carrier. Ideally, the coated particle suspension should be devoid of free, soluble antibody, which can compete for binding of the ligand. Our laboratory has had consistent success with the AuroProbe reagent series (GE Healthcare, Little Chalfont, UK), but high-quality conjugates are available from other sources as well. Particle concentration can be expressed either as number of particles (usually in the range of 10^{12}–10^{13} per ml in typical commercial preparations) or optical density. For a given particle concentration, the latter parameter increases with particle diameter; typical optical densities (measured at 520 nm) for commercial preparations of 5, 15, and 30 nm particles from one major supplier are 2.5, 3.5, and 7.0, respectively. High-quality colloidal gold particles are also supplied with an estimate of particle size distribution (e.g., coefficient of variation, ideally less than 10% for particles measuring 10 nm and larger) and information on the relative paucity of particle clumps. The latter are a particular hazard for ultrastructural immunolabeling, as their nonspecific sticking to tissues can mimic clusters of gold particles bound to a focal concentration of antigen.

12.8.2 OTHER ELECTRON-DENSE LABELS

In addition to colloidal gold, reagents for transmission EM can be conjugated with ferritin, a 750-kDa protein that derives its electron density from large quantities of bound iron. Other iron-containing compounds, including iron dextran (Imposil) and iron mannan, can also be used as ultrastructural immunolabels. These polymers appear as distinctive oblong particles in the electron microscope, facilitating double-label applications.[40,41]

Reagents conjugated with HRP, a popular reagent for light microscopic immunohistochemistry, can also be used for IEM and reportedly offer an advantage in specimen permeability.[42] HRP itself is not electron dense, but the oxidized precipitate formed by its action on a common substrate, 3,3-diaminobenzidine, can be stained with osmium or gold chloride. Other HRP substrates with greater affinity for osmium have also been described recently.[43] Some cells, particularly hematopoietic cells, contain endogenous peroxidases. Depending on the method of fixation and embedment, these cellular enzymes may retain catalytic activity and produce a 3,3-diaminobenzidine reaction product. This nonspecific reaction must be blocked (e.g., by incubation for 30 min in 0.3% H_2O_2 in distilled water followed by rinsing with phosphate-buffered saline) or controlled for in peroxidase labeling experiments. Both iron-containing labels and the reaction products of HRP also have the disadvantage of producing a somewhat indistinct, "smudgy" signal in the electron microscope, as compared with the crisp opacity of gold particles.

Immunolabeling is also occasionally performed on scanning EM specimens, usually for ligands on the surfaces of the specimens. Colloidal gold particles, particularly those of larger diameter, are a common choice of label for this application. Other uniform particles visible in the scanning EM, such as virus particles (e.g., tobacco mosaic virus, bacteriophages) are also occasionally conjugated to antibodies or secondary reagents and used as labels.

12.8.3 PITFALLS AND HINTS FOR VISUALIZING HEAVY METAL LABELS

Though most ultrastructural metal immunolabels have a fairly distinctive appearance, detecting some and distinguishing them from background features can occasionally be tricky. Small gold particles (5 nm and less) cannot be identified reliably at microscope magnifications less than ×25,000, higher than the range in which many microscopists are accustomed to working. Seeing them also requires that the microscope be properly stigmated at high magnifications. Unlike larger gold particles, which are absolutely opaque and dense black, the smaller ones have a less distinctive grey appearance.

Sections labeled for IEM are typically contrasted with soluble heavy metal salts (e.g., uranyl acetate) to allow visualization of tissue ultrastructure, in a manner analogous to the hematoxylin counterstain often used for light microscopic immu-noperoxidase staining. Small gold particles and other heavy metal tags such as ferritin are difficult to pick out in images of resin sections contrasted with large amounts of heavy metal stains. This problem is exacerbated if precipitates are present in the heavy metal solution or form after it is applied. Uranyl acetate tends to precipitate at high concentrations, particularly if exposed to phosphate buffers or neutral to high pH. This problem can be minimized by use of a relatively low concentration of uranyl acetate (e.g., 1–2% for 5–12 min) for contrasting and thor-ough washing of grids with water before its application to remove phosphates from reagent diluents. Methylcellulose, often applied in a solution with uranyl acetate in the final embedment step for ultracryosections (see the following section), can form clumps with entrapped stain; these can be removed by microcentrifugation of the reagent before use. Copper grids react with some buffers and form precipitates; this problem can be avoided by using nickel grids.

12.8.4 MULTIPLE LABELING APPLICATIONS

Simultaneous ultrastructural localization of more than one antigen in tissue speci-mens can be accomplished using colloidal gold particles of various sizes or with gold and other electron-dense tags. Combinations of immunolabeling and other detection modalities, such as *in situ* hybridization,[5,44] ultrastructural autoradiography, or uptake experiments with cationized ferritin, are also possible. For any multiple label experiment, it is advisable to work out the staining conditions for each label singly before attempting the combination. Similarly, it is important to rule out "cross-talk" between reagent systems (e.g., to make sure that a goat anti-rabbit secondary antibody used in an experiment with one rabbit and one mouse primary does not cross-react with mouse immunoglobulin).

There are myriad ways to associate disparate labels with distinct antigens, too numerous to catalog here. Briefly, the simplest form of experiment employs primary antibodies directly conjugated to different tags (e.g., 5- and 10-nm gold particles). More sensitive, but also more complex, experiments employ multiple primary antibodies without heavy metal tags that can be detected selectively by differently tagged secondary reagents (e.g., primaries raised in different species, primaries of different isotypes from one species, biotinylated and nonbiotinylated primaries). In many instances, multiple reagents can be applied synchronously. In others, sequential application in a carefully chosen order is required (e.g., rabbit primary antibody, goat anti-rabbit immunoglobulin conjugated with 5-nm gold, biotinylated rabbit primary antibody, avidin conjugated with 10-nm gold). It is even possible to utilize pairs of primary antibodies that are formally indistinguish-able with proper sequences of reagents (e.g., rabbit primary antibody, protein A conjugated with 5-nm gold, unlabeled protein A to block any residual binding sites, second rabbit primary antibody, protein A conjugated with 10-nm gold). As noted previously, an interposed silver enhancement step is another possible approach to this problem. For resin sections, double labeling reportedly can also be accomplished by staining the two sides of a grid (and thus of the section) with two series of reagents.[45]

12.8.5 Quantitation of Labeling

Quantitative IEM requires attention to detail at several steps.[2] A wide variety of factors that influence the efficiency and specificity with which primary antibodies bind to their target ligands must be taken into consideration, including antigenic preservation after fixation, avidity of antibody for antigen, type of section (cryosection vs. resin embedment, epoxy vs. acrylic resin), plane of section through the organelle of interest, steric hindrance to antibody binding, movement of soluble antigens within the specimen, and background staining. Assuming that the primary antibody finds its target in a manner that bears direct relation to antigen density, quantitation of binding is best performed using directly labeled primary antibodies or indirect methods with a secondary reagent, such as protein A or protein G, that binds to the primary with one-to-one stoichiometry. Enumeration of binding is also facilitated by use of colloidal gold as a label, because its particles can be counted with relative ease, an advantage not shared by amorphous labels such as peroxidase reaction products.

12.9 TISSUE SUBSTRATES FOR LABELING

12.9.1 Fixation

In general, fixation of tissues for IEM should be accomplished with the gentlest suitable fixative at the minimum concentration and duration of fixation necessary to achieve the desired level of ultrastructural preservation. Paraformaldehyde (2–8%) in suitable buffer (e.g., phosphate, PIPES, HEPES) is a common choice. This can be supplemented with small quantities of glutaraldehyde (0.1–0.5%) to improve fixation. Fixation (and subsequent embedment and staining) can be enhanced and accelerated by microwave irradiation.[6,46]

Osmium tetroxide (OsO_4) is often employed in processing of specimens for conventional EM, where it serves as both a fixative and an en bloc heavy metal stain. It should usually be avoided in immunostaining experiments, because it causes extensive oxidative damage to a majority of antigens. On occasion, however, supplementation of the fixative in such experiments with small quantities of OsO_4 may be necessary to reveal ultrastructural detail in the tissue sections (see the following section). In such cases, antigenicity can sometimes be restored by treatment of sections with sodium metaperiodate.[47]

12.9.2 Preembedment Labeling

It is possible to label suspensions of cells or thin slices of aldehyde-fixed tissue produced with a vibrating microtome with antibodies before embedment, with subsequent embedment, sectioning, and examination by conventional methods (Figure 12.1). Preembedment labeling has the advantage of preserving epitopes that are often lost or rendered inaccessible in the embedment process. Because the tissues can be osmicated after immunolabeling, ultrastructural detail is also easy to see in the finished product. Osmium tetroxide can oxidize metallic silver and may cause degradation of signal if used after silver-enhanced gold labeling. In such cases,

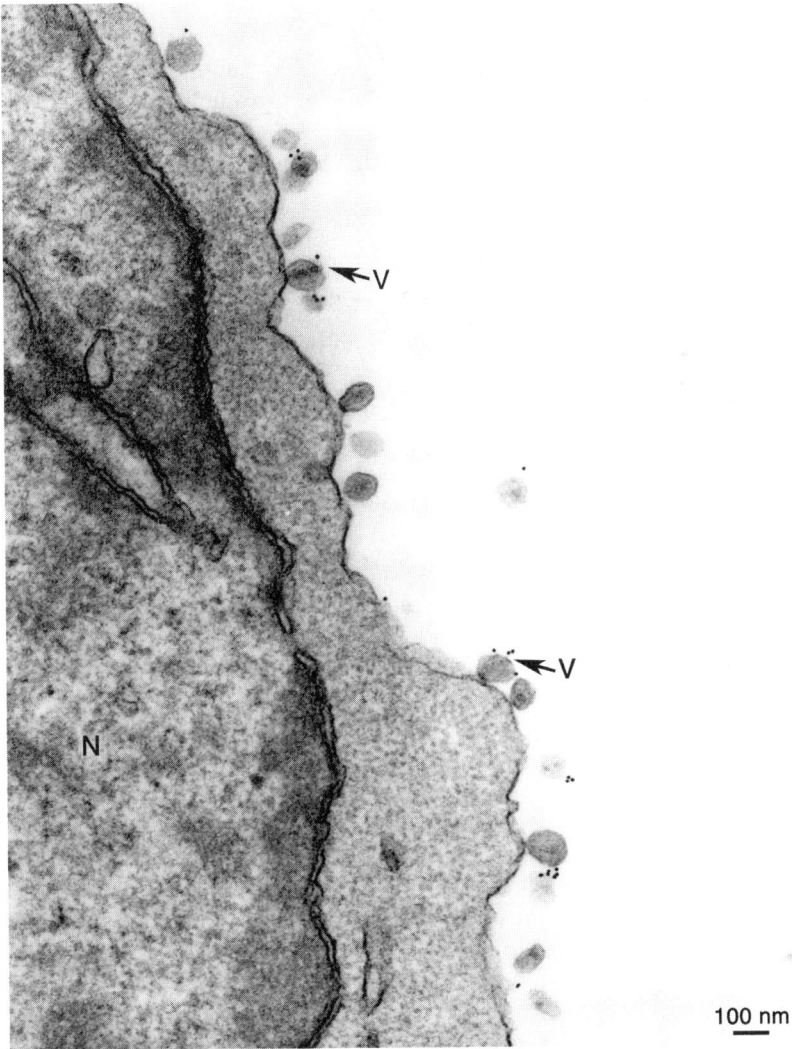

FIGURE 12.1 Preembedment immunolabeling. Cells infected with human immunodeficiency virus were immunolabeled in suspension on ice with a rabbit antibody against a protein on the virus envelope. After washing by centrifugation and resuspension in buffer three times, goat anti-rabbit antibody conjugated to 10-nm colloidal gold was added. After incubation, the cells were again washed and then fixed in glutaraldehyde, osmium, and uranyl acetate for routine thin sectioning. Note the normal-appearing, darkly stained membranes. N = nucleus; V = virus labeled with gold.

preservation of labeling can often be achieved by slight overenhancement coupled with a limited osmium exposure (e.g., 1% OsO_4 for 15 min).

The primary disadvantage of preembedment labeling methods is that only sites on the surfaces of the cells or tissue slices are available to the labeling reagents. Serial ultrathin sections, the substrate of choice for performance of negative control staining, are also perforce unavailable. Permeability can be improved somewhat be treating the cells or tissues to be labeled with methanol or nonionic detergents (e.g., saponin, Triton). These permeabilizing reagents, which must be washed out before application of antibodies, allow penetration of labels to a limited depth of 8–9 μm. Unfortunately, they also destroy cell membranes and compromise ultrastructure to varying degrees and may allow relocation of antigenic determinants during the subsequent staining process.

12.9.3 PARTICULATE SPECIMENS

In some cases, small particulate specimens, such as viruses (Figure 12.2), micelles, and membrane preparations, can be immunolabeled on a film-coated grid or in suspension and then negatively stained,[48–51] avoiding the need for embedment altogether. Particles can be applied to grids coated with support films by any of a number of methods, including simple placement of the grid on a drop of suspension, centrifugation techniques, and concentration methods such as the agar diffusion method.[52] In the latter, a drop of suspension is placed onto solid 1% agarose and a grid is floated film-side-down on the surface of the drop. The fluid diffuses into the agar, concentrating the particles onto the surface of the support film. After the particles are attached to the grid, immunostaining can be performed in much the same manner used for thin sections, followed by negative staining to provide contrast. Colloidal gold is the label of choice for such experiments.

Viruses and viral antigens can also be immunolabeled in suspension, then deposited on film-coated grids and examined. This approach is limited by the lack of a straightforward method for washing between reagent applications. Reaction of antibodies with particles in suspension can be measured indirectly by particle clumping. In some cases, unlabeled antibodies bound to particles can also be identified as a "fuzzy coat" at high magnification. If an indirect staining method is required, use of a high-avidity primary antibody at low concentration is recommended to minimize competition of unbound primary for the secondary reagent.

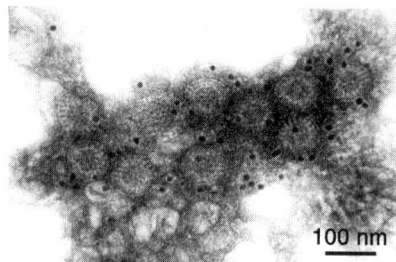

FIGURE 12.2 On-grid immunolabeling of viruses. Rotavirus was adsorbed to a Formvar and carbon-coated grid. The grid was then floated on monoclonal antiviral antibody, washed in buffer five times, and floated on goat anti-mouse immunoglobulin conjugated to 10-nm colloidal gold. After three water washes, the virus was negatively stained with uranyl acetate.

12.9.4 Replica Labeling

Replica labeling is another method for examining antigens on specimen surfaces without embedment, particularly useful for studying antigens associated with cell membranes.[53,54] For antigens on the cell surface, immunogold labeling can be performed on intact cells. The inner face of the membrane can be studied by sticking the outer cell surfaces to a substrate (e.g., a plastic film) and peeling off the attached membranes, exposing their undersides, or by freeze-fracture techniques. In either case, the specimen is dried after immunolabeling (e.g., by freeze drying or critical-point drying), and a carbon replica is made in a vacuum evaporator or freeze etch apparatus. The gold particles are embedded in the replica. After replication, the cell material is digested away with sodium hypochlorite (undiluted commercial bleach) or acid (e.g., 10% hydrofluoric or 60% sulfuric acid). Carbon replicas provide an en face view of the cell membrane, as opposed to conventional sections, in which membranes are usually seen in cross-section.

12.9.5 Labeling of Frozen Sections

Immunostaining of cryosections produced from frozen tissue blocks, often referred to as the "Tokuyasu method" in recognition or its initial developer, is the best and most general technique for assuring preservation of antigenic epitopes,[55-57] but has some drawbacks. The ultrastructural staining pattern seen in cryosections (light membranes against a dark cytosol) is the reverse of that produced with conventional fixation/embedment methods (Figure 12.3); interpretation of cryosection images requires a certain amount of experience. The equipment needed for producing cryosections is expensive and somewhat complicated to operate. Size limitations for frozen tissue blocks are even more severe than those for conventional EM blocks, and their long-term storage requires an ultra-low temperature or liquid nitrogen freezer. These limitations notwithstanding, cryosectioning is a powerful method for ultrastructural immunolabeling.

In most cases, tissue is fixed lightly with paraformaldehyde (sometimes supplemented with a low percentage of glutaraldehyde, but without OsO_4), infiltrated with 2.1–2.3 M sucrose, and snap frozen. Infiltration can also be performed with a mixture of 1.15 M sucrose and 10% polyvinyl pyrrolidone in 0.1 M phosphate buffer.[58] Unlike specimens for freeze substitution techniques, which require special freezing conditions (see the following section), sucrose-cryoprotected specimens for ultracryotomy can be frozen by simple immersion in liquid nitrogen. Sectioning is performed on a liquid nitrogen–cooled ultracryomicrotome at a temperature of –80° to –110°C. The sucrose concentration and sectioning temperature can be varied to accommodate specimens of different plasticity; in general, the higher the concentration of sucrose used, the softer the frozen block and the colder the temperature required for cutting.[57]

Dry sections are picked up with wire loops (1.5–2.0 mm diameter) containing drops of 2.3 M sucrose, brought out of the cryomicrotome chamber to thaw, and touched to fresh grids coated with a support film.[52] The grids are then placed, section side down, on top of drops of reagent; these can be deployed conveniently in the

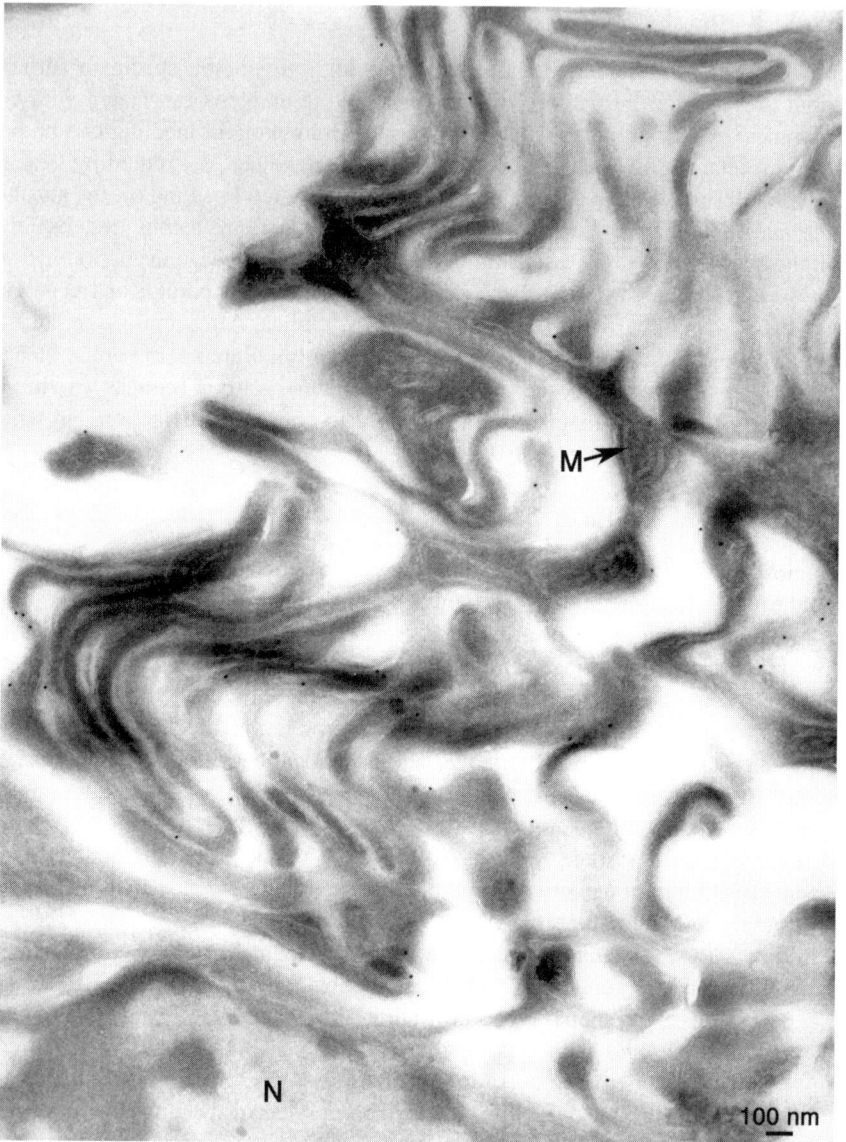

FIGURE 12.3 Labeling of ultrathin cryosections. Ocular ciliary body was lightly fixed in paraformaldehyde, cryoprotected in sucrose, and frozen in liquid nitrogen. Ultrathin cryosections (60–80 nm) were cut and placed onto a grid containing a Formvar and carbon support film. They were immunostained with an antibody against a membrane protein, washed in buffer five times, and reacted with a secondary antibody conjugated to 10-nm colloidal gold. Finally, they were washed in water, contrasted with uranyl acetate in methyl cellulose, and air-dried in a dust-free chamber. Note that the image is a "reverse" image with light membranes against a dark cytosol. N = nucleus; M = mitochondrion.

required sequence on the surface of a sheet of paraffin film. An example of a staining sequence, usually performed at room temperature, is as follows:

Reagent	Duration (min)
Phosphate-buffered saline (PBS) washes (×3)	5 each
50 mM glycine (block reactive aldehydes)	10
Primary antibody	30–60
PBS washes (×5)	5 each
Gold-conjugated secondary reagent	30
Water washes (×3)	5 each
2% Methylcellulose/0.3% uranyl acetate	20

The last solution embeds and stabilizes the sections (methylcellulose) and provides a limited amount of ultrastructural staining/contrast. Several variations on the final staining-embedding step have been described.[2] The grids are transferred from drop to drop with a wire loop (3–4 mm), never allowing them to dry until the final step. They should be suspended on the reagent drops by surface tension, and never allowed to sink beneath the reagent surface, as wetting of their backs can create high background staining. After the final step, excess fluid is wicked away using filter paper, and the grids are allowed to air dry while still suspended in their loops, held in place by a thin methylcellulose membrane.

12.9.6 POSTEMBEDMENT LABELING OF NONFROZEN TISSUES

Sections of tissue embedded in polar acrylic resins (LR White, LR Gold, Lowicryl K4M or K11M) are a viable alternative to frozen sections for immunostaining in many cases (Figure 12.4).[59] Though antigenic preservation and accessibility with these resins does not match that of frozen tissue, they are superior in several aspects to epoxy resins while providing ultrastructural preservation similar to that afforded by epoxy embedment. The hydrophilic nature of polar acrylic resins facilitates some penetration of aqueous reagents, allowing immunostaining without the harsh etching steps usually required for epoxy-embedded tissues (see the following section). They are miscible with water to some extent during polymerization, obviating the need for complete specimen dehydration. Acrylic resins that maintain low viscosity at low temperatures and can be polymerized with ultraviolet light have been developed[60]; avoidance of the high-temperature curing required for epoxy resins preserves many antigens. The favorable permeability provided by acrylic resins can be augmented by use of the smallest available immunoprobes (e.g., peroxidase-conjugated antibodies, Nanogold or undecagold conjugates) to increase penetration.

Epoxy-embedded specimens (or specimens embedded in non-polar acrylic resins such as Lowicryl H20 or H23) are generally a less desirable substrate for immuno-labeling than those embedded in polar acrylic resins. Epoxy embedment destroys many antigenic epitopes, and the hydrophobicity of epoxy is a barrier to the penetration of aqueous reagent solutions. Because epoxy is the favored embedment medium for clinical EM laboratories, however, specimens embedded in epoxy may be the only available option in some cases, particularly for studies of archival clinical material.

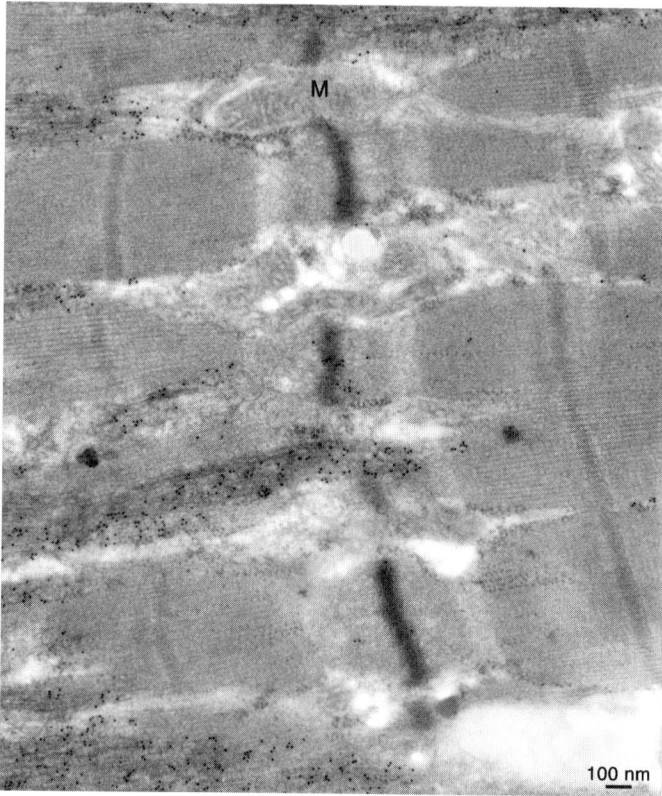

FIGURE 12.4 Postembedment immunolabeling. Mouse muscle was fixed lightly in paraformaldehyde and embedded in LR White, an acrylic resin, without en bloc heavy metal staining. Ultrathin sections on grids were floated on drops of primary antibody against a mutant muscle protein, washed in buffer five times, and then incubated with secondary antibody conjugated to 10-nm colloidal gold. After three water washes, the grids were poststained with uranyl acetate and lead citrate. Note that, although ultrastructure is identifiable, the membranes are unstained because the tissue is not osmicated. M = mitochondrion.

Successful labeling can occasionally be achieved without special treatment of sections if the antigen in question is unusually resilient or present in high concentration. More often, techniques to expose epitopes, including chemical etching with oxidizing agents and heat-induced epitope retrieval, are necessary to allow successful staining.[4]

Both acrylic and epoxy resins remain permanently in the sections they are used to embed. An alternative is provided by polyethylene glycol (Carbowax), which can be used to embed tissues for ultracryotomy and subsequently removed from the sections in a manner analogous to paraffin embedment and deparaffinization of routine histologic sections.[61-63] Tissues embedded in polyethylene glycol are brittle and difficult to cut, but can be sectioned on standard instruments. The sections are cut with a dry knife and manipulated with wire loops in much the same manner as ultracryosections. After removal of the polyethylene glycol, immunostaining reagents have good access to antigenic epitopes.

12.9.7 FREEZE SUBSTITUTION

Both frozen sectioning and conventional EM embedment techniques allow spatial dislocation of antigens that may be significant at the ultrastructural level. Rapid freezing immobilizes molecules, but they can move to a considerable degree when ultracryosections are allowed to thaw. By the same token, a certain amount of molecular movement can occur during conventional slow chemical fixation with aldehydes. Freeze substitution is a hybrid technique designed to immobilize antigens in their *in vivo* locations.[64,65] It can provide excellent ultrastructural detail and immunolabeling, but is labor intensive and requires special equipment for optimal results. Keys to success include very small sample size (<0.5 mm), rapid freezing, and long infiltration times (up to several days per solution) with high reagent volumes (1,000× specimen volume).

Tissue fragments are frozen under conditions that promote vitrification but prevent formation of ice crystals. This can be accomplished by immersion in a slurry of liquid ethane or propane cooled in liquid nitrogen. (Direct immersion of the specimen in liquid nitrogen leads to formation of insulating nitrogen bubbles on the specimen surface.) Freezing can be accelerated with special instruments by forced impact ("slamming") against a cooled metal block or exposure to a high-pressure liquid propane jet. Specimens are kept at very low temperatures, usually by transfer to a dedicated freeze substitution instrument that allows precise temperature control. (A conventional –80°C freezer is a less-expensive alternative, but does not afford consistent temperature regulation during warming steps.) In most protocols, the specimens are infiltrated with fixatives (aldehydes and sometimes osmium) in an organic solvent (e.g., acetone). The constant low temperature minimizes migration of antigenic molecules from their original locations. Depending on the method employed, the tissues are then either infiltrated and cured in the cold using a resin designed for low-temperature manipulations and ultraviolet light polymerization (e.g., Lowicryl) or warmed to room temperature for conventional resin infiltration. Immunostaining can then be performed by any method appropriate for ultrathin resin-embedded sections.

12.10 CORRELATIVE IMMUNOMICROSCOPY

This chapter has hopefully provided a clear view of the power and range of ultrastructural immunolabeling techniques. Though many (perhaps most) readers of this text will undoubtedly be more familiar with light microscopic immunostaining methods, it is hoped that this introduction will encourage them to extend their studies into the ultrastructural realm if the need or opportunity arises. In closing, it is worth noting that IEM is most often employed in correlative fashion with light-microscopic immunolabeling modalities, providing a natural lead-in for investigators more familiar with the latter methods. The rationales for correlative immunomicroscopy are manifold. The utility of testing reagents at the light microscopic level before applying them to IEM has already been stressed. Beyond that, however, the use of light microscopic methods to provide an overall context followed by EM to flesh out the details at high resolution is a tried and true formula that usually provides ample rewards for the extra effort.

Though correlative immunomicroscopy often involves employment of two or more entirely disparate detection/staining systems, it should be noted that many of the ultrastructural labels discussed in this chapter can be detected with a variety of microscopes. Colloidal gold particles can be imaged at the light microscopic level with darkfield illumination, epi-illumination, or video-enhanced microscopes equipped for differential interference contrast, and by a range of electron microscopes, including conventional transmission and scanning microscopes and high-voltage instruments, the latter of which can be used to image whole cells without sectioning.[66] Silver-enhanced gold labels can be visualized using a conventional light microscope. As noted previously, conjugates employing both gold and fluorescent tags for combined immunofluorescence and electron microscopic detection have also been developed, and immunoperoxidase reaction products can be rendered visible in the electron microscope by heavy metal staining. The reader is encouraged to pick one of these methods, find an electron microscope, and have a look.

REFERENCES

1. Polak, J.M. and Varndell, I.M., Eds., *Immunolabelling for Electron Microscopy*, Elsevier, New York, 1984.
2. Griffiths, G., *Fine Structure Immunocytochemistry*, Springer-Verlag, Berlin, 1993.
3. Hayat, M.A., Ed., *Colloidal Gold: Principles, Methods, and Applications*, vol. 1–3, Academic Press, San Diego, 1989.
4. Hayat, M.A., *Microscopy, Immunohistochemistry, and Antigen Retrieval Methods for Light and Electron Microscopy*, Kluwer Academic/Plenum Publishers, New York, 2002.
5. Morel, G., Ed., *Hybridization Techniques for Electron Microscopy*, CRC Press, Boca Raton, 1993.
6. Kok, L.P. and Boon, M.E., *Microwave Cookbook for Microscopists. Art and Science of Visualization*, 3rd ed. Coulomb Press Leyden, Leiden, 1992.
7. Jasani, B. et al., Dinitrophenyl (DNP) hapten sandwich staining (DHSS) procedure. A 10 year review of its principle reagents and applications, *J. Immunol. Methods*, 150, 193, 1992.
8. Pathak, R.K. and Anderson, R.G.W., Use of dinitrophenyl IgG conjugates: immunogold labeling of cellular antigens on thin sections of osmicated and Epon-embedded specimens, in *Colloidal Gold. Principles, Methods, and Applications*, vol. 3, Hayat, M. A., Ed., Academic Press, San Diego, 1991.
9. Amato, P.A. and Taylor, D.L., Probing the mechanism of incorporation of fluorescently labeled actin into stress fibers, *J. Cell Biol.*, 102, 1074, 1986.
10. Ishida-Yamamoto, A. et al., Electron microscopic *in situ* DNA nick end-labeling in combination with immunoelectron microscopy, *J. Histochem. Cytochem.*, 47, 711, 1999.
11. Cenacchi, G. et al., *In situ* hybridization at the ultrastructural level: localization of cytomegalovirus DNA using digoxigenin labelled probes, *J. Submicrosc. Cytol. Pathol.*, 25, 341, 1993.
12. Muller-Hocker, J. *et al.*, The application of a biotin-anti-biotin gold technique providing a significant signal intensification in electron microscopic immunocytochemistry: a comparison with the ultrasmall immunogold silver staining procedure, *Histochem. Cell Biol.*, 109, 119, 1998.

13. Childs, G.V. and Unabia, G., Application of a rapid avidin-biotin-peroxidase complex (ABC) technique to the localization of pituitary hormones at the electron microscopic level, *J. Histochem. Cytochem.,* 30, 1320, 1982.

14. Sternberger, L.A. et al., The unlabeled antibody enzyme method of immunohistochemistry: preparation and properties of soluble antigen-antibody complex (horseradish peroxidase-antihorseradish peroxidase) and its use in identification of spirochetes, *J. Histochem. Cytochem.,* 18, 315, 1970.

15. Slot, J.W. and Geuze, H.J., A new method of preparing gold probes for multiple-labeling cytochemistry, *Eur. J. Cell Biol.,* 38, 87, 1985.

16. Horisberger, M. and Rosset, J., Colloidal gold, a useful marker for transmission and scanning electron microscopy, *J. Histochem. Cytochem.,* 25, 295, 1977.

17. Slot, J.W. and Geuze, H.J., Sizing of protein A-colloidal gold probes for immuno-electron microscopy, *J. Cell Biol.,* 90, 533, 1981.

18. Horisberger, M., Colloidal gold: a cytochemical marker for light and fluorescent microscopy and for transmission and scanning electron microscopy, in *Scanning Electron Microscopy,* vol. 2, Johari, O., Ed., SEM Inc., AMF O'Hare, Chicago, 1981.

19. Horisberger, M., Electron-opaque markers: a review, in *Immunolabeling for Electron Microscopy,* Polak, J.M. and Varndell, I. M., Eds., Elsevier, Amsterdam, 1984.

20. Handley, D.A., Methods of synthesis of colloidal gold, in *Colloidal Gold. Principles, Methods, and Applications,* vol. 1, Hayat, M. A., Ed., Academic Press, San Diego, 1989.

21. Molday, R.S. and Maher, P., A review of cell surface markers and labelling techniques for scanning electron microscopy, *Histochem. J.,* 12, 273, 1980.

22. Albrecht, R.M. and Hodges, G.M., Eds., *Biotechnology and Bioapplications of Colloidal Gold,* Scanning Microscopy International, Chicago, 1987.

23. Bullock, G.R. and Petrusz, P., Eds., *Techniques in Immunocytochemistry,* vol. 3, Academic Press, New York, 1983.

24. Namork, E., Double labeling of antigenic sites on cell surfaces imaged with backscattered electrons, in *Colloidal Gold. Principles, Methods, and Applications,* vol. 3, Hayat, M. A., Ed., Academic Press, New York, 1991.

25. Eskelinen, S. and Peura, R., Location and identification of colloidal gold particles on the cell surface with a scanning electron microscope and energy dispersive analyzer, *Scanning Microsc.,* 2, 1765, 1988.

26. Thimonier, J. et al., Thy-1 immunolabeled thymocyte microdomains studied with the atomic force microscope and the electron microscope, *Biophys. J.,* 73, 1627, 1997.

27. Avci, R. et al., Comparison of antibody-antigen interactions on collagen measured by conventional immunological techniques and atomic force microscopy, *Langmuir,* 20, 11053, 2004.

28. Baschong, W. and Stierhof, Y.D., Preparation, use, and enlargement of ultrasmall gold particles in immunoelectron microscopy, *Microsc. Res. Tech.,* 42, 66, 1998.

29. Hainfeld, J.F. and Furuya, F.R., A 1.4-nm gold cluster covalently attached to antibodies improves immunolabeling, *J. Histochem. Cytochem.,* 40, 177, 1992.

30. Hainfeld, J.F. and Powell, R.D., New frontiers in gold labeling, *J. Histochem. Cytochem.,* 48, 471, 2000.

31. Hainfeld, J.F., A small gold-conjugated antibody label: improved resolution for electron microscopy, *Science,* 236, 450, 1987.

32. Malecki, M. *et al.*, Molecular immunolabeling with recombinant single-chain variable fragment (scFv) antibodies designed with metal-binding domains, *Proc. Natl. Acad. Sci. U. S. A.,* 99, 213, 2002.

33. Burry, R.W., Vandre, D.D., and Hayes, D.M., Silver enhancement of gold antibody probes in pre-embedding electron microscopic immunocytochemistry, *J. Histochem. Cytochem.*, 40, 1849, 1992.

34. Danscher, G. et al., Trends in autometallographic silver amplification of colloidal gold particles, in *Immunogold Silver Staining: Principles, Methods and Applications*, Hayat, M. A., Ed., CRC Press, New York, 1995.

35. Holgate, C.S. et al., Immunogold-silver staining: new method of immunostaining with enhanced sensitivity, *J. Histochem. Cytochem.*, 31, 938, 1983.

36. Bienz, K., Egger, D., and Pasamontes, L., Electron microscopic immunocytochemistry. Silver enhancement of colloidal gold marker allows double labeling with the same primary antibody, *J. Histochem. Cytochem.*, 34, 1337, 1986.

37. Scopsi, L. et al., Silver-enhanced colloidal gold probes as markers for scanning electron microscopy, *Histochemistry*, 86, 35, 1986.

38. Goode, D. and Maugel, T.K., Backscattered electron imaging of immunogold-labeled and silver-enhanced microtubules in cultured mammalian cells, *J. Electron Microsc. Tech.*, 5, 263, 1987.

39. Van de Kant, H.J. et al., A rapid immunogold-silver staining for detection of bromodeoxyuridine in large numbers of plastic sections, using microwave irradiation, *Histochem. J.*, 22, 321, 1990.

40. Dutton, A.H., Tokuyasu, K.T., and Singer, S.J., Iron-dextran antibody conjugates: general method for simultaneous staining of two components in high-resolution immunoelectron microscopy, *Proc. Natl. Acad. Sci. U. S. A.*, 76, 3392, 1979.

41. Geiger, B. et al., Immunoelectron microscope studies of membrane-microfilament interactions: distributions of alpha-actinin, tropomyosin, and vinculin in intestinal epithelial brush border and chicken gizzard smooth muscle cells, *J. Cell Biol.*, 91, 614, 1981.

42. Romano, E.L. and Romano, M., Historical aspects, in *Immunolabeling for Electron Microscopy*, Polak, J.M. and Varndell, I.M., Eds., Elsevier, Amsterdam, 1984.

43. Krieg, R. and Halbhuber, K.J., Recent advances in catalytic peroxidase histochemistry, *Cell. Mol. Biol. (Noisy-le-grand)*, 49, 547, 2003.

44. Chevalier, J. et al., Biotin and digoxigenin as labels for light and electron microscopy *in situ* hybridization probes: where do we stand?, *J. Histochem. Cytochem.*, 45, 481, 1997.

45. Bendayan, M., Double immunocytochemical labeling applying the protein A-gold technique, *J. Histochem. Cytochem.*, 30, 81, 1982.

46. Zondervan, P.E. et al., Microwave-stimulated incubation in immunoelectron microscopy: a quantitative study, *Histochem. J.*, 20, 359, 1988.

47. Bendayan, M. and Zollinger, M., Ultrastructural localization of antigenic sites on osmium-fixed tissues applying the protein A-gold technique, *J. Histochem. Cytochem.*, 31, 101, 1983.

48. Spehner, D. et al., Enveloped virus is the major virus form produced during productive infection with the modified vaccinia virus Ankara strain, *Virology*, 273, 9, 2000.

49. Hopley, J.F. and Doane, F.W., Development of a sensitive protein A-gold immunoelectron microscopy method for detecting viral antigens in fluid specimens, *J. Virol. Methods*, 12, 135, 1985.

50. Kjeldsberg, E., Demonstration of calicivirus in human faeces by immunosorbent and immunogold-labelling electron microscopy methods, *J. Virol. Methods*, 14, 321, 1986.

51. Muller, G. and Baigent, C.L., Antigen controlled immuno diagnosis—'ACID test', *J. Immunol. Methods*, 37, 185, 1980.

52. Hayat, M.A. and Miller, S.E., *Negative Staining: Applications and Methods*, McGraw-Hill, New York, 1990.
53. Nicol, A. et al., Labeling of structural elements at the ventral plasma membrane of fibroblasts with the immunogold technique, *J. Histochem. Cytochem.*, 35, 499, 1987.
54. Rutter, G. et al., Demonstration of antigens at both sides of plasma membranes in one coincident electron microscopic image: a double-immunogold replica study of virus-infected cells, *J. Histochem. Cytochem.*, 36, 1015, 1988.
55. Tokuyasu, K.T., Immunochemistry on ultrathin frozen sections, *Histochem. J.*, 12, 381, 1980.
56. Griffiths, G. et al., Immunoelectron microscopy using thin, frozen sections: application to studies of the intracellular transport of Semliki Forest virus spike glycoproteins, *Methods Enzymol.*, 96, 466, 1983.
57. Tokuyasu, K.T., Immuno-cryoultramicrotomy, in *Immunolabeling for Electron Microscopy*, Polak, J.M. and Varndell, I.M., Eds., Elsevier, Amsterdam, 1984.
58. Tokuyasu, K.T., Use of poly(vinylpyrrolidone) and poly(vinyl alcohol) for cryoultramicrotomy, *Histochem. J.*, 21, 163, 1989.
59. Carlemalm, E., Garavito, R.M., and Villiger, W., Resin development for electron microscopy and an analysis of embedding at low temperature, *J. Microsc.*, 126, 123, 1982.
60. Carlemalm, E. and Villiger, W., Low temperature embedding, in *Techniques in Immunocytochemistry*, vol. 4, Bullock, G.R. and Petrusz, P., Eds., Academic Press, New York, 1989.
61. Wolosewick, J.J., The application of polyethylene glycol (PEG) to electron microscopy, *J. Cell Biol.*, 86, 675, 1980.
62. Wolosewick, J.J., Cell fine structure and protein antigenicity after polyethylene glycol processing, in *The Science of Biological Specimen Preparation for Microscopy and Microanalysis*, Barnard, T., Revel, J.P., and Hagg, G., Eds., SEM Inc., AMF O'Hare, Chicago, 1984.
63. Kondo, H., Ultrastructural localization of actin in the intermediate lobe of rat hypophysis, *Biol. Cell*, 60, 57, 1987.
64. Acetarin, J.D., Carlemalm, E., and Villiger, W., Developments of new Lowicryl resins for embedding biological specimens at even lower temperatures, *J. Microsc.*, 143, 81, 1986.
65. Nicolas, M.T., Bassot, J.M., and Nicolas, G., Immunogold labeling of luciferase in the luminous bacterium Vibrio harveyi after fast-freeze fixation and different freeze-substitution and embedding procedures, *J. Histochem. Cytochem.*, 37, 663, 1989.
66. Goodman, S.L., Park, K., and Albrecht, R.M., A correlative approach to colloidal gold labeling with video-enhanced light microscopy, low-voltage scanning electron microscopy and high-voltage electron microscopy, in *Colloidal Gold. Principles, Methods, and Applications*, vol. 3, Hayat, M.A., Ed., Academic Press, New York, 1991.

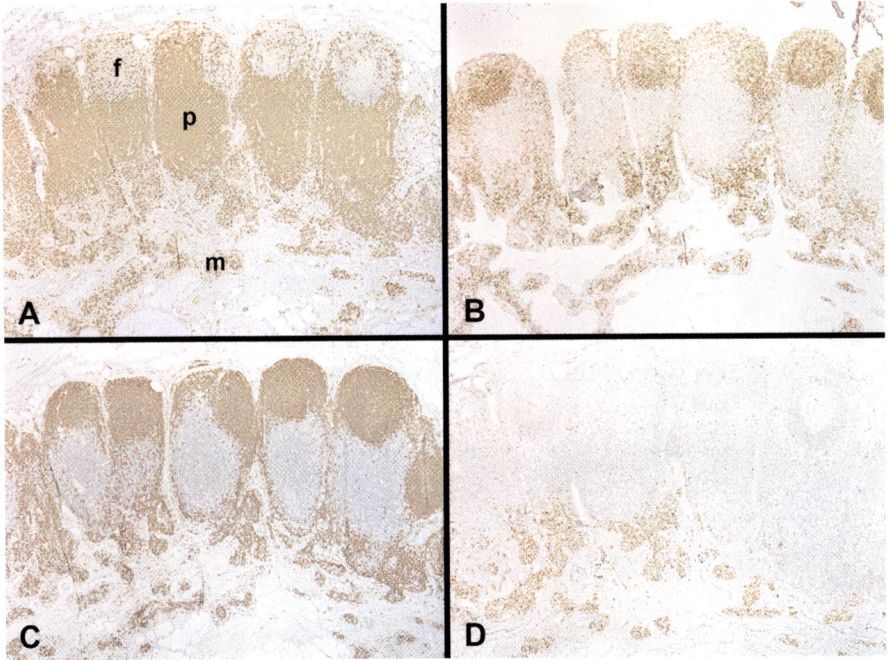

FIGURE 11.1 Knowing the location of the antigen of interest is paramount to determining the quality of the immunostaining. In this lymph node, four antigens to CD3 (A), CD79a (B), CD20 (C), and MUM1 (D) were compared. The follicular areas (f) are mostly positive for CD79a and CD20 with a few positive cells for CD3 and fewer for MUM1. The paracortex (p) is mostly populated by CD3-positive cells with few positive cells for the other markers. The medulla (m) contains mostly CD79a, CD20, and MUM1-positive cells with fewer T cells. Immunoperoxidase-DAB, hematoxylin counterstain.

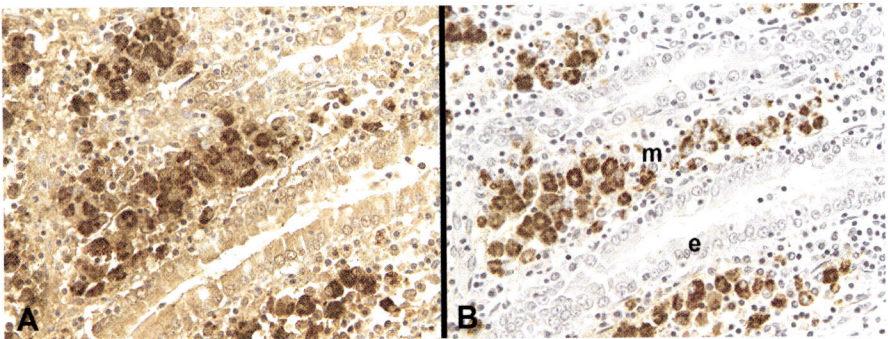

FIGURE 11.2 From the intestine of a cow with paratuberculosis produced by *Mycobacterium avium* subsp. *paratuberculosis*. Optimal dilution of the primary antibody is critical. Insufficiently diluted primary antibody (A) results in nonspecific staining in intestinal epithelium and lamina propria. The optimal dilution (B) detects only antigen in infected macrophages (m) without staining epithelium (e). Immunoperoxidase-DAB, hematoxylin counterstain.

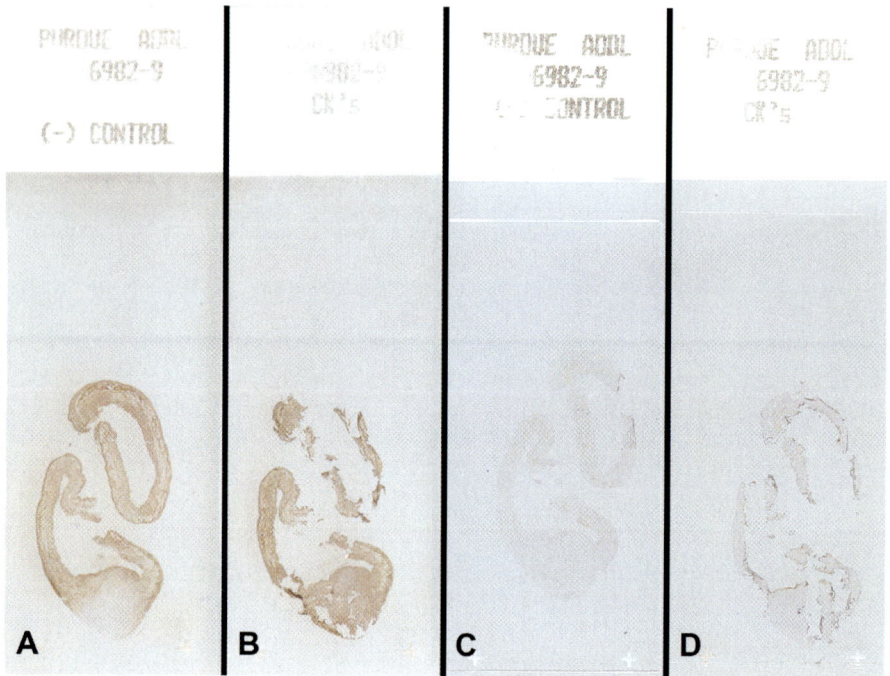

FIGURE 11.3 Multiple-step methods are more sensitive than direct methods but may cause unexpected reactions. The urinary bladder from a goat was stained with monoclonal antibody to cytokeratins. Figures A and B were stained with LSAB+ link antibody that recognizes rabbit, mouse, and goat primary antibodies. The identical diffuse staining in the section with the negative reagent control section (A) and the anti-cytokeratin antibody (B) is caused by nonspecific binding of the link antibody to endogenous (tissue) goat immunoglobulins. To avoid this problem a detection system that does not recognize goat immunoglobulins (ENVISION+) was used in images C and D. The negative reagent control (C) does not show staining, and the positive reagent tissue (D) has distinct staining in the expected areas of the tissue. Immunoperoxidase-DAB, hematoxylin counterstain.

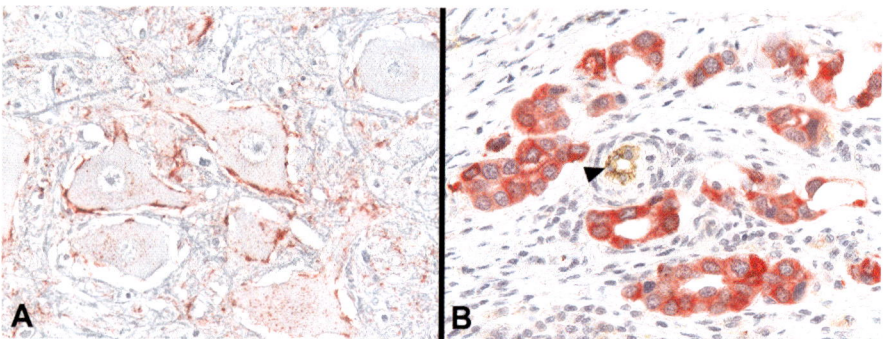

FIGURE 11.4 (A) Immunoalkaline phosphatase staining for prion protein in the brainstem of a sheep. The staining is fuchsia against a blue counterstain. (B) Double immunostaining for COX-2 (fuchsia) and CD31 (brown). Transitional cell carcinoma cells expressing COX-2 are detected with alkaline phosphatase, and endothelial cells expressing CD31 (arrowhead) are detected with peroxidase.

13 Flow Cytometry

Kristi R. Harkins and M. Elaine Kunze

CONTENTS

13.1 INTRODUCTION

Flow cytometry is a process in which light-based measurements are made while cells in suspension pass single file through a measuring apparatus focused onto a fluid stream. This measured light can be scattered by the cell or induced fluorescence from the cell as it passes through the fluid stream and past the illumination source.[1] Antibody detection is one of the most popular applications for flow cytometry. The antibodies are covalently labeled with a signal moiety, such as a fluorescing dye molecule. Antibodies can be purchased ready to use or flourescently labeled by the end user. A variety of antibody labeling kits are on the market today as well as documented approaches to labeling cells with antibody for discerning functional or phenotypic information from homogeneous and heterogeneous populations. The method of choice will depend upon the availability of commercial antibodies, the concentration of the antigen being detected, and the capabilities of the cytometric technology available to the end user.

This chapter is written with the understanding that the individual has generated or obtained an antibody and characterized it for protein binding (e.g., whole-cell assay, enzyme-linked immunosorbent assay, or blotting assay). This introduction is organized to provide the reader with a thorough understanding of the issues to consider in designing a flow cytometric assay before performing the experiments. A number of well-characterized commercial kits provide reliable assay platforms. This chapter will focus on methods that work equally well for noncommercialized antibody use in flow cytometry. These are the critical issues to be considered in developing the final assay method.

13.1.1 AVAILABLE TECHNOLOGY

Flow cytometric technology within hospital and clinical pathology labs, corporations (e.g., pharmaceutical companies), and academic institutions is typically found in a core laboratory. These core labs provide users with access to trained operators who can offer suggestions on fluorochromes that are compatible with the systems on site and advice on the proper controls. Several bench-top instruments are on the market and differ primarily in the number of detection parameters, excitation and emission detection options available to the user. Some of them are listed in Table 13.1.

Some of the available dyes (fluorochromes) that are compatible with most flow cytometry excitation and emission optical setups are defined in Table 13.2. Many vendors offer these fluorochromes in convenient conjugation kit formats or as pure products. Fluorochromes can have inherent properties that affect use and brightness (intensity of the emitted photons). A good fluorochrome has strong absorbance at the exciting wavelength, a well-separated emission spectra (Stokes

TABLE 13.1
Benchtop Cytometers Typically Available in Core Labs

Vendor	Device	Excitation Wavelength	# of Parameters	Wavelength Range
Beckman-Coulter	XL	488 nm	4	Green to far red
	FC500	488, 633	5	Green to far red
	Cell Lab Quanta	488 nm, arc 546 nm or 365 nm	2	Violet to far red
Beckton Dickinson	Facs Calibur	488, 633	5	Green to far red
	Facs Canto	488, 633	6	Green to far red
Dako Cytomation	Cyan	488, 635	9	
Guava		546 nm	2	Red to far red
Partec	Cyflow Space	488 nm, 635	8	Green to far red
	CyflowSL	Single laser choice	1–5	Variable

TABLE 13.2
Common Dyes Used in Flow Cytometry

488 nm Excitation				633 nm Excitation
530 nm	575 nm	605 nm	675 nm	>633 nm
Fluorescein[2]	Phycoerythrin[3]	PE-Texas Red[5]	PE-Cy5[5]	Allophycocyanine[3]
(Fitc)	(PE)	(ECD)	(Cychrome)	(APC)
Alexa 488[1]			PerCP[3]	Cy5[4]
Cy2[4]			PE-Cy7[5]	APC-Cy7[5]
				(PharRed)
			PE-Cy5.5[5]	Alexa 660[1]
				Alexa 680[1]

[1]Invitrogen, [2]Sigma, [3]Prozyme, Inc, [4]Amersham Life Science, [5]These tandem dyes can be produced using the following protocol: M. Roederer, Conjugation of monoclonal antibodies (August, 2004). http://www.drmr.com/adcon/

shift), high quantum yield, and photostability. Always try to select the brightest fluorochrome for cells that express antigens at low intensity. Choose them in order of brightness:

PE > PE-Cy5 tandems > PE-Texas Red tandems > APC > FITC > PerCp

Refer to the indicated references here and to Chapters 9 and 11 for the appropriate conjugation methods.[2,3] A web site (http://www.drmr.com/abcon/) with spectra and conjugation methods that are outlined for monoclonal antibodies with flow cytometric applications in mind.

13.1.2 ANTIBODY SELECTION

Both monoclonal and polyclonal antibodies can be used in flow cytometry. Monoclonal antibodies are prepared from a single B-cell clone fused with a melanoma cell line to form hybridoma cells and recognize a single epitope of the antigen. While they are expensive to produce, they provide a renewable source of antibody with reproducible specificity and affinity from batch to batch. They can be of a specific isotype (e.g., immunoglobulin [Ig]G$_1$) and allow the selection of valid isotype controls to standardize the assay performance.

Polyclonal antibodies also have clear advantages for use. They are much less expensive to produce than monoclonal antibodies. Large quantities of high-affinity polyclonal antibodies (~10 mg/ml) can be produced from the serum of an immunized animal 2–3 months after the initial immunization. Because polyclonal antibodies contain an entire antigen-specific antibody population, they are polyvalent having more binding sites per protein. The greater number of binding sites can yield a brighter flow cytometric signal. This same property can also raise issues with the specificity of the antibody that can be alleviated by affinity purifying the antibody before use. The ideal negative control here is preimmune serum taken from the same animal used in the immunization process before immunization.

13.1.3 STARTING CONSIDERATIONS

Some basic rules must be followed for optimal results. First, it is critical that the number of cells used in the staining process be between 10^5 and 10^6 cells. This provides sufficient cells for analysis and insures that the antibody concentration does not become limiting because the number of cells becomes too high. Second, the cell-labeling process can be performed in a variety of tube or plate formats. Make sure the appropriate centrifuge with adapters operating at the ideal g force is available to handle the cell centrifugation during the washing steps.[4]

Three approaches must be considered when labeling cells for flow cytometric analysis. If the antibody is directly conjugated to the fluorochrome, it can be added to the cell mixture, followed by incubation, a wash step (optional), and analysis. This method provides the advantage of speed and added versatility when combining mouse monoclonal antibodies each with a different color and antigen specificity in one test tube for flow cytometry. If the antibody is conjugated to biotin, then after incubation and a wash step, a streptavidin-dye conjugate is added to label the bound primary antibody. This method requires a wash step, but allows amplification of the fluorescent signal with the multiple streptavidin binding potential of each biotin. If the antibody is used in an unconjugated form, then a secondary antibody must be added after incubation and washing to label the primary antibody in a species-specific manner (e.g., goat anti-rabbit fluorescein conjugate is used to label an antigen-specific rabbit polyclonal antibody). This method provides the advantage of economy (no need to conjugate the antibody) and the amplification of fluorescence with multiple primary binding sites for the secondary. Also, a single secondary antibody conjugated to a specific fluorochrome can be used with a variety of different primary antibodies increasing the flexibility of dye combinations for future experiments.

Antibody concentration, time, and temperature are three critical factors of the staining process. Antibody labeling will occur faster at higher antibody concentrations and reach saturation faster at higher temperatures. However, most labeling procedures are performed at 4°C for 30 min. A process known as capping or internalization of extracellular bound antibodies can be reduced by labeling live cells at lower temperatures and including sodium azide (0.1% w/v) in the labeling solution.[5]

Cells are often preserved or fixed after staining in a formaldehyde solution (1–4% w/v). Fixation cross-links proteins and provides stability of the antibody binding and cell structure, allowing the researcher to collect and prepare samples in time-course studies, and also provides some flexibility for appointments on usually busy core lab flow cytometers. Samples can typically be held for several days at 4°C in the dark before analysis with no effect on the analysis results.

13.1.4 CONTROLS

Labeling and experimental controls are vital to the proper interpretation of the flow cytometric data. Examples of labeling controls include: (1) cells that do not receive antibody are used to determine if the cells exhibit autofluorescence; (2) for indirect labeling methods, cells exposed to the secondary fluorochrome-conjugated reagent are used to determine the degree of nonspecific binding of the secondary antibody, and (3) for direct labeling methods, cells exposed to a fluorochrome-conjugated monoclonal antibody of similar species and isotype but raised to an antigen (typically a hapten) that should not be present in or on the cells of interest. Isotype antibodies should be used at the same protein concentration and have a similar number of fluorochrome molecules per protein ratio as the antibody of interest in the assay. Typically in a flow cytometry experiment the detector sensitivity is adjusted to place these controls in the lowest decade of the fluorescence histogram plot.

For experiments involving multicolor multiple antigen detection schemes, single-color stained negative and positive controls (i.e., cells that are known to express the antigen of interest) should be included to enable proper electronic compensation of the flow cytometer.[6] Dr. Mario Roederer at the National Institute of Health has created a web site covering compensation in great detail (see http://www.drmr.com/compensation/index.html).

13.1.5 SPECIFICITY

The specificity of the antibody is critical to the success of the assay. When working with whole cells as in flow cytometry, the antigen will be presented to the antibody in a specific format (e.g., membrane-bound or -embedded, intracellular, extracellular). The selection of an antibody should include, whenever possible, an antigen-positive and antigen-negative population during the titration process.

Another factor that affects specificity is the presence of Fc receptors on the surface of cells.

Monocytes (as a subset of cells in whole blood) bind many antibodies through the non-antigen-specific Fc domain. The use of Fab or $F(ab)_2'$ fragments in the assay is one way to prevent this form of nonspecific binding. Also blocking agents can be included in the cell solutions (animal sera or commercial Fc block) to block these sites before the addition of the assay antibody. Commercial blocking agents bind to the Fc receptors before primary antibody staining can minimize this problem (such as Becton Dickinson catalog number 553141/553142). In the case of indirect labeling methods, it is equally important to verify that the secondary reagents do not bind to any of the Fc blocking agents.

Finally, dead cells bind antibodies nonspecifically and can cause misinterpretation of data and potential false positive detection. There are three very common methods for eliminating dead cells from flow cytometry data. For very uniform cell populations, light scatter alone may be enough to eliminate the dead cells from the distribution. Dead cells generally have a lower level of forward scatter and a slightly higher level of side scatter. In less uniform populations, propidium iodide can be added. This dye passes through compromised cell membranes, and the dead cells, which fluoresce red, can be eliminated by excluding these red cells from the analysis.[7] Cells cannot be fixed after labeling for this red dye exclusion assay. A related dye, ethidium monoazide, covalently links the azide group to nuclear histones by light exposure, and excess dye can then be removed before antibody labeling. In contrast to propidium iodide, the cell suspension can then be fixed without fear of excluding the cells that were viable before fixation.[8]

13.1.6 INSTRUMENT DISPLAY AND QUALITY CONTROL

The operator should routinely analyze a set of standard beads that typically contains multiple intensity levels of a specific fluorochrome (see Table 13.1). The operator uses this information to standardize and monitor the device performance from a qualitative and quantitative viewpoint. Note the histogram in Figure 13.1 showing data output on a flow cytometer with a base 10-log scale of increasing fluorescence intensity on the X-axis and particle count on the Y-axis. This histogram was collected while using an excitation light of 488 nm and an emission of 525 nm and accumulated for a total count of 5000 events. Each peak on the histogram represents a bead population with a negative (lower left) or positive number of molecules of equivalent soluble fluorochrome (MESF) concentration. Software programs allow the user to define analysis gates (e.g., M1) that have a lower and upper limit and can be drawn around each peak. Statistics include a median channel of fluorescence intensity as well a number and percentage of particles or cells found within the gated region. The operator then plots the median channel for each peak versus the MESF value provided by the manufacturer to determine the lower limit of detection as well as the linearity of the output of the flow cytometer.[9] Also note the graph in Figure 13.2.

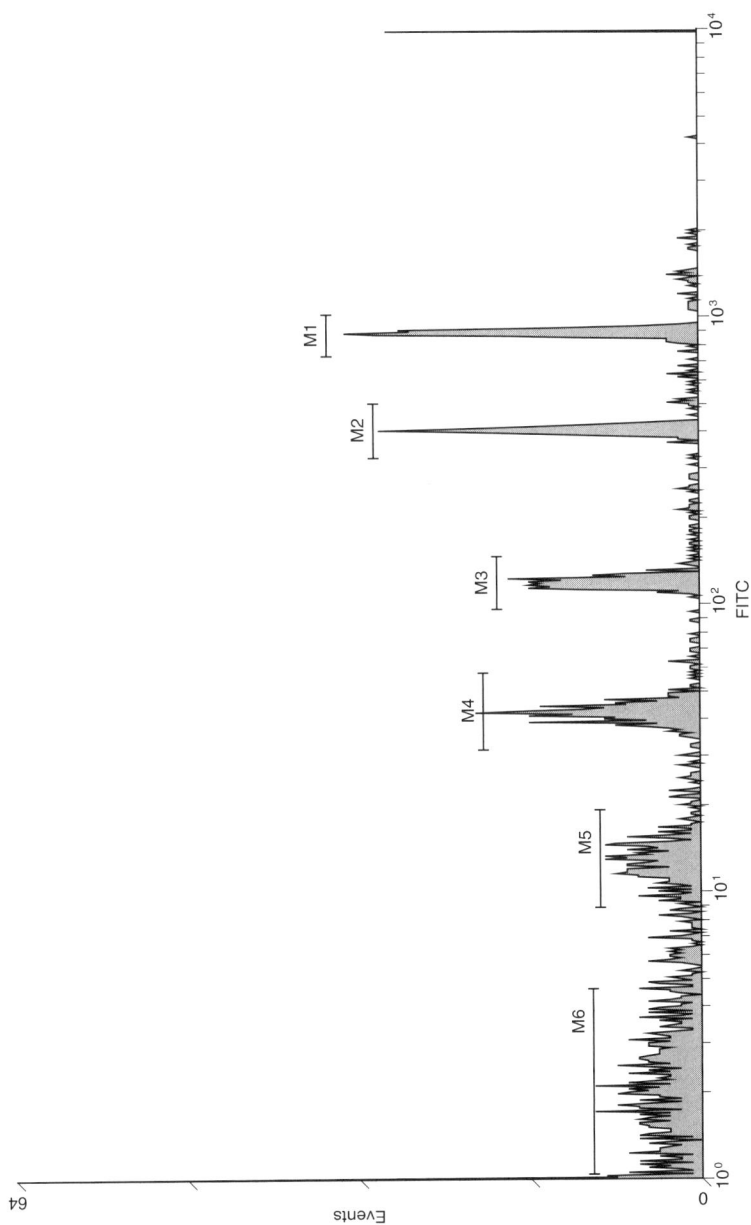

FIGURE 13.1 A mixture of six fluorospheres (Rainbow Calibration Particles RCP-30–5, Spherotech, Inc. Libertyville, IL) each with different amounts of dye. Analyzing this mixture under the same conditions as samples ensures consistency across flow cytometry platforms and laboratories. Each gate (labeled M1–M6) provides analysis to determine median channel number for fluorescence brightness.

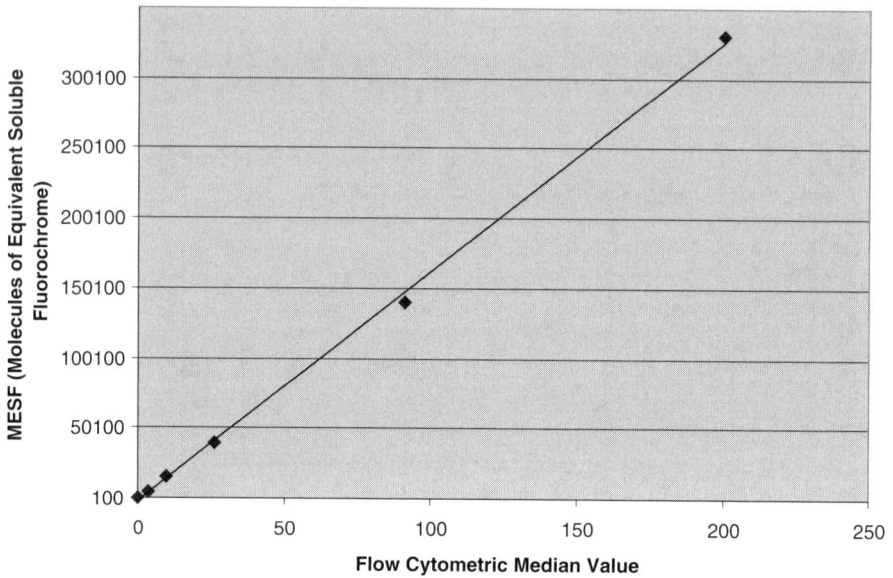

FIGURE 13.2 A typical calibration curve for the Rainbow particle shown in Figure 13.1. Each population of spheres consists of uniform beads with known number of fluorochromes attached. By plotting the median from each population of beads vs. the molecules of equivalent soluble fluorochrome (MESF in this case is for fluorescein), calibration is possible for most standard fluorescent dyes.

13.2 IMMUNOPHENOTYPING

As noted in the introduction to this chapter, the researcher should now have selected the antibody and fluorochrome to be used that is compatible with the available flow cytometer, determined the appropriate antigen-positive and antigen-negative cells for analysis, and identified the negative controls to include in the study. The researcher should also be aware of the negative effect dead cells can have on the assay results and other potential causes of high fluorescence in the negative-control samples.

13.2.1 ANTIBODY TITRATION

Titration is the first critical step in the process of setting up a flow cytometry assay to ensure that the antibody concentration is not limiting or too high in the labeling process. Proper antibody concentration enables the researcher to obtain potential quantitative information of the antigen level within a population and provides optimal separation of a subset of cells expressing a specific antigen within a heterogenous population. Antibody titration should be performed for every new antibody lot and for new batches of previously titered antibodies. Make sure to titer the antibodies at the same temperature and for the same amount of time as planned in your experimental design.

Typically, 3 μg of antibody per 100 μl of 10^6 cells is the concentration at which nonspecific binding of all IgG initially becomes detectable, and so this concentration is a convenient starting point for titration. Antibody concentration should be high enough to achieve effective equilibrium between bound and unbound antibody in a reasonable time. Variation in the staining volume is not critical within the accuracy range of the pipetting technology used. Experimental costs can be reduced by keeping the staining volume low so less antibody is used.

13.2.1.1 Materials Required

Target cells that express the epitope
Control cells that do not express the epitope
Specific primary antibody
Isotype control exactly matched in concentration to the specific antibody
Phosphate-buffered saline (PBS), pH 7.2 (equilibrated at 4°C)
PBS+ (0.1% [w/v] sodium azide and 1% [v/v] fetal bovine serum or 0.1% [w/v] bovine serum albumin [BSA], or 200 µg/ml goat Ig) in PBS, filtered through a 0.2–2 µm filter and equilibrated at 4°C. The addition of sodium azide is to reduce the capping phenomenon and the serum or purified goat Ig fraction provides unlabeled and nonspecific IgG to bind and block any open Fc receptors. The excess proteins present in the serum and BSA are used to reduce the potential nonspecific binding that may occur in these assay types.

13.2.1.2 Time Required

60–90 min

13.2.1.3 Special Equipment

Centrifuge
Cell counting device (e.g., Coulter counter)
Micropipettors (accurate volume handling from 10 to 200 µl and 1000 µl)
Flow cytometer

13.2.1.4 Procedure

1. Determine the concentration of specific primary antibody and control antibody in the stock solutions and centrifuge each stock for 10 min at 15,000 g at 4°C. Leave the aggregated antibody in the pellet and collect the supernatant. (Antibodies have a tendency to form aggregates during storage that can cause problems with the labeling process. These aggregates should be removed by centrifugation prior to dilution for use.)
2. Prepare 30 µl containing 9 µg of primary antibody (300 µg/ml) in PBS. Prepare six tubes with 1:4 serial dilutions (10 µl antibody plus 30 µl PBS) for the specific antibody.
3. Prepare 30 µl containing 9 µg of control antibody (300 µg/ml) in PBS. Prepare a single 1:4 serial dilution (10 µl antibody plus 30 µl PBS) for the isotype antibody.

4. Prepare a target cell suspension (a mix of epitope-positive and -negative cells) containing 5×10^6 cells per ml in PBS+.
5. Add 10 µl of each specific antibody dilution to 100 µl aliquots of cell suspension in separate 12 × 75–mm tubes.
6. Add 10 µl the isotype control antibody dilution to a 100 µl aliquot of cell suspension in a 12 × 75–mm tube.
7. Prepare a control tube containing only a cell suspension (100 µl aliquot of cells) plus 10 µl of PBS+.
8. Incubate all samples for 15 min at 4^0C.
9. Add 1 ml of cold PBS. Vortex.
10. Centrifuge at 1500g for 3 min. Use a pipet to remove the liquid. Be careful not to disturb the pellet.
11. Add 0.5 ml of PBS and vortex.
12. Add 0.5 ml of formaldehyde fixative (2% v/v). Store fixed samples at 4°C and protect them from light until the analysis is complete ($\sim 5 \times 10^5$ cells).
13. Acquire 5000 gated cells for each sample using a flow cytometer.

This method defines titration for using a directly conjugated monoclonal antibody. When using a nonconjugated primary of a polyclonal nature, use the manufacturers recommended final titer plus 6× for the labeled secondary (substituting this dilution in step 3). Perform steps 1–10 as defined in the procedure above, substituting PBS for the isotype control antibody in step 6. The control sample will be a secondary only antibody control. After step 10, there is an additional step added with the addition 50 µl of PBS to all the cell pellets followed by 10 µl of the diluted sixfold recommended titer secondary antibody. Follow steps 8–13 as outlined previously.

13.2.1.5 Results

Regardless of the origin of the cells used for immunofluorescence labeling, the operator will measure how the cells within the sample scatter the excitation light. This scattered light is measured in two directions relative to the illumination source.[6] Forward light scatter is measured in front of the source and provides a relative measure of cell size. Orthogonal light scatter (also called side scatter or 90° light scatter) is measured at right angles or 90° to the light source and can provide information on the intracellular structure or granularity of the cells.

Accumulated light scatter data is presented in a two-parameter histogram in Figure 13.3 with increasing 90° scatter intensity on the x-axis and increasing forward scatter intensity on the y-axis. The z-axis defines cell number which in this dot density histogram can be interpreted by difference in the density of the dots on the graph. This data can be very useful in further defining the fluorescence data output of the system. Note in Figure 13.3, the presence of several populations with different scatter characteristics and the circle or lymphocyte gate drawn around the primary population of cells in the histogram. (Light scatter signals can shift with formaldehyde fixed cells. Also, dead cells may be distinguishable from live cells with some cell types becoming lower in forward scatter and higher in 90 degree scatter.) This gate defines the cells that will be analyzed for fluorescence in the graphs shown in Figure 13.4.

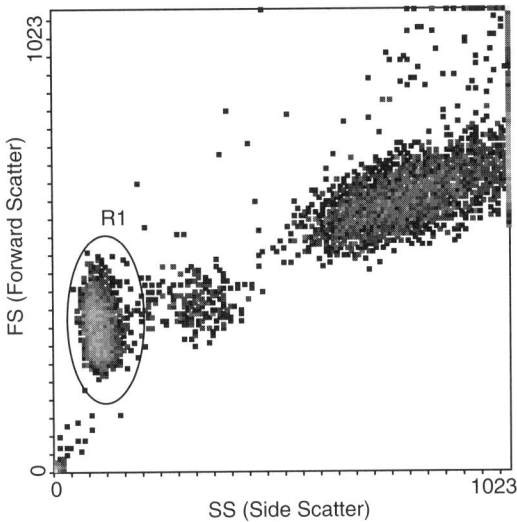

FIGURE 13.3 A typical two parameter scatter plot of mouse white blood cells showing a gate for subsequent analysis. This scattergram plots 10,000 events with 90° light scatter on the x-axis (cellular complexity) and forward angle light scatter on the y-axis (size and refractive index). A gate R1 is placed around lymphocytes.

Most flow cytometry data are collected as individual bits of stored data for each parameter measured for each trigger event (usually based on the presence of a cell) in a format known as listmode. This allows the researcher to replay the file data, set up gates, and combine parameters in one- and two-parameter histograms with analysis windows offline long after the cells have been analyzed on the flow cytometer. Instrument manufacturers provide software for analysis, and freeware packages are also available (e.g., WinMDI available as a download from http://facs.scripps.edu/software.html and used to generate the histogram for these figures).

The sample data in Figure 13.4 represent the titration of a fluorescein-labeled antibody specific to CD4 using a heterogeneous lymphocyte population of cells. The data in the histogram for the isotype control or the unstained sample are used to generate the analysis gates defined as negative and positive in the top control graph so that 99% of the negative cells fall in the negative gate and <1% in the positive. Remember that only cells within the lymphocyte scatter gate in Figure 13.3 are analyzed for fluorescence in Figure 13.4.

To determine the signal-to-noise ratio, the software analysis program determines the median channel fluorescence intensity for both positive (signal) cells and negative (noise) cells from the histograms for each of the six antibody dilution samples. Compute the signal-to-noise ratio by dividing the median channel fluorescence intensity value for the positive cells by that for the negative cells. Plot these values as a function of antibody dilution (Figure 13.5). The optimal titer is the one that generates the highest signal-to-noise ratio, because this provides the greatest

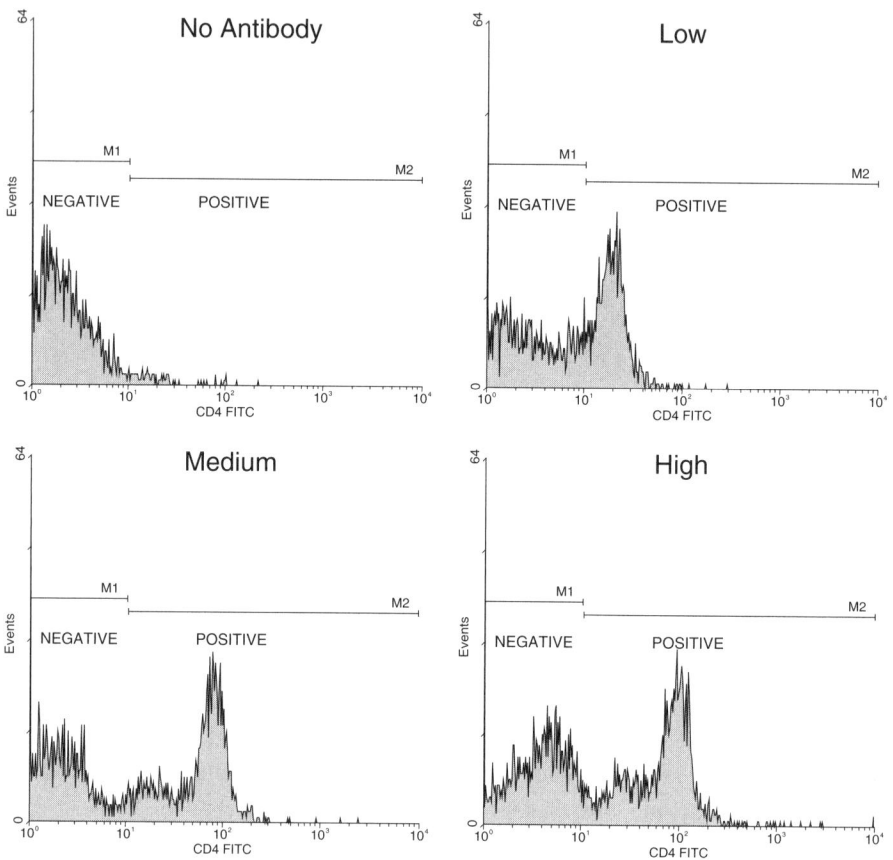

FIGURE 13.4 Titration of a fluorescein-labeled antibody specific to CD4, gated on lymphocytes. The top left graph shows control (negative cells) and the three remaining graphs depict increasing antibody concentrations. Note that in the highest concentration the negative cells are beginning to pick up a significant amount of fluorescence, a certain indication of nonspecific binding caused by excessive antibody.

discrimination between positive and negative cells, regardless of the absolute value of the fluorescence intensity.

13.2.1.6 Problems and Troubleshooting Tips

Variations in epitope density, cell concentrations, and staining procedures can affect the validity of antibody titrations. It is best to set up titration studies using very healthy viable cells that have the highest level of antigen density, make sure the cell concentration is determined with a measure of accuracy for each experiment and control the labeling method (time, temperature, and controls) in a consistent manner. Avoid using inappropriate target cell (cell lines, frozen cells, or fixed cells) during the titration process if they are not indicative of the cells to be used in the final

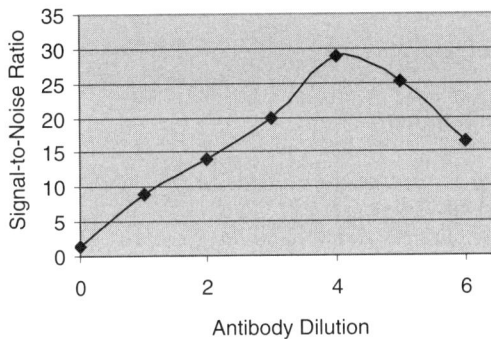

FIGURE 13.5 An Excel plot of the six CD8 antibody dilutions described in 13.2.1.4. The signal-to-noise ratio was computed by dividing the median channel fluorescence intensity for the positive cells (M2) by the median channel fluorescence intensity for the negative cells (M1). The optimal antibody concentration is found where the signal-to-noise ratio is the highest (in this case, dilution 4).

experiment. Low fluorescence intensity suggests a low number of antigen sites, so consider using a brighter fluorochrome if available.

When working with indirect labeling, use F(ab′) or F(ab) fragments for secondary antibodies. These should be affinity purified and should not react with the cellular epitope. By eliminating the Fc domain on these polyclonal antibodies, much of the background fluorescence associated with negative controls can be removed.

13.2.2 DIRECT LABELING METHOD

When performing multicolor labeling, directly conjugated antibodies can be added simultaneously, rather than sequentially. The following sample method demonstrates a direct three-color labeling scenario targeting T-lymphocyte subsets from a population of white blood cells purified from contaminating red blood cells. Each antibody should have been previously titered.

13.2.2.1 Materials Required

White blood cell suspension (determined cell count and gradient purified)
Anti-CD3 phycoerythrin (PE)
Anti-CD8 fluorescein (FITC)
Anti-CD4 CyChrome (Cy5)
Isotype control conjugate for PE, FITC, Cy5 exactly matched in protein concentration to the specific antibody
PBS, pH 7.2, equilibrated to 4°C
Wash buffer (PBS + 0.1% sodium azide and 1% fetal bovine serum or BSA) filtered to 0.22 μm and equilibrated to 4°C.
Formaldehyde (Polysciences, Warrington, PA; Cat. No. 18814, methanol free 16% w/v)

13.2.2.2 Time Required

60–90 min

13.2.2.3 Special Equipment

Centrifuge
Cell counting device
Micropipettors (accurate volume handling from 10–200 µl and 1000 µl)
Flow cytometer

13.2.2.4 Procedures

1. Harvest or isolate the cells to ensure a single-cell suspension in final suspension of cold wash buffer at an adjusted cell concentration of 5×10^5/ml.
2. Place 1 ml of the cell suspension into each of the 12×75 tubes preparing a tube for each single antibody (primary and isotype), a tube for all the antibodies, and a tube to be used as a negative control.
3. Centrifuge at 1500 g for 3 min. Use a pipet to remove the liquid. Be careful not to disturb the pellet. A slight amount of liquid can remain (~100 µl).
4. Add the appropriate amount of antibody conjugate or control sera or isotype as previously by titration.
5. Vortex and keep the tubes on ice for 15 min in the dark.
6. Add 1 ml of cold PBS. Vortex.
7. Centrifuge at 1500 g for 3 min. Use a pipet to remove the liquid. Be careful not to disturb the pellet.
8. Add 0.5 ml of PBS and vortex.
9. Add 0.5 ml of formaldehyde fixative (2% v/v). Store fixed samples at 4°C and protect them from light until analysis complete.
10. Acquire 10,000 gated cells for each sample using a flow cytometer.

13.2.2.5 Results

The flow cytometric data shown in Figure 13.6 are presented in the form of two-parameter histograms with independent emissions on the x- and y-axes. The z-axis defines cell count which is depicted as scaled changes in the dot density of each plot. Each sample is analyzed for FITC vs. PE, PE vs. CyChrome, and FITC vs. CyChrome. The single color samples are used to verify that fluorescence is present in the expected labeled population and to adjust electronic compensation for any of that color signal detected on the other color axes. The data in the histogram for the isotype control or the unstained sample are used to generate the analysis gates defined as negative and positive in the top control graph so that 99% of the negative cells fall in the negative gate and <1% in the positive. Note the lines that cross the x- and y-axes providing data on the number of cells within each of the four quadrants of the graph. The bottom three graphs provide the data of interest with information on cells that express only one of the three epitopes (upper left or lower right quadrants) and the cells that express two epitopes (upper right quadrant).

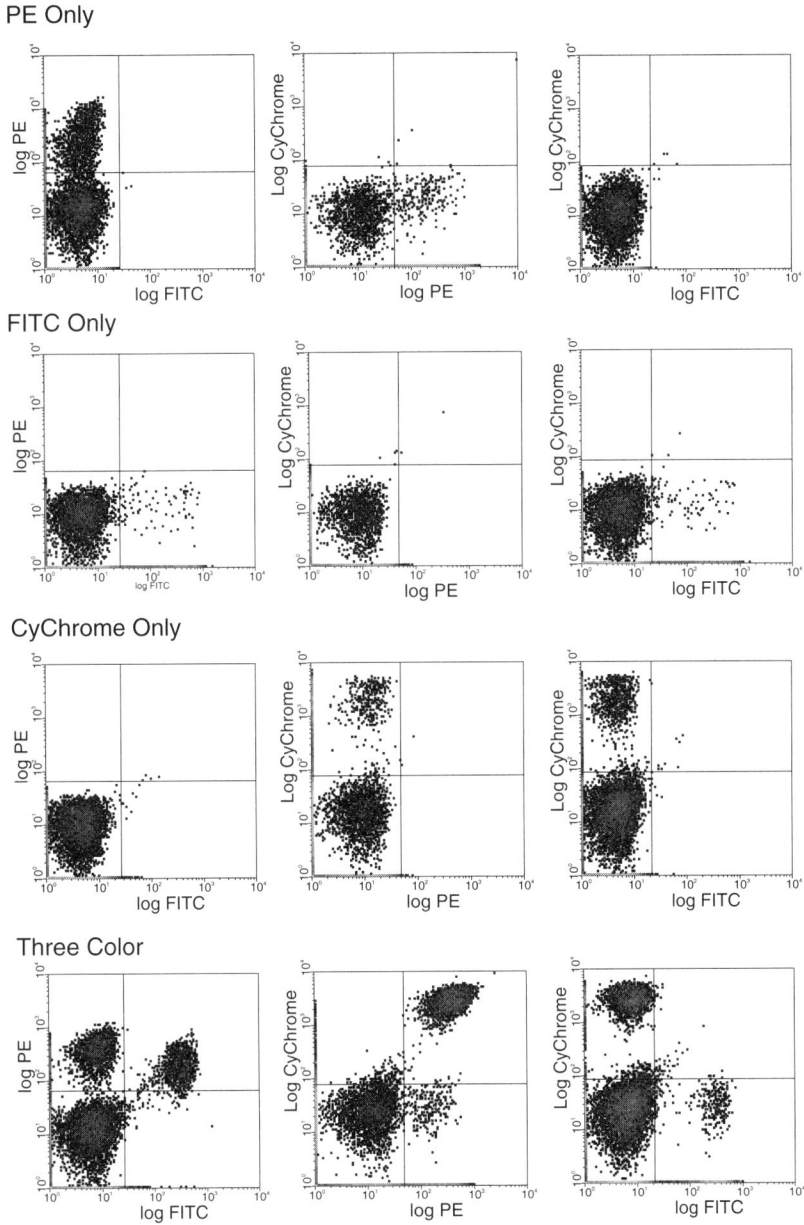

FIGURE 13.6 A series of two parameter histograms with each fluorochrome in combination with the two others. Single color staining is necessary to assure that positive cells fall only into the quadrants appropriately. The single stained samples are also used for compensation which can now be accomplished electronically or with subsequent reanalysis of the data in several software packages (FCS Express and FloJo).

13.2.2.6 Problems and Troubleshooting Tips

A negative cell population exhibiting a shift up in fluorescence can indicate cross-reactivity of the labeling antibodies with the cell-surface proteins. Blocking agents should be carefully considered (e.g., sera, commercial Fc block, purified Ig fraction) to identify the one that results in a low and consistent negative population. As noted previously, the presence of dead cells within the assay can lead to false positive or higher negative cell fluorescence. Consider performing a trypan blue dye exclusion assay to determine the viability of the cells before labeling. Use the scatter gating to attempt to exclude dead cells from the assay. Dead cells typically have a higher brighter 90 scatter signal than live cells on the forward scatter vs. 90° scatter histogram.

13.2.3 INDIRECT LABELING METHOD

The following sample method demonstrates an indirect two-color labeling scenario targeting T-lymphocyte subsets from a population of white blood cells purified from contaminating red blood cells. The primary antibodies are mouse monoclonal antibodies, each of a unique isotype allowing the selection of rat anti-mouse isotype-specific secondary antibodies with unique fluorochrome conjugates. This assay format works with an unlabeled isotypic antibody as a control, and this type of labeling can also be performed with polyclonal sera from unique animal systems. For example a goat-specific antibody for the first primary can be labeled with an anti-goat secondary conjugate to fluorochrome 1 and a rabbit-specific antibody for the second primary can be labeled with an anti-rabbit secondary conjugated to fluorochrome 2. One can imagine how complicated this series can become when additional and species unique primaries are added for multicolor labeling. Plus there is the added concern that species-specific polyclonal antibodies must be adsorbed against other species to remove any chance of cross-labeling the wrong species.

13.2.3.1 Materials Required

All the antibodies listed in this section can be obtained from Southern Biotech, Birmingham, AL. Recommended working dilution information is provided with each antibody. Secondary antibody conjugates are typically used at 1 $\mu g/10^6$ cells.

> Lymphocyte cell suspension (determined cell count and gradient purified)
> Anti-CD3 IgG1 mouse monoclonal antibody
> Anti-CD8 IgG2a mouse monoclonal antibody
> Isotype antibody for mouse IgG1 and IgG2a
> Rat anti-mouse IgG1 affinity-purified antibody with PE conjugate
> Rat anti-mouse IgG2a affinity-purified antibody with FITC conjugate
> PBS, pH 7.2, equilibrated at 4°C

Wash buffer (PBS + 0.1% sodium azide and 1% fetal bovine serum or 0.1% BSA, or 200 µg/ml goat Ig), filtered to 0.22 µm

Formaldehyde (Cat. No. 18814, methanol free 16% w/v, Polysciences, Warrington, PA)

13.2.3.2 Time Required

90–120 min

13.2.3.3 Special Equipment

Centrifuge
Cell counting device
Micropipettors (accurate volume handling for 10–200 µl and 1000 µl)
Flow cytometer

13.2.3.4 Procedures

1. Harvest or isolate the cells to ensure a single cell suspension in final suspension of cold wash buffer at an adjusted cell concentration of 5×10^5/ml.
2. Place 1 ml of the cell suspension into each of the 12×75 tubes preparing four tubes, one for each single antibody (unconjugated primary and isotype), a tube for all the antibodies, and a tube to be used as a secondary only control tube.
3. Centrifuge at 1500 g for 3 min. Use a pipette to remove the liquid. Be careful not to disturb the pellet. A slight amount of liquid can remain (~100 µl).
4. Add the appropriate amount of antibody (primary or isotype) as previously by titration.
5. Vortex and keep the tubes on ice for 15 min in the dark.
6. Add 1 ml of cold wash buffer. Vortex.
7. Centrifuge at 1500 g for 3 min. Use a pipette to remove the liquid. Be careful not to disturb the pellet. A slight amount of liquid can remain (~100 µl).
8. Add the vendor-recommended amount of secondary antibody conjugate to each tube.
9. Vortex and keep the tubes on ice for 15 min in the dark.
10. Add 1 ml of cold PBS. Vortex.
11. Centrifuge at 1500 g for 3 min. Use a pipette to remove the liquid. Be careful not to disturb the pellet.
12. Add 0.5 ml of PBS and vortex.
13. Add 0.5 ml of formaldehyde fixative (2% v/v). Store fixed samples at 4°C and protect them from light until analysis is complete.
14. Acquire data from 10,000 gated cells for each sample using a flow cytometer.

13.2.3.5 Results

The results anticipated in an indirect labeling would be similar in appearance to the FITC vs. PE combinations shown in Figure 13.6 with the added advantage of better signal-to-noise between the negatives in the lower left quadrant and the positives in the other three quadrants of the figure. This is due to the amplification potential of multiple secondary antibodies binding to one unlabeled primary antibody. Data were analyzed in the same manner as discussed previously.

13.2.3.6 Problems and Troubleshooting Tips

In situations where the fluorescence intensity is still low and the secondary antibody labeling system and brightest fluorochrome has been used to amplify the signal for very low-density cell-surface markers (e.g., cytokine receptors), this method can be modified to a three-step protocol that may amplify the staining: purified primary antibody, followed by a biotinylated anti-Ig, and then a fluorochrome-conjugated avidin or streptavidin as the final step with centrifugation and a wash step added between each addition to remove the excess nonbound antibodies and reagents.

13.2.4 Intracellular Labeling

Protein detection inside a cell can take the form of a secreted product (e.g., cytokine), an architecture material (e.g., tubulin), or an incorporated product (bromodeoxyuridine or virus protein). In the case of a secreted product, some form of cell activation may be under study. The use of agents that block the secretory pathway with protein accumulation in the Golgi (e.g., monensin or brefeldin A; both available from Sigma-Aldrich) can be useful in setting up the assay during the antibody titration step and to verify that the antibody is binding to the correct antigen.[6] Fluorescence should be higher in monensin-treated versus nontreated cells. Additionally, to further verify that the staining is intracellular only in nature, preincubating cells with the unconjugated antibody before fixation should eliminate the detection of any surface antigen when the fluorochrome antibody has been bound.

For intracellular staining, cells must first be fixed and permeabilized. The fixed cells are permeabilized, allowing conjugated antibodies access to proteins within the cell. The common fixation is paraformaldehyde in PBS that cross-links proteins; however, this fixation method may limit antigen accessibility. The final concentration of fixative can vary from 0.25% to 4% (w/v). Other tissue fixatives (e.g., methanol) may affect the staining properties of the antibody by denaturing the antigen structure. Some antibodies are sensitive to the conformational structure of their target epitope. For permeabilization, 0.1% saponin in a balanced salt solution is effective to facilitate antibody entry into cells.[10] Other permeabilization agents may be used (e.g., 0.02% Tween 20). Because this permeabilization can be reversible, it is necessary to keep saponin in all buffers used (both staining and washing steps).

The use of a directly conjugated monoclonal (versus an indirect labeling of a polyclonal antibody) will minimize nonspecific binding. This also allows for use of an isotype control to further validate the specificity of the antibody binding. For storage of samples longer than overnight, the sample should be resuspended in formaldehyde solution after completing the internal staining procedure.

13.2.4.1 Material Required

Formaldehyde (Cat. No. 18814, methanol-free 16% [w/v] Polysciences, Warrington, PA) diluted 1:4 to form a final concentration of 4% (w/v) in PBS
PBS (pH 7.2)
Saponin buffer, 0.1% (w/v) saponin with 0.05% (w/v) NaN_3 in Hanks' balanced salt solution. Stable for 1 month at room temperature
Anti-intracellular antigen monoclonal antibody conjugated to Alexa 488
Conjugated isotype antibody

13.2.4.2 Time Required

90–120 min

13.2.4.3 Special Equipment

Centrifuge
Cell counting device
Micropipettors (accurate volume handling for 10–200 μl and 1000 μl)
Flow cytometer

13.2.4.4 Procedures

In some assay scenarios, the researcher may wish to label both intracellular and extracellular antigens. If this is the case the extracellular antigens are labeled first, before proceeding with the steps of this method (fixation, permeabilization, and intracellular labeling).

1. Centrifuge 5×10^5 cells at 1500 g for 3 min removing the supernatant with a pipette. Add 1 ml PBS, vortex and resuspend, creating a single cell suspension.
2. Add the cells in 1 ml of cold 4% formaldehyde fixative with incubation at room temperature for 10 min. (Cells are vortexed at minute intervals in order to maintain a single cell suspension.)
3. Centrifuge at 1500 g for 3 min. Use a pipet to remove the liquid. Be careful not to disturb the pellet. Resuspend the cell pellet in 1 ml of PBS.
4. Centrifuge at 1500 g for 3 min. Use a pipet to remove the liquid. Be careful not to disturb the pellet. Resuspend the cell pellet in 1 ml of Saponin buffer.
5. Centrifuge at 1500 g for 3 min. Use a pipet to remove the liquid. Be careful not to disturb the pellet.
6. Resuspend the cell pellet in 100 μl of Saponin buffer.
7. Add the appropriate amount of antibody (primary or isotype).

8. Vortex the tubes briefly and incubate for 30 min in the dark at room temperature.
9. Repeat steps 4 and 5.
10. Add 1 ml of PBS, gently mix and centrifuge at 1500 g for 3 min.
11. Add 0.5 ml of PBS and gently mix.
12. Acquire data from 10,000 gated cells for each sample with a flow cytometer

13.2.4.5 Results

The sample data shown in Figure 13.7 depict increasing Alexa 488 fluorescence on the x-axis vs. cell count on the y-axis, for a negative control (A) and a positively labeled intracellular antigen (B). Data analysis of a homogeneous population of cells as shown here uses the entire population in the software analysis program to determine the median channel fluorescence intensity for both positive (signal) cells and negative (noise) cells from the histograms. On a well-standardized flow cytometer (as described in the Introduction and as discussed in the Instrument Display and Quality Control section), the operator analyzes daily a multiple intensity set of specific color beads with known MESF values. In this case the beads should be representative of the number of fluorescent dye molecules. The relative MESF values are plotted relative to the mean channel software values and a linear standard curve is generated. Researchers that want to be able to quantitate the number of MESF values must tell the operator, so that the instrument settings can be fixed to those used to assay the multiple intensity beads before the accumulation of the sample data. Thus it will be possible to plot the mean channel of the labeled population and determine the intensity level. This type of experiment may be useful when the researcher is looking at comparing modulation in the expression level of the antigen (intracellular or extracellular) from day to day, animal to animal, or between treatment conditions in the experimental design.

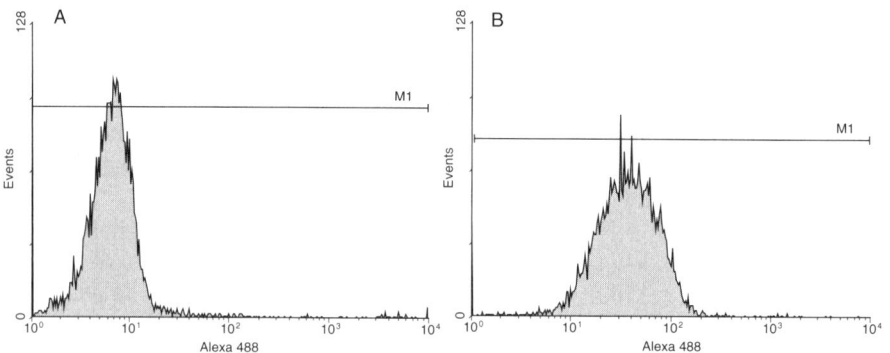

FIGURE 13.7 Sample data depicting intracellular staining. Analysis of the control cells (isotype antibody) is shown on the left and the positive (with primary antibody) cells on the right.

13.2.4.6 Problems and Troubleshooting Tips

Verification of antibody specificity and quantitative binding nature of the assay will require careful thought about the positive and negative controls. In the case of secretory products, use of a secretory inhibitor should allow accumulation and increase fluorescence detection if the antibody is specific. In the case of the structural protein, consider using unlabeled antibody to demonstrate loss of fluorescence by blocking the antigen binding site. In the case of incorporated products (e.g., bromo-deoxyuridine), use untreated samples as the negative control to establish a base line of fluorescence. Because removal of the fixative early on in this process and the use of permeabilizing agents, these cells cannot be stored for long periods of time before flow cytometric analysis and should be analyzed on the cytometer within 4 h of labeling.

REFERENCES

1. Shapiro, H., *Practical Flow Cytometry*, 4th ed., John Wiley and Sons, New York, 2003.
2. Brinkley, M., A brief survey of methods for preparing protein conjugates with dyes, haptens and cross-linking reagents, *Bioconjugate Chem.* 3, 2, 1992.
3. Hermanson, G.T. *Bioconjugate Techniques*, Hermanson, G.T., ed., Academic Press, San Diego, 1996.
4. Givan, A.L., The basics of staining for cell surface proteins, in *In Living Color: Protocols in Flow Cytometry and Cell Sorting*, Diamond, R. and Demaggio, S., eds., Springer, Berlin, 2000.
5. Rao, P.E., Talle, M.A., Kung, P.C., and Goldstein, G., Five epitopes of a differentiation antigen on human inducer T cells distinguished by monoclonal antibodies. *Cell. Immunol.*, 80, 310, 1983.
6. Givan, A.L., *Flow Cytometry: First Principles*, John Wiley and Sons, New York, 2001.
7. Bagwell, C.B. and Adams, E.G., Fluorescence spectral overlap compensation for any number of flow cytometry parameters, *Ann. N. Y. Acad. Sci.*, 677, 167, 1993.
8. Riedy, M.C. et al., The use of a photobleaching technique to identify nonviable cells in fixed homogeneous or heterologous cell populations, *Cytometry* 12, 133, 1991.
9. Schwartz, A. et al., Standardizing flow cytometry: a classification of fluorescence standards use for flow cytometry, *Cytometry*, 33, 106, 1998.
10. Bauer, K.D. and Jacobberger, J.W., Analysis of intracellular protein, in *Methods in Cell Biology: Vol. 41 Flow Cytometry*, Darzykiewicz, Z., Robinson, J.P. and Crissman, H., Eds., 2nd ed., Academic Press, San Diego, 1994.

14 ELISAs

John Chen and Gary C. Howard

CONTENTS

14.1 INTRODUCTION

Enzyme-linked immunosorbent assays (ELISAs) are a fundamental tool of modern biology and a multimillion dollar market in diagnostics and analysis. This powerful technique combines the sensitivity and specificity of antibodies and other binding partners in a format that is easily automated for high-throughput assays.

ELISAs are extraordinary adaptable and take advantage of many combinations of ligand pairs, methods of measuring signal (radioactive, fluorescent, or enzymatic), and experimental strategies. The presence or concentration of an antigen, antibody, or other target can be determined in solution with excellent limits of detection, generally in the ng/ml to pg/ml.

The two basic types of ELISA are sandwich and competitive, but there are many variations. Each has its own set of advantages and limitations in any given experimental setting. Many excellent ELISAs for specific purposes are commercially available. In this chapter, we will give some general guidelines for all ELISAs and describe each method in some detail and provide a step-by-step procedure for each.

14.2 GETTING READY

There are two key factors to consider: finding the right antibody pair and stability of the solid phase antibody.

14.2.1 ANTIBODIES

Identifying a proper antibody pair is the most crucial component of any ELISA. Having two antibodies that are directed at the target of interest is no guarantee that they can be used to make a functional ELISA. For example, in the sandwich ELISA, the most common application, both antibodies must bind the same antigen molecule. If the two epitopes overlap, the binding might be stearically inhibited. This physical blocking is most problematic with polyclonal antibodies (pAbs).

Monoclonal antibodies (mAbs) and pAbs have different strengths and limitations in ELISAs. Because pAbs bind to a wide array of epitopes, they can be particularly valuable as the labeled second antibody in a simple ELISA. mAbs bind to a single epitope. Their high specificities and avidities are very useful, but care must still be exercised to ensure that the critical epitopes do not overlap. In addition, the use of mAbs in commercial ELISAs may be subject to patent protection.

Purity of the antibodies is important. In a simple ELISA, the capture antibody is first bound to the solid phase. The number of protein binding sites in each well is limited. Any other proteins in the solution (e.g., from serum, ascites fluid, cell supernatants, bacterial cells) will compete with the antibody molecules for those binding sites. The resulting ELISA will be less robust than it might have been with pure antibody.

Antibodies can be purified by ammonium sulfate precipitation, diethyl-aminoethyl (DEAE) chromatography, or affinity chromatography. Purification is covered in detail in Chapter 7. The reader is encouraged to carefully review that chapter.

Antisera can be easily tested by a simple Ouchterlony or immunodiffusion method. An agarose gel is formed with a center well and several other wells arranged around the center. The antigen is placed in the center well, and serial dilutions of the antisera are put in the outer wells. Incubate the slide overnight at room temperature or 37°C. A white band will form between the inner and outer wells in those reactions of useful dilutions. A white line for dilutions of greater than 64-fold indicates a promising antiserum.

Cost is another consideration. pAbs tend to be a fraction of the cost of mAbs. Any purification steps will also add time and cost.

14.2.1.1 Attaching Proteins to the Plates

There are several choices for a solid support. Polyvinylchloride microtiter plates with 96 wells are commonly used. The volume and shape of the wells are also important. Flat-bottomed wells are best for ELISAs, especially when some sort of plate reader will be used to determine the signal in each well.

Binding of the capture antibody to the plate is a critical element of an ELISA. Antibody solution is added to each well of the plate, and allowed to attach, most likely through hydrophobic interactions. By diluting or concentrating the antibody solution, the amount of the bound capture antibody can be adjusted easily. Poly-vinylchloride plates bind approximately 100 ng of protein per well. We suggest using at least 1 μg/well for maximal binding. The binding reaction is fairly fast, but most protocols call for leaving the antibody in the well for 15–60 min at room temperature or overnight at 4°C.

After the unbound antibody has been washed away with two to three exchanges of buffer, the remaining unbound protein sites in the wells must be blocked. The best choices for blocking are bovine serum albumin or a control serum (ideally from the species of any secondary antibody).

14.2.1.1.1 Stability

For most research applications, stability is only a minor concern: the ELISA plates are used quickly. However, for commercial plates, stability is a major issue. Unfor-tunately, stability results from the interaction of many variables, and each must be determined empirically. Each antibody is different, but immunoglobulin Gs tend to be more stable whether dried or not, than immunoglobulin Ms.

The stability of the many antibodies bound to the dried plates can be improved with careful attention to production details. Drying by lyophilization is ideal. If the plates are to be dried, the buffer solution in the wells should be removed: as the plate dries, the microenvironment of the bound antibodies will be subjected to very high salt concentrations. Sugars (e.g., sucrose, trehalose) can sometimes help with the stability of dried plates.

14.2.1.1.2 Choice of Ligand Pairs

Many ELISA assays are based on antibody interactions with a target molecule. However, others can be used. Lectins lack the binding strength of most antibodies. Nevertheless, their specificity for carbohydrates makes them valuable components in selected applications.

Biotin-avidin (or streptavidin) binding is often used in the secondary reactions. The very high binding affinity ($>10^{15}$ M^{-1}) of avidin and biotin makes them a highly desirable ligand pair for many assays. The specificity and strength of the binding reaction allow reaction times to be shorter and allow the use of conditions that reduce nonspecific binding. Many biotinylating reagents are commercially available, and some offer features that may be useful in particular applications.

14.2.2 DETECTING THE REACTION

ELISA reactions can be visualized by eye or quantitated by several methods, includ-ing radioactivity, colorimetry, fluorescence, and chemiluminescence. Fortunately, antibodies typically have numerous moieties that can be used for binding labels, including α and ε aminos, carboxyls, sulfhydrals, and carbohydrates. An excellent description of the many methods for labeling antibodies can be found in Chapter 9.

14.2.2.1 Radioactivity

^{14}C, ^{3}H, and especially ^{125}I have been used to quantitate immune assays. The very small radioactive moieties usually have little, if any, effect on the antibody's activity. However, the rather short half-life of ^{125}I in particular and the regulatory constraints and disposal problems make the use of radioactively labeled antibodies problematic.

14.2.2.2 Enzyme Colorimetric Substrates

An attractive alternative to radioactivity is to introduce an enzymatic reaction by attaching an enzyme to a secondary antibody or by using avidin–biotin binding or another lectin pair. The sensitivity of these reactions is comparable to that of radioactive detection. In both cases, the color developed after a given time is measured by optical density at a specific wavelength and compared to a standard curve to determine the amount of antibody/antigen present (Table 14.1).

Horseradish peroxidase (HRP) and alkaline phosphatase (AP) are the most commonly used enzymes as active catalyst. HRP reacts with hydrogen peroxide to release an oxygen free radical, which in turn reacts with a colometric substrate. One commonly used substrate is 2,2′-azino-*bis*(3-ethylbenzthiazoline-6-sulfonic acid). This product is soluble in aqueous solutions and its reproducibility and high extinction coefficient make it an attractive option. The resulting green color can be measured at 405 nm with conventional plate readers. Another HRP substrate is 3,3′,5,5′-tetramethylbenzidine. The products of this reaction can be read at 650 nm or, after stopping the enzymatic reaction with acid, read the optical densities at 450 nm.

AP removes a phosphate from a substrate that, in turn, reacts with another component to yield a colored product. *p*-Nitrophenylphosphate yields a soluble bright yellow product that can be quantitated at 405–420 nm.

14.2.2.3 Fluorescence

Fluorescent labels can also be used and offer excellent sensitivity. The fluorescent molecule can be a small chemical (e.g., flouresceine, Texas red, rhodamine) or certain proteins (e.g., phycoerytherin).

TABLE 14.1
Methods of Detection in ELISAs

Chromogen	Color	Read (nm)
Horseradish peroxidase (all with hydrogen peroxide as substrate)		
2,2-azo-*bis*(3-ethylbenzthiazoline-6-sulfonic acid) (ABTS)	Green	405
o-phenylenediamine (OPD)	Orange	450
3,3,5,5- tetramethylbenzidine base (TMB)	Blue	650
Alkaline phosphatase		
p-nitrophenylphosphate (pNPP)	Yellow	405–420

"Sandwich" ELISA

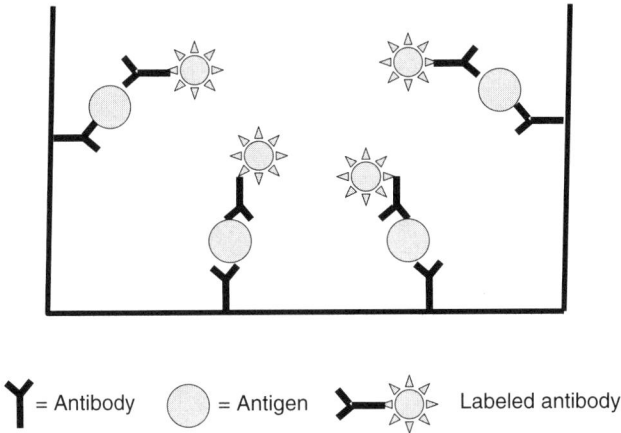

Y = Antibody   ◯ = Antigen   Y—✳ Labeled antibody

FIGURE 14.1 In the "sandwich" enzyme-linked immunosorbent assay, bound antibody holds antigen, and the amount of bound labeled second antibody is proportional to the amount of bound antigen.

14.2.2.4 Chemiluminescence

Commercially available chemiluminescence detection kits are a very good option. The sensitivity and speed are outstanding. In general, they also are based on the action of alkaline phosphatase or horseradish peroxidase.

14.3 SAFETY

As in all laboratory work, safety is an important consideration. Ensure that all personnel are adequately trained. Practice good general laboratory safety. Some antigens and samples may be biohazards. They may require special handling and disposal. Detection methods can also be of concern. Radioactive isotopes require special handling and disposal. Some chromogens are carcinogenic or toxic. Secure the material safety data sheets for all components of the assays. Be sure you understand and obey all applicable rules and regulations regarding the training, use, storage, and disposal of radioactive material and biohazards.

14.3.1 Sandwich ELISA Assays

The "sandwich" assay is one of the simplest and most useful immunoassays and is used to determine the concentration of an antigen in a sample (Figure 14.1). One antibody is attached to the microtiter plate. When a solution containing an antigen of interest is added to the well, the bound antibody captures the antigen, and careful washings remove any unbound antigen. A second antibody that recognizes a separate epitope binds only to the antigen that is already bound to the solid phase by the primary antibody. Excess secondary antibody is removed by washing. The secondary antibody usually is labeled to facilitate quantitation.

In addition to being quick and easy, the sandwich ELISA is reasonably accurate and reproducible. The antigen does not need to be pure.

Sensitivity depends on several factors: the capture antibody bound to the well, the avidity of the antibodies, and the "specific activity" of the second antibody.

14.3.1.1 Procedure

1. Coat the capture antibody onto the plate by adding 50 µl of the antibody solution to each well (20 µg/ml in phosphate-buffered saline [PBS]). Each assay is different, but 1 µg/well of antibody is a very good starting point. Incubate for 1 h at room temperature or overnight at 4°C.
2. Wash the wells twice with PBS. This can be done with a multichannel pipetter or a squirt bottle.
3. Block the rest of the protein binding sites with blocking buffer. A solution of bovine serum albumin (3%) or nonfat dry milk in PBS or Tris-buffered saline (TBS) is commonly used. Incubate for 1 h. (See 14.3.1.2a.)
4. Wash wells twice with PBS. (See 14.3.1.2b.)
5. Add 50 µl of the antigen solution to the wells (the antigen solution should be titrated). All dilutions should be done in the blocking buffer (3% bovine serum albumin/PBS). Incubate for at least 2 h at room temperature in a humid atmosphere.
6. Wash the plate four times with PBS.
7. Add the labeled second antibody. The amount to be added can be determined in preliminary experiments. For accurate quantitation, the second antibody should be used in excess. All dilutions should be done in the blocking buffer.
8. Incubate for 2 h or more at room temperature in a humid atmosphere.
9. Wash with several changes of PBS.
10. Add substrate as indicated by manufacturer. After suggested incubation time has elapsed, optical densities at target wavelengths can be measured on an ELISA plate reader. (See 14.3.1.2c.)

For quantitative results, compare signal of unknown samples against those of a standard curve. Standards must be run with each assay to ensure accuracy.

14.3.1.2 Procedural Notes

(a) Sodium azide is an inhibitor or horseradish peroxidase. Do not include sodium azide in buffers or wash solutions if an HRP-labeled antibody will be used for detection.
(b) If the plates are for future use, they must be prepared for storage. Cool storage, freeze-drying, and the addition of sugars or glycerol to the coating solution have all been used. The optimal method for any given antibody plate must be determined empirically.

(c)   Some enzyme substrates are considered hazardous, due to potential carcinogenicity. Handle with care and refer to material safety data sheets for proper handling precautions.

Note that azide is extremely toxic.

14.3.2 COMPETITIVE ELISA ASSAYS

Competitive ELISA require a little more work than the sandwich assay, but it is the method of choice when the target antigen is very small (Figure 14.2). In addition, impure primary antibodies can be used successfully.

In this assay, one of the reagents must be conjugated to a detection molecule, such as ^{125}I, HRP, or a fluorescent molecule. In the protocol described here, a labeled immunogen is used as the competitor.

As in the sandwich assay, a purified antibody is bound to the wells of a 96-well microtiter plate. Unbound antibody is removed by washing and any other protein binding sites in the well are tied up by incubation with a blocking reagent (often a nonfat dry milk solution). The wells with the bound primary antibody are then incubated with known amounts of unlabeled standard and unknown solutions. After this reaction has reached equilibrium, a labeled immunogen is added and binds to the primary antibody that is not already blocked by the unlabeled immunogen. The label is then quantitated, and the amount of labeled antigen that binds is inversely proportional to the amount of bound unlabeled antigen.

"Competitive" ELISA

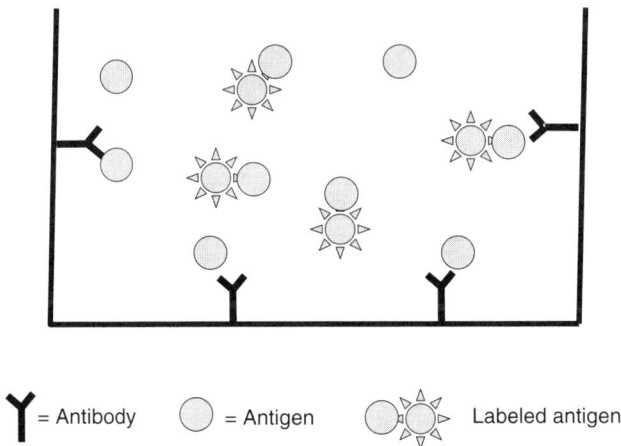

FIGURE 14.2 In the "competitive" enzyme-linked immunosorbent assay, a known amount of labeled antigen competes with the unlabeled antigen in an experimental sample. The amount of labeled antigen detected is inversely proportional to the amount of antigen in the sample.

TABLE 14.2
Suppliers of ELISA Reagents

Supplier	Location	Products
BD Biosciences	San Jose, CA	P, M, E, S
BioSearch International	Camarillo, CA	P, M, E
Chemicon	Temecula, CA	P, M, E
Sigma-Aldrich	St. Louis, MO	P, M, L, BA, S, E
Vector Laboratories	Burlingame, CA	P, M, L, BA, S, E

BA, biotin-avidin reagents; E, enzyme conjugates; L, lectins; M, monoclonal antibodies; P, polyclonal antibodies; S, substrates.

14.3.2.1 Procedure

1. Add 50 μl of diluted capture antibody solution to each well for 4 h at room temperature or 4°C overnight.
2. Block any unbound sites in the well with unlabeled second antibody (50 μl of a 20 μg/ml antibody solution per well). Incubate the wells overnight at 4°C to ensure complete blocking.
3. Wash the wells twice with PBS.
4. Block any remaining protein binding sites in the microtiter wells with blocking buffer by filling the wells to the top with 3% bovine serum albumin in buffer. (See 14.3.2.2a.) Incubate for 2 hours to overnight at room temperature.
5. Wash wells twice with PBS.
6. Add 50 μl of standard or sample solution to the wells. (See 14.3.2.2b.) All dilutions should be done in the blocking solution. (See 14.3.2.2c.)
7. Add 50 μl of the antigen-conjugate solution to the wells. All dilutions should be done in the blocking solution. Incubate 2 h at room temperature.
8. Wash the plate twice times.
9. Measure the bound labeled with the appropriate method (e.g., counting in a gamma counter, measure fluorescence or enzymatic color changes spectroscopically). (See 14.3.2.2d.)

14.3.2.2 Procedural Notes

(a) Detergent (e.g., 0.05% Tween 20) can be added if nonspecific binding is a problem.
(b) Flip out the remaining excess liquid to leave the wells empty before adding the standards and samples to the wells.
(c) Sodium azide is an inhibitor of horseradish peroxidase. Do not use it in buffers or wash solutions, if an HRP-labeled conjugate will be used for detection.
(d) Remember that this type of assay will yield an inverse relationship.

14.3.3 TROUBLESHOOTING ELISAs

1. Rerun the test to ensure that a simple mistake was not made.
2. Make sure the plates are still good. Ensure that the primary antibody is still stable on the plates.
3. Check the controls.

   a. If the negative controls are giving positive results, one of the solutions may be contaminated. Examine the substrate solution, enzyme-labeled antibody, or the controls themselves.
   b. If the positive controls are negative, examine all the reagents for integrity.

4. Check the enzyme reaction. For example, hydrogen peroxide, the substrate in the HRP reaction, is very labile. Use a fresh solution or check the stock supply for the strength of the hydrogen peroxide using an extinction coefficient of 43.6 at 240 nm.
5. Ensure that the controls are working.

   a. If the negative controls are positive, one of the reaction components may be contaminated.
   b. If the positive controls are negative, one of the components may be bad. Check the expirations dates for all the reagents for dating. Then check their concentrations.
   c. If only a small amount of color appears in the positive controls and the test samples, the dilution of the enzyme solution or a reactant may be off. Check them all.
   d. Check the wavelength settings for the plate reader.

6. When you are troubleshooting an assay, change only one experimental factor at a time.

REFERENCES

1. Bensadoun, A., Sandwich immunoassay for measurement of human hepatic lipase, *Methods Enzymol.*, 263, 333, 1996.
2. Butler, J.E., Enzyme-linked immunosorbent assays. In Howard, G.C., ed., *Methods in Non-radioactive Detection*, Appleton & Lange, Norwalk, CT, 1993.
3. Butler, J.E., ed., *Immunochemistry of Solid-Phase Immunoassay*, CRC Press, Boca Raton, FL, 1991.
4. Butler, J.E., Ni L., Brown W.R., Rosenberg B., Chang J., and Voss, E.W., Jr., The immunochemistry of sandwich ELISA. VI. Greater than ninety percent of monoclonal and seventy-five percent of polyclonal antifluorescyl capture antibodies (CAbs) are denatured by passive adsorption, *Mol. Immunol.*, 30, 1165, 1993.
5. Daly, D.S., White, A.M., Varnum, S.M., Anderson, K.K., and Zangar, R.C. Evaluating concentration estimation errors in ELISA microarray experiments, *BMC Bioinform.*, 6, 17, 2004.

6. Giltinan, D.M. and Davidian, M. Assays for recombinant proteins: a problem in non-linear calibration, *Stat. Med.*, 13, 1165, 1994.

7. Jones, G., Wortberg, M., Kreissig, S.B., Hammock, B.D., and Rocke, D.M., Sources of experimental variation in calibration curves for enzyme-linked immunosorbent assay, *Anal. Chim. Acta,* 313, 197, 1995.

8. Joshi, K.S., Hoffmann, L.G., and Butler, J.E., The immunochemistry of sandwich ELISAs. V. The capture antibody performance of polyclonal antibody-enriched fractions prepared by various methods, *Mol. Immunol.*, 29, 971, 1992.

9. Mashishi, T. and Gray, C.M., The ELISPOT assay: an easily transferable method for measuring cellular responses and identifying T cell epitopes. *Clin. Chem. Lab. Med.,* 40, 903, 2002.

10. Sittampalm, G.S., Smith, W.C., Miyakawa, T.W., Smith, D.R., and McMorris, C., Application of experimental design techniques to optimize a competitive ELISA, *J. Immunol. Methods*, 190, 151, 1996.

11. Vann, W.F., Sutton, A., and Schneerson, R., Enzyme-linked immunosorbent assay, *Methods Enzymol.* 184, 537, 1990.

12. Varnum, S.M., Woodbury, R.L., and Zangar, R.C., A protein microarray ELISA for screening biological fluids. *Methods Mol. Biol.*, 264, 2004.

13. Wong, C.H. and Bryan, M.C., Sugar arrays in microtiter plates, *Methods Enzymol.*, 362, 225, 2003.

14. Ziouti, N., Triantaphyllidou, I.E., Assouti, M., Papageorgakopoulou, N., Kyriakopoulou, D., Anagnostides, S.T., and Vynios, D.H., Solid phase assays in glycoconjugate research: applications to the analysis of proteoglycans, glycosaminoglycans and metalloproteinases, *J. Pharm. Biomed. Anal.,* 34, 771, 2004.

15 Antibodies in the Future: Challenges and Opportunities

Matthew R. Kaser and Gary C. Howard

CONTENTS

15.1 OVERVIEW

In the more than 200 years since Jenner's time, antibodies have become an indispensable tool in the study of biology and medicine. In biology, they have been a key component of the surge in fundamental knowledge that has occurred in the last quarter century. In the practice of medicine, multiple vaccines have led to the control (at least in the developed world) of many infectious diseases (e.g., polio, mumps, measles, chicken pox). They have also been envisioned to deliver drugs and other compounds to the site of disease.

In the future, they certainly will continue to be incredibly versatile and powerful agents. It is impossible to know where new discoveries might take biomedical science in the future. However, much of the future of antibodies is already known. They are so powerful that their current uses—enzyme-linked immunosorbent assays (ELISAs),

Western blotting, immunohistochemistry, and flow cytometry—will certainly remain important in the future. Nevertheless, new methods for making and using antibodies will be found. Here we will point out a few exciting areas. The reader is encouraged to review the references at the end of the chapter for further information.

15.2 NEW ANTIBODY SOURCES

15.2.1 HUMAN ANTIBODIES FROM TRANSGENIC ANIMALS

The immunogenicity of antibodies made in other species for use in humans may be solved, in part, by the use of transgenic animals.[1] In this way, the mouse (for example) might see a normal human antibody as "self" rather than nonself.

15.2.2 GENERATION OF ANTIBODIES BY GENETIC IMMUNIZATION

To elicit an immune response, a protein is usually injected into an animal. Tang et al. demonstrated that the process of isolating sufficient amounts of protein can be avoided by inoculating the animal with the gene of the antigen instead.[2] They coated gold microbeads with the DNA encoding the protein and injected them into animals. The system essentially mimics the actions of infective virus particles that usurp the cell's own expression mechanisms to produce new virus. This overall strategy has been used by numerous groups.[3]

15.2.2.1 Antibodies from Plants

Plants offer many advantages for producing antibodies and other peptides. They are cheap, transform easily, and have few of the ethical problems associated with the use of animals. Purification does have some challenges.[4–6]

Some of the original experiments to produce antibodies from plants were done in transgenic tobacco plants.[7] Plants were generated to express a mouse monoclonal antibody kappa chain, a hybrid heavy chain, and joining chain and a rabbit secretory component. When the plants were appropriately crossed, all four protein chains were expressed and assembled into a functional immunoglobulin molecule.

15.2.2.2 Humanizing Antibodies

Antibodies for therapy of malignancy and autoimmune diseases are being developed and approved for use in treatment, such as adalimumab and rituximab.[8,9] Although therapeutic antibodies have great promise, they are also immunogens themselves. The patient's own immune system will attempt to clear the therapeutic antibody, thus reducing its half-life in the patient and its effectiveness. Developing antibodies in another organism that are not, in turn, recognized by the foreign by the patient is a critical step.

Strategies for humanizing mouse or other antibodies are based on comparisons of the various antibody domains or framework regions.[10–12] These are conserved to a greater or lesser extent between the species, and finding those with the greatest degree of conservation is important. By changing perhaps only a small number of amino acids, the antigenicity of the engineered antibody may be greatly reduced.[13,14]

15.2.3 AVIMERS: ALTERNATIVES TO ANTIBODIES?

Although they have been the molecules of choice for some time, antibodies have several characteristics that limit their applications. Making them in mammalian systems is expensive. Bacterial systems lack the ability to glycosylate antibodies properly. Furthermore, it is difficult to engineer antibodies so they can cross the blood-brain barrier or other boundaries.

One answer may be another class of binding proteins called avimers.[15] Avimers are engineered protein molecules that combine several human A-domains of cell-surface receptors in a single molecule. The binding sites can recognize the same or multiple epitopes.

15.3 NEW WAYS TO USE ANTIBODIES

15.3.1 ANTIBODY FRAGMENTS AND SINGLE DOMAINS

Antibodies' size has always limited their use. For some applications, a smaller binding molecule is desirable. Thus, genetically engineered antibody fragments (e.g., Fab) and single domain molecules are of great interest, particularly for their ability to access previously inaccessible antigen, such as cavities (see reviews by Hudson and Souriau and Holliger and Hudson[16,17]). The fragments retain their binding affinities but traditionally have been difficult to produce in useful quantities. Recombinant DNA methods have solved many of the production challenges.

Several types of smaller fragments are available. Early efforts took advantage of enzymatic cleavage to produce Fab and Fc fragments. Although the Fc region aids in stability of the overall antibody molecule and interacts with complement and receptors, in many applications, these functions are not needed, and the Fc domain can be removed. More recently, the Fv fragments have become options. Some single-domain molecules combine multiple Fv domains into one construct with increases in binding affinity and other properties.

15.3.2 ANTIBODIES AND PROTEOMICS

Antibodies have been employed in arrays to do proteomics studies.[18-20] However, proteins have limitations in arrays, and some researchers have used small DNA or RNA oligomers called aptamers as antibody mimics.[21] Furthermore, aptamers can be genetically engineered to contain reporter moieties, and they work very well in array technologies as well as in ELISAs and Western blots.

15.3.3 HYBRID AND CHIMERIC ANTIBODY CONSTRUCTS

Recombinant technology enables us to now create hybrid or chimeric molecules that can combine both antigen (or ligand) binding activity of an antibody with a portion of another molecule, for example a G-protein-coupled receptor (GPCR) transmembrane fragment, to produce a molecule that could gate ions or drugs into the cell in response to an endogenous soluble or infused antigen or ligand.

We also expect that the fields of biotechnology, microelectronics, and nanoelectronics will become increasingly more intertwined, such as coupling Fab-type domains to an electrical circuit. We anticipate that Ag- or ligand-binding fragments of antibodies will be used as molecular switches in response to ligand concentrations for use in micropumps for delivery of therapeutic drugs. Similarly, such a system might be used for diagnostics, *in vitro* or even *in vivo*, to detect increases in a soluble antigen.

Recent research at several major institutions is showing promise in the field of diagnostics, for example, linking antibodies, antibody fragments, or antigens to quantum dots or other nanoparticles that may be used to detect the presence of a molecular marker.

15.3.4 DNA VACCINES

In DNA vaccines, gold particles or other microbeads are coated with circular DNA that encodes an antigen of interest and control sequences that allow antigen expression in mammalian cells. Adenovirus has also been suggested as a delivery vehicle for the antigen gene.[22]

Interesting results have been observed using this technique for many diseases, including various cancers, HIV, and Alzheimer's disease.[23–26]

Other subtle, but significant, challenges remain. Cancer is caused both by normal proteins whose expression has become dissociated from normal regulatory mechanism and by mutations that cause recombination between the oncogene or tumor suppressor) and another unrelated gene that is differentially regulated by the cell. To combat these proteins, antibodies against those same normal proteins will have to be made and their activity controlled. Similarly, autoimmune diseases occur when the body begins to see self as nonself, and the immune system begins to attack and degrade cells and their surface molecules. Understanding the nature of those epitopes or antigen involved before such immunologic changes will help us understand how to design therapies that may prevent such an event in susceptible individuals.

15.3.5 OTHER APPLICATIONS

Immunoglobulins, and more generally, immunoglobulin domains, are unique in nature in that they have the property of being able to bind to probably any epitope found in almost any environment. Their ability to bind with high affinity to a molecule can allow them to act as molecular switches using the "lock-and-key" analogy, which might have many potential uses in nanotechnological applications.

The very property of somatic recombination allows the immunoglobulin variable domain to define a polypeptide that can theoretically fill every probably space and thereby provide a surface that can interact with any, repeat, *any*, molecule that is presented to it. Such a robust molecular tool will undoubtedly have many future uses as yet undreamed, many perhaps in the field of detoxification of fluids. For example, gene shuffling can produce vast libraries of immunoglobulins that can be recombined with signal transduction molecules that are, in turn, linked to an electronic substrate. Such biosilicon chips could be used to screen molecules with binding activity to known toxins or beneficial compounds and thereby remove (purify) them from a carrier medium. Other transduction systems could include an electron transfer system that results in a color change of a plant flavanoid. Such biochips could have uses in detecting airborne pathogens and reacting at least as fast as a conventional ELISA, but having the combined advantage of miniaturization and sensitivity. Government agencies and the plant biotech industry take note.

Indeed, if the immunoglobulins were produced inexpensively using bacterial systems in a pseudo-monoclonal approach, the product of each clone could be selectively bound to an individual site on a silicon chip, thereby removing the need to separately purify and characterize each clone. Those clones that do not react with a target could either be disposed or retained for another screen.

Improvements in protein engineering and production capacity may enable vast amounts of antibodies to be produced very inexpensively. Several applications spring to mind if antibodies, monoclonal, polyclonal, or synthetic, could be produced in industrial quantities (kilograms to tons). For example, an antibody raised against "red tide" toxins could be using spray technology already developed for crop dusting, thereby binding to and neutralizing (or even eliminating) red tide that threatens so many of the worlds fisheries and coral reefs.

15.4   THE FUTURE

What of the future? Biomedical science has come a long way since Jenner's pioneering experiments for treating smallpox with an attenuated virus. Although he did not understand the molecular mechanisms, his early vaccinations created the "magic bullet" that prevented the dreaded pox. Today antibodies are still our magic bullet. Jenner had no inkling of the many uses in which antibodies are used today. In like manner, it is probably just as impossible that we are able to foresee tomorrow's uses. Certainly, many of the current techniques discussed will still be in use for some time. We hope that this book will be helpful for today's biomedical researchers to use with those future techniques.

REFERENCES

1. Lonberg, N., Human antibodies from transgenic animals, *Nat. Biotechnol.*, 23, 1117, 2005.
2. Tang, D.C., DeVit, M., and Johnston, S.A., Genetic immunization is a simple method for eliciting an immune response, *Nature* 356, 152, 1992.
3. Chambers, R.S. and Johnston, S.A., High-level generation of polyclonal antibodies by genetic immunization, *Nat. Biotechnol.*, 21, 1088, 2003.
4. Hiatt, A., Antibodies produced in plants, *Nature,* 334, 469, 1990.
5. Giddings, G., Allison, G., Brooks, D., and Carter, A., Transgenic plants as factories for biopharmaceuticals, *Nat. Biotechnol.,* 18, 1151, 2000.
6. Arntzen, C., Plotkin, S., and Dodet, B., Plant-derived vaccines and antibodies: potential and limitations, *Vaccine,* 23, 1753, 2005.
7. Ma, J.K., Hiatt, A., Hein, M., Vine, N.D., Wang, F., Stabila, P., van Dolleweerd, C., Mostov, K., and Lehner, T., Generation and assembly of secretory antibodies in plants, *Science,* 268, 716, 1995.

8. Rau, R., Adalimumab (a fully human anti-tumour necrosis factor α monoclonal antibody) in the treatment of active rheumatoid arthritis: the initial results of five trials, *Ann. Rheum. Dis.*, 61, ii70, 2002.

9. Rizvi, S.A. and Bashir, K., Other therapy options and future strategies for treating patients with multiple sclerosis, *Neurology*, 63, S47, 2004.

10. Bucher, P., Morel, P., and Bühler, L.H., Xenotransplantation: an update on recent progress and future perspectives, *Transplant Int.*, 18, 894,

11. Krauss, J., Arndt, M.A.E., Martin, A.C.R., Liu, H., and Rybak, S.M., Specificity grafting of human antibody frameworks selected from a phage display library: generation of a highly stable humanized anti-CD22 single-chain Fv fragment, *Prot. Engineer*, 16, 753, 2003.

12. Hwang, W.Y., Almagro, J.C., Buss, T.N., Tan, P., and Foote, J., Use of human germline genes in a CDR homology-based approach to antibody humanization, *Methods* 36:35, 2005.

13. Ross, J.S., et al., Antibody-based therapeutics: focus on prostate cancer, *Cancer Metastasis Rev.*, 24, 521, 2005.

14. Villamor, N., Montserrat, E., and Colomer, D., Mechanism of action and resistance to monoclonal antibody therapy, *Semin. Oncol.*, 30, 424, 2003.

15. Jiong, K.J., Mabry, R., and Georgiou, G., Avimers hold their own. *Nat. Biotechnol.*, 23, 1493, 2005.

16. Hudson, P.J. and Souriau, C., Engineered antibodies. *Nat. Med.*, 9, 129, 2003.

17. Holliger, P. and Hudson, P.J., Engineered antibody fragments and the rise of single domains, *Nat. Biotechnol.*, 23, 1126, 2005.

18. Michaud, G.A., Salcius, M., Zhou, F., Bangham, R., Bonin, J., Guo, H., Snyder, M., Predki, P.G., and Schweitzer, B.I., Analyzing antibody specificity with whole proteome microarrays, *Nat. Biotechnol.*, 21, 1509, 2003.

19. Bradbury, A.R., Velappan, N., Verzillo, V., Ovecka, M., Marzari, R., Sblattero, D., Chasteen, L., Siegel, R., and Pavlik, P., Antibodies in proteomics, *Methods Mol. Biol.*, 248, 519, 2004.

20. Uhlen, M. and Ponten, F., Antibody-based proteomics for human tissue profiling, *Mol. Cell Proteomics*, 4, 384, 2005.

21. Hamaguchi, N., Ellington, A., and Stanton, M., Aptamer beacons for the direct detection of proteins, *Anal Biochem.*, 294, 126,

22. Liu, M.A., DNA vaccines: a review, *J. Intern. Med.*, 253, 402, 2003.

23. Schellekens, H., Immunogenicity of therapeutic proteins: clinical implications and future prospects, *Clin. Therap.*, 24, 1720, 2002.

24. Brekke, O.H. and Sandlie, I., Therapeutic antibodies for human diseases at the dawn of the twenty-first century, *Nat. Rev. Drug Discov.*, 2, 52, 2003.

25. Presta, L., Antibody engineering for therapeutics, *Curr. Opin. Struct. Biol.*, 13, 519, 2003.

26. Nabel, G.J., Genetic, cellular and immune approaches to disease therapy: past and future, *Nat. Med.*, 10, 135, 2004.

27. Baker, M., Upping the ante on antibodies, *Nat. Biotech.*, 23, 1065, 2005.

Index

Page numbers followed by the letter *f* refer to figures; page numbers followed by the letter *t* refer to tables.